Corrosion Control
by
Organic Coatings

Corrosion Control by Organic Coatings

Henry Leidheiser, Jr.—Editor

National Association of Corrosion Engineers

Published by
National Association of Corrosion Engineers
1440 South Creek Drive
Houston, Texas 77084

**Library of Congress
Catalog Number
81-84733**

Neither the National Association of Corrosion Engineers, its officers, directors, or members thereof accept any responsibility for the use of the methods and materials discussed herein. The information is advisory only and the use of the materials and methods is solely at the risk of the user.

Printed in the United States of America. All Rights reserved.
This book, or parts thereof, may not be reproduced in any form without permission of the copyright owners.

Copyright 1981
National Association of Corrosion Engineers

Foreword

Over the years, the National Association of Corrosion Engineers (NACE) has sponsored a number of special conferences related to localized corrosion, corrosion/erosion of coal conversion system materials, stress corrosion cracking, hydrogen embrittlement of iron base alloys, high temperature, high pressure electrochemistry in aqueous solution, and passivity in iron and iron base alloys. All of the proceedings of these conferences were published by NACE. In 1978, the Research Committee decided to formalize these conferences, and won approval from the Board of Directors to draw up a plan covering the ten year period, 1980 to 1989. I was asked to appoint an organizing committee and to arrange the first of these conferences, in 1980, to be concerned with "Corrosion Control by Organic Coatings." The charge included the mandate to make the Conference truly international in character. The Conference was held at Lehigh University, Bethlehem, Pennsylvania, August 10 to 15, 1980.

The international character of the conference was assured by the selection of a committee of outstanding workers in the field of corrosion, namely

    Dr. Werner Funke — West Germany
    Dr. H. Jullien — France
    Dr. John D. Keane — U. S. A.
    Dr. K. Meguro — Japan
    Dr. A. N. McKelvie — England
    Dr. J. E. O. Mayne — England
    Dr. Ing. M. Svoboda — Czechoslovakia
    Dr. Joseph Yahalom — Israel
    Dr. M. Yaseen — India

Special thanks go to W. Funke, H. Jullien, and K. Meguro, who publicized the conference widely in their own countries.

The 161 attendees represented 15 countries, and of the 41 papers presented in written form, 17 originated from England, France, India, Belgium, Hungary, Germany, Japan, Saudi Arabia, and Israel.

The papers included in this book range from the highly theoretical to the very practical, with the majority falling in the region where the distinction between basic and applied science is not sharp. Most of the authors appear to be motivated by the desire to understand corrosion phenomena, so that better coating systems can be developed. In some instances, the editor has used a sharp scissors on the submitted papers to minimize proprietary product advocacy, or promotion of an organization. Several authors make strong statements about generic coatings that should, or should not, be used under certain circumstances. These recommendations are the sole responsibility of the authors. NACE, the Organizing Committee, and the editor do not endorse the recommendations by virtue of acceptance of the manuscript for publication.

The papers, viewed in their entirety, illustrate very well the complexity of a full understanding of corrosion control by organic coatings. The major problem appears to be to relate the observations at the atomic and molecular level to the observations of the system as a whole, as exemplified by the electrical properties and the behavior of the system in an accelerated corrosion test.

Henry Leidheiser, Jr.
Lehigh University
November, 1980

Contents

Enhancing Polymer Adhesion to Iron Surfaces by
Acid Base Interaction
    F. M. Fowkes, C-Y Sun, and S. T. Joslin 1

The States of Water in Polymer Films
    *G. E. Johnson, H. E. Bair, and
    E. W. Anderson* 4

The Inhomogeneous Nature of Polymer Films and
Its Effect on Resistance Inhibition
    D. J. Mills and J. E. O. Mayne 12

Polyurethane Films as Electrochemical Membranes:
How to Progress in Anticorrosion Protection
    H. Jullien 18

Permeation Properties of Organic Coatings in the
Control of Metallic Corrosion
    M. Yaseen 24

Studies on the Subcoating Environment of Coated
Iron Using Qualitative Ellipsometric and
Electrochemical Techniques
    J. J. Ritter and J. Kruger 28

Electrochemical Values—Their Significance When
Applied to a Coated Substrate
    M. Piens and R. Verbist 32

Properties and Behavior of Corrosion Protective
Organic Coatings as Determined by Electrical
Impedance Measurements
    J. V. Standish and H. Leidheiser, Jr. 38

An AC Impedance Probe as an Indicator of Corrosion
and Defects in Polymer/Metal Substrate Systems
    M. C. Hughes and J. M. Parks 45

Impedance Techniques for the Study of Organic
Coatings
    J. D. Scantlebury and G. A. M. Sussex 51

Dielectric Study of Paint and Lacquer Coatings
    J. Devay, L. Meszaros, and F. Janászik 56

The Transport Properties of Polyurethane Paint
    R. T. Ruggeri and T. R. Beck 62

Rate Controlling Steps in the Cathodic Delamination
of 10 to 40 μm Thick Polybutadiene and Epoxy Polyamide
Coatings from Metallic Substrates
    H. Leidheiser, Jr. and W. Wang 70

Cathodic Disbondment of Well Characterized
Steel/Coating Interfaces
  J. E. Castle and J. F. Watts 78

Underfilm Corrosion Currents as the Cause of
Failure of Protective Organic Coatings
  E. L. Koehler 87

Blistering of Paint Films
  W. Funke 97

Adhesion Loss of Organic Coatings—Causes and
Consequences for Corrosion Protection
  W. Schwenk 103

Some Aspects of Cathodic Electrodeposition of
Epoxy Latexes as Corrosion Resistant Coatings
  C. C. Ho, A. Hymayun, M. S. El-Aasser, and
  J. W. Vanderhoff 111

Can Failures Still Occur When the Correct Coating
(For a Given Environment) is Selected and Applied
Properly?
  K. B. Tator 122

Film Application Method as Related to Corrosion
Control
  H. J. Schmidt, Jr. 128

Corrosion Protective Properties of Paint Films
  I. Sekine 130

Performance of Marine and Industrial Coatings in
the Arabian Gulf
  K. I. Rhodes and E. M. Moore, Jr. 138

Sulfonate Based Coatings—Their Chemical, Physical
and Performance Properties
  L. S. Cech, J. W. Forsberg and W. A. Higgins 144

Water Displacing, Corrosion Preventive Compounds
  C. R. Hegedus 150

Corrosion Control under Thermal Insulation
and Fireproofing
  J. F. Delahunt 158

Corrosion Caused by Drying Paint Film
  J. H. White, Z. Kielmanson, and P. Letai 165

An Anomalous Effect of Limited Drying Time on the
Performance of a Vinyl Coating
  J. H. White and W. Rothschild 172

Evaluation of Linings for SO_2 Scrubber Service
  D. M. Berger, R. J. Trewalla, and C. J. Wummer 178

The Impact of Environmental Restraints on Corrosion
Resistant Coatings
  R. N. Washburne 186

Influence of Pigments on the Effectiveness of
Anticorrosive Primers
P. Kresse, V. Szadkowski, and R. H. Odenthal 197

Study of the Use of Inhibitors in Coatings to Control
Stress Corrosion Cracking of Line-Pipe Steel
E. W. Brooman, D. M. Lineman, W. E. Berry,
and R. R. Fessler 203

The Paintability of High Strength Cold Rolled
Steels
J. A. Kargol and D. L. Jordan 211

Relation of Steel Surface Profile to Coating
Performance
L. K. Schwab and R. W. Drisko 222

Evaluation of Surface Pretreatment Methods for
Application of Organic Coatings
F. Mansfeld, J. B. Lumsden, S. L. Jeanjaquet,
and S. Tsai 227

The Interaction of Chlorinated Rubber Based Lacquers
with Abraded Mild Steel Substrates
D. H. Smelt 238

Is the Salt Fog Test an Effective Method to Evaluate
Corrosion Resistant Coatings?
T. Liu 247

Comparative Investigations of Corrosion
Performance of Coating Systems for Automobiles
by Different Methods of Accelerated Weathering
W. Goering, E. Koesters, and R. Muenster 255

Corrosion Inhibitor Test Method
G. A. Salensky 263

A Modified Sequential Sampling Plan for Painting
Inspection
R. Tooke, Jr. 267

High Temperature, Long Term Performance of
Temperature Indicating Paints
J. F. Delahunt 280

Painting "O"-Ring Sealing Surfaces to Prevent
Corrosion
C. J. Sandwith 285

Index 295

Enhancing Polymer Adhesion to Iron Surfaces By Acid-Base Interaction

*Frederick M. Fowkes, Chen-Yu Sun, and Sara T. Joslin**

Introduction

The adhesion of organic polymers to inorganic oxides such as those on steel surfaces has been shown earlier to result entirely from acid-base interactions between the polymer and the oxide.[1] Hydrogen bonds were shown to be a subset of acid-base interactions, and dipole-dipole interactions were shown to be so negligibly small that no experimental evidence for them has yet been found.[2] It therefore becomes important to find a method of measuring and predicting acid-base interactions in the bonding of organic polymers to metal oxide surfaces, especially the oxide layers on steel.

In earlier studies, we have made good use of the "C and E" equation of Drago[2] for correlating and summarizing heats of acid-base interactions:

$$-\Delta H_{ab} = C_A C_B + E_A E_B \quad (1)$$

in which the acidic strength of the acid is fully characterized by the constants C_A and E_A, and the basic strength of the base is fully characterized by the constants C_B and E_B. We now propose to test a hypothesis that the heat of adsorption of bases on metal oxide powders is precisely equal to the heat of acid-base interaction of these bases with the acidic sites of the oxide and that the acid strength of these oxides sites can be fully characterized by C_A and E_A parameters.

In this work, we shall determine the heats of adsorption from the temperature coefficient of adsorption isotherms. If the equilibrium surface concentration Γ of adsorbed base is determined at equilibrium bulk concentration C, these are related by an equilibrium constant K for Langmuirian adsorption:

$$K = \frac{\theta}{(1-\theta)C} = \frac{\Gamma}{(\Gamma_m - \Gamma)C} \quad (2)$$

which can be rearranged to give:

$$\frac{C}{\Gamma} = \frac{1}{\Gamma_m K} + \frac{C}{\Gamma_m} \quad (3)$$

where Γ_m is the surface concentration for a complete monolayer. If the adsorbed molecules are not too tightly packed, Equation (3) predicts a straight line plot of C/Γ vs C, with intercept $1/\Gamma_m K$, from which the equilibrium constant K of Equation (2) may be determined. If this equilibrium constant is determined at two or more temperatures, the heat of adsorption may be calculated from:

$$\Delta H_{ads} = \frac{RT_1 T_2}{T_2 - T_1} \ln \frac{(K \text{ at } T_2)}{(K \text{ at } T_1)} \quad (4)$$

Experimental Details

Fe_3O_4 powder was used for most of these studies. Its surface area, determined by the BET method with argon gas at liquid nitrogen temperatures, was 7.0 M²/g. Some Fe_2O_3 powder from the same source was also used; its surface was 12.0 M²/g.

The powders were dried for 1 hour at 120 C in a vacuum oven at 1×10^{-4} torr. This amount of drying is not believed to be enough to remove all of the adsorbed water. The slurries were stirred and tumbled with glass beads in cyclohexane solutions of the model bases.

The triethylamine and pyridine were obtained from another source. The pyridine was 99+% Gold Label grade and triethylamine was 99% purity.

Discussion of Results

Figure 1 shows the adsorption isotherms obtained with triethylamine, pyridine, ethyl ether, and ethyl sulfide. The amounts of pyridine adsorbed per unit area were about twice those of triethylamine; consequently, for the pyridine curves the ordinates have been reduced by a factor of two to allow them to fit on the same graph. The ether and sulfide data are too inaccurate at this time to be reported in more detail, but the triethylamine and pyridine adsorption data were reproducible and Langmuirian in all details; the graphs of these data according to Equation (3) are shown in Figures 2 and 3. The straightness of the lines indicates that the isotherms are indeed Langmuir isotherms; that is, the equilibrium constant K is constant at all surface concentrations, suggesting that all

---

*Department of Chemistry, Lehigh University, Bethlehem, Pennsylvania

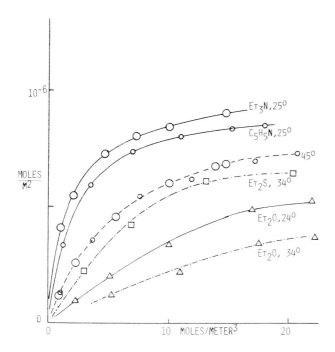

FIGURE 1 — Adsorption isotherms for pyridine, triethylamine, ethyl sulfide and ethyl ether adsorbing from cyclohexene solutions on iron oxide powder (7 m²/g).

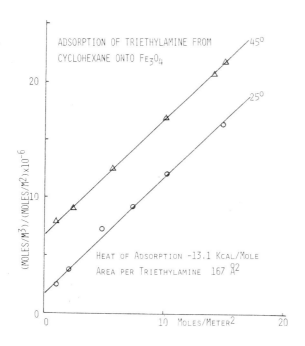

FIGURE 2 — Langmuir isotherm for adsorption of triethylamine from cyclohexane onto iron oxide powder.

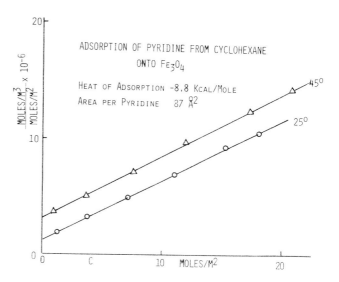

FIGURE 3 — Langmuir isotherm for adsorption of pyridine from cyclohexane onto iron oxide powder.

the acid sites are of equal strength and that the adsorbed bases are not tightly crowded together at full coverage.

Adsorption of triethylamine and of pyridine was checked on three different iron oxide samples and nearly identical results were obtained. Even on Fe_2O_3 the adsorption of pyridine was the same. Examination of the Fe_3O_4 then showed that it was not pure Fe_3O_4 but was about 20% oxidized to Fe_2O_3, probably on the surface. Therefore, the results really apply only to Fe_2O_3 surfaces.

The heats of adsorption were -13.1 ± 0.3 Kcal/mole for triethylamine and -8.8 ± 0.3 Kcal/mole for pyridine on Fe_2O_3 surfaces. These data are the minimum information necessary to determine the C_A and E_A for the acidic sites of iron oxide. Since we know that triethylamine has $C_B = 11.09$ and $E_B = 0.99$, the relation of C_A to E_A can be determined from:

$$C_A = -(H_{ads} + E_A E_B)/C_B$$

or

$$C_A = -(-13.1 - 11.09 E_A)/0.99$$

Figure 4 shows a graph of C_A vs E_A for the triethylamine and pyridine experiments. The two lines differ appreciably in slope and the values of C_A and E_A at the intersection are believed to be the correct values: $C_A = 1.0$, and $E_A = 2.04$. These values give a C/E ratio of 0.5, indicating that the acid sites on iron oxide are quite soft, though not as soft as iodine (which has a C/E ratio of unity). This evidence suggests the sites to be Fe^{3+} ions rather than FeOH acid sites, which would be expected to be much harder, and have a C/E ratio nearer to 0.1.

The softness of iron acid sites suggests that adhesives and coatings for iron and steel should feature basic groups (electron donors) which are strong soft bases, for these might best withstand the incursion of water into the interface.

Conclusions

A method of interpreting heats of adsorption as acid-base interactions predictable by the Drago equa-

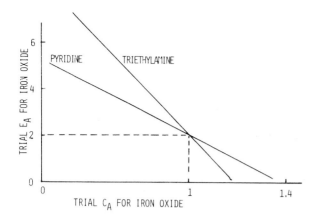

FIGURE 4 — Trial E_A vs C_A plot for iron oxide surfaces: $E_A = (-H_{ads} - C_A D_B)E_B$; C_B and E_B values taken from Drago.[1]

tion is presented. It appears that the dominant acidic adsorption sites can best be characterized by their C_A and E_A values of 1.0 and 2.04, respectively. This information, plus the measured surface density of sites, should allow prediction of the adsorption of any bases at any temperature, and should predict the energy of adhesion of any basic coating to such surfaces. Further work is needed to verify these predictions.

Acknowledgement

Appreciation is expressed to the Office of Naval Research which provided support for this research.

References

1. R.S. Drago, G.C. Vogel, and T.E. Needham, *J. Am. Chem. Soc.*, **93**, p. 6014 (1971); R.S. Drago, L.B. Parr, and C.S. Chamberlain, *J. Am. Chem. Soc.*, **99**, p. 3203 (1977).
2. F.M. Fowkes and M.A. Mostafa, *Ind. Eng. Chem.*, Prod. R&D. **17**, p. 3 (1978); F.M. Fowkes, Adhesion and Adsorption of Polymers, A. pp. 43-52 (Plenum, 1980) ed. L.H. Lee.

The States of Water in Polymer Films

G.E. Johnson, H.E. Bair, and E.W. Anderson*

Introduction

Recently we reported a way to measure the amount of water which has associated to form microscopic water-filled cavities (clusters) in polyethylene at a level of 10 ppm and greater.[1,2] By combining this calorimetric technique with a coulometric method, it was possible to differentiate between clustered water and the total water sorbed by the polyethylene. It was found that clusters are formed when polyethylene is saturated with water at an elevated temperature and is rapidly cooled to room temperature. During cooling, the solubility of water in polymers is lowered and some water condenses in the form of microscopic water filled cavities, provided the internal pressure (which is generated by the excess water) exceeds the strength of the polymer. Figure 1 shows 2-micron clusters, formed in polyethylene, quenched from the melt in the presence of water.

In sorption studies of polycarbonate,[3] it was learned that this polymer absorbs water in two stages. In the initial period of absorption at an elevated temperature, but below T_g, all of the water was identified in a bound state when cooled to room temperature. In the second stage, at later times, most of the water gained by the polymer was identified in a separate liquid phase (clustered water). In addition, after the polymer was saturated with water at a temperature above T_g and cooled, its solubility was lowered, and water condensed in the form of microscopic water filled cavities. Below T_g, the clusters were formed only after the polycarbonate's strength ($M_w = 26,600$) was decreased by hydrolysis, whereas above T_g clusters were formed without degradation.

Early investigations of the effect of water on the low-temperature relaxations of several aromatic polymers including polycarbonate, polyamides, and a polyurethane have shown several low temperature anomalies.[4,5] In the case of a water-saturated polysulfone polymer which exhibited a doublet in its β-loss process, Jackson suggested that the secondary peak may be due to water filled cavities. We have employed the DSC technique for water cluster analysis, along with the total water content measurements, to elucidate the water sorption behavior of polysulfone and poly(vinyl acetate) (PVAc).[6] In addition, DSC and dielectric methods were used cooperatively to understand the T_g behavior of polymers in the presence of water.

Experimental

Dielectric Measurements

Dielectric measurements[7] on polyethylene in the MHz region used an HP-4243A Q-meter and a General Radio 1690 micrometer-electrode cell. Also used at 200 MHz, was a re-entrant cavity. At room temperature, liquid displacement measurements were made at 30 MHz using a Boonton Electronics Admittance Bridge 33B. When feasible, measuring setups were calibrated with a conductance standard to obtain absoulute accuracy.

Dielectric measurements on polyethylene, polycarbonate, and poly-sulfone were conducted at 10^2, 10^3, and 10^4 Hz. The data were obtained by combining a Princeton Applied Research 124 lock-in amplifier and

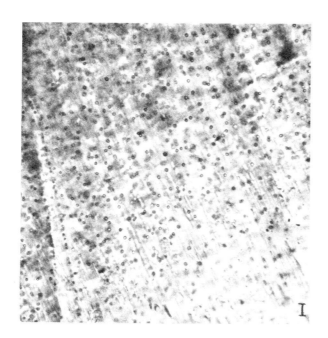

FIGURE 1 — 2 micron water clusters in polyethylene.

*Bell Laboratories, Murray Hill, New Jersey

a General Radio 1615A capacitance bridge. The bridge was connected to a Balsbaugh LD3 research cell inside a test chamber. After the test chamber was equilibrated at −160 C for one hour, measurements were made at ten to twenty degree intervals, with a fifteen minute waiting period between each discrete change in temperature, until room temperature was reached.

Dielectric measurements on poly(vinyl acetate) were obtained using a Fourier transform dielectric spectrometer developed in our laboratory.[8] A voltage step pulse was applied to the sample, and the time dependent integrated current response, Q(t), was collected by computer. The frequency dependent dielectric properties, ε' and ε'' were then obtained from the Fourier transform of the integrated current. For this study, frequency dependent dielectric data in the 10^0 and 10^4 Hz range were obtained isothermally. A sample cell with low thermal mass and copper screening allowed rapid equilibration between temperatures. The frequency dependence at several temperatures was measured while minimal water was lost from the samples.

Total Water Measurements

The duPont Moisture Analyzer (26-321A) was used to determine the total amount of water in a sample. This instrument used a coulometric technique to measure water quantitatively. In order to analyze samples with 50 ppm or less, we enclosed our instrument in a glove bag or a dry box which was continuously flushed with dry nitrogen. The samples were heated for 15 minutes above T_g to drive off the water. The water vapor then passed through an electrolytic cell composed of phosphorus pentoxide-coated platinum wires. The cell absorbed water vapor and quantitatively electrolyzed it.

Clustered Water Measurement

The determination of clustered water was done calorimetrically, using a Perkin-Elmer Differential Scanning Calorimeter (DSC-2). Samples were placed in a nitrogen-flushed dry box before they entered the DSC sample holder. Most experimental runs were made at 20 C/minute. The sample was placed in the calorimeter and cooled to −55 C. Subsequently, the sample was heated to 30 C, and a first order transition detected at 0 C with a heat of transition, ΔH_{tr}. We attribute the 0 C transition to the melting of ice crystals which formed from the clustered water in the polymer sample.

If the concentration of clustered water in ppm is represented by C, then:

$$C = \frac{\Delta H_{tr}}{M \Delta H_f} \times 10^6 \qquad (1)$$

where M is the weight of the sample in milligrams and ΔH_f is the heat of fusion of water, 79.7 cal/g.[9]

Results and Discussion

Polyethylene

The nature of the water in polyethylene was studied both dielectrically and calorimetrically. When studies of the saturation content of water in polyethylene were undertaken, the water was found to exist in two states in samples which had been saturated at an elevated temperature and then quenched in 23 C water. Figure 2 shows the saturation water content

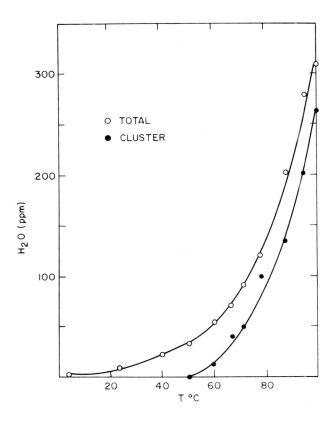

FIGURE 2 — Saturation water content and cluster formation in SG polyethylene.

from 4 to 100 C and the amount of that water found in the clustered state upon quenching. When clustered water was detected, a Maxwell-Wagner loss peak (Figure 3) was found dielectrically. This dielectric loss peak had a room temperature maximum frequency of a few MHz, comparing favorably with literature data[10] on the effect of water spheres.

Though the clustered water had been seen calorimetrically and dielectrically, initial attempts to find it microscopically failed. A cooperative effort with Phelps Dodge using high pressure steam and 23 C water quenching, yielded samples with up to 6000 ppm of water, which was detectable microscopically as the 2 micron spheres of Figure 1. Coulometric (DuPont moisture analyzer) and calorimetric results showed all water within experimental error was in the clustered state.

To look at the water in the clustered state dielectrically in various polyethylenes, a loss measuring

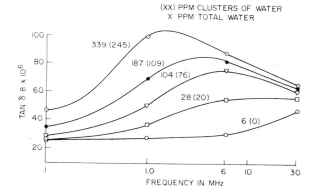

FIGURE 3 — Clustered water's effect on the dielectric loss of SG polyethylene.

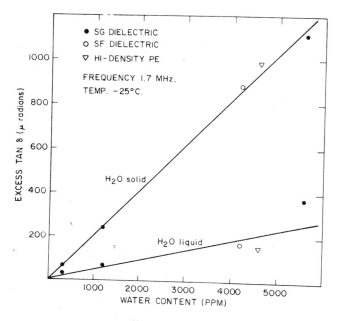

FIGURE 4 — Excess dielectric loss as a function of water content for polyethylenes.

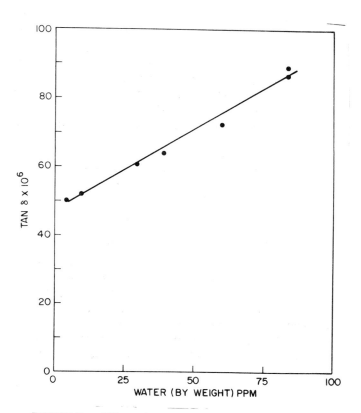

FIGURE 5 — Effect of bound water on 30M Hz loss of polyethylene.

setup[7] was maintained at −25 C and 1.7 MHz. The samples containing up to 6000 ppm water were quenched in liquid nitrogen and then placed in the −25 C measuring setup. After the tanδ was measured, the sample was warmed to about room temperature and remeasured at −25 C. The data of Figure 4 show no dependence on the excess dielectric loss at −25 C due to the different polyethylenes, but a dramatic effect due to the difference in thermal history. The water in clusters at room temperature must be brought significantly below −25 C to change the liquid to solid within the time scale of these experiments. Thus, by using different thermal histories both solid and liquid water are observed at −25 C.

When less than 100 ppm of water were present in polyethylene, a different dielectric behavior was observed. It was noted that the plaques showing 25 microradians excess loss also showed an initial water content of 60 ppm. A linear relationship, between water content up to 80 ppm and 30 MHz dielectric loss, was established, and the slope showed an increase of 1 microradian for 2.5 ppm additional water (Figure 5). The water which showed this behavior was found to be in a nonfreezable state (bound). Dielectric measurements on these samples between 6 and 200 MHz showed proportional loss increases due to water. These measurements showed that the excess loss could be associated with the polyethylene's γ mechanism, since the increased loss appears to be an enhancement of the γ mechanism. The excess loss is then based on the correlated motion of the water dipoles according to the dynamics of the polyethylene molecules. Work by S. Matsuoka, et al,[11] shows that each mole percent chlorine in chlorinated polyethylene contributes 2500 microradians to the polymer's dielectric loss at 30 MHz and 23 C. If the water dipole, which has approximately the same dipole moment as that of the C-Cl (2 Debye), acted in a manner similar to chlorine, then 3 ppm of water would give a 1 microradian increase of 30 MHz. This value is close to the 2.5 ppm per microradian experimentally observed. Stein's dielectric studies[12] on oxidized polyethylene are in close agreement with chlorinated polyethylene when carbonyl groups are considered.

Polycarbonate

Molded polycarbonate bars were immersed in water at a constant temperature until they reached equilibrium. A typical plot of polycarbonate's water content, as a function of time at 97 C, is shown in Figure 6. Polycarbonate became saturated rapidly with about 0.62 Wt% water. Then the total water con-

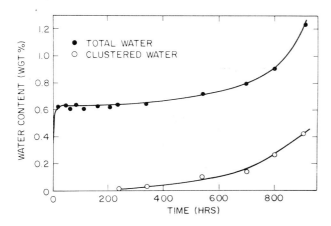

FIGURE 6 — Time dependent water sorption of polycarbonate at 97 C.

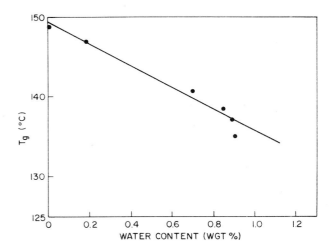

FIGURE 7 — Effect of water on polycarbonate's glass transition.

tent, as measured by the Coulometric method, remained constant for about 200 hours. However, after 240 hours, the polycarbonate sample began to absorb more water. This unexpected gain in total water content was accompanied by the onset of haziness in the previously clear bar. This second state of water sorption, which began after the sample had been at apparent equilibrium, caused a doubling of the sample's total water content to 1.2 Wt% after 900 hours at 97 C.

Calorimetric examination of the polycarbonate samples which had spent 240 hours or more in the 97 C water bath revealed these specimens had water clusters present. The quantity of clustered water was observed to increase with increasing time in the water bath (open circles, Figure 6). Within experimental error, all of the additional water gained after 240 hours was in the form of clusters until 900 hours. This increase in the level of bound water after 900 hours at 97 C is due to an increased solubility in the hydrolyzing polycarbonate.

Samples were also prepared by placing them in an autoclave at 125 C. With the larger amounts of bound water (up to 1%), it was possible to follow by DSC the T_g dependence on water as shown in Figure 7. The bound water plasticized the glass transition.

Since a spectrum of water contents and types could be produced, dielectric measurements were undertaken as a function of water content to further increase our understanding of water-polymer interactions. The dry sample showed a broad absorption region amplitude at 1 kHz between -140 and 20 C; the maxima in tanδ occurred near -50 C. The sample of intermediate wetness (0.35 Wt%) yielded a stronger absorption at low temperatures, with the maxima at -50 C. The sample with 0.65 Wt% water had the strongest absorption, with the maxima still at -50 C. The increase in the area under the loss curve of the samples, which contained 0.35 and 0.65 Wt% water, was proportional to the increase in the amount of water in the two samples. This dielectric loss increase of the polymer's β mechanism can be explained based on the correlated motion of the water dipoles according to the dynamics of the polymer molecules. A

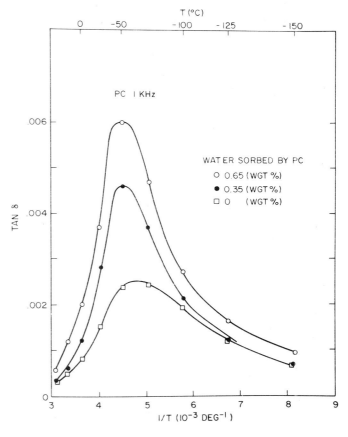

FIGURE 8 — Effect of bound water on polycarbonate's secondary dielectric loss transition (β).

calculation of the added dipolar contribution of the water was made using the curves of Figure 8.

$$\Delta\varepsilon = \frac{2\Delta H}{\pi R} \int_0^\infty \varepsilon'' d\frac{1}{T} \qquad (2)$$

$$N\mu^2 = \frac{3KT}{4\pi} \frac{(2\varepsilon_R + \varepsilon_u)}{3\varepsilon_R} \left[\frac{3}{\varepsilon_u + 2}\right] \Delta\varepsilon \qquad (3)$$

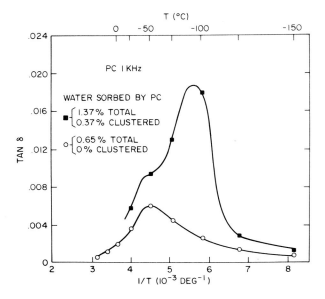

FIGURE 9 — Effect of clustered water on polycarbonate's low temperature dielectric loss.

The value of $N\mu^2$ [(dipoles/unit volume) × (dipole moment)^2] was one third the value that would be obtained if the water dipoles participated completely.

The effect of clustered water on the low temperature loss behavior of polycarbonate is shown in Figure 9. This polycarbonate plaque contained 0.37 Wt% clustered water and a total water content of 1.37%. Note that the amount of bound water has increased to 1.00%. There was a doublet which appeared in the loss of the polycarbonate at low temperatures. A narrow peak occurred at −90 C with a shoulder at about −50 C. The β mechanism of the polymer, enhanced by the 1% bound water, accounted for the shoulder at −50 C. The loss mechanism at −90 C was assigned qualitatively to the clustered water present as ice crystals. However, this peak was not found to be directly proportional to the amount of clustered water. This was based upon the dielectric data from two polycarbonate samples which contained 0.13 and 0.37 Wt% clustered water.

Polysulfone

Compression molded samples of polysulfone were immersed in water at 100 C and below until they came to equilibrium. Above 100 C, an autoclave was used. As with polycarbonate, a mild temperature dependence in equilibrium absorption was noted. The amount of nonfreezable water at saturation went from 0.8% at 23 C to 1.2% at 132 C. No clustered water appeared in samples exposed below T_g (190 C). When the polymer was exposed to steam at 208 C for 1 minute, and quenched to 23 C, cluster formation was noted. There was, however, only 0.04% water found in the clustered state. Five minute exposure to steam increased the clustered water content to 0.16%. Microscopic analysis of cross sections of dielectric specimens showed a nonuniformity of cluster size and distribution. Two DSC peaks (−34 C and −42 C) were observed on cooling the sample at 20 C/minute

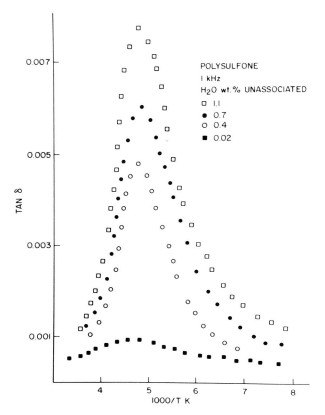

FIGURE 10 — Effect of bound water on polysulfone's β transition.

and a broadened melting peak starting at 0 C on heating at the same rate. The dielectric loss behavior of polysulfone samples was measured below 23 C as a function of bound water content. Figure 10 shows such data taken at 1 kHz. The activation energy of the process is 11.4 kcal/mole (1.1% H_2O) from the plot of the logarithm of frequency of dielectric loss maxima vs reciprocal absolute temperature of Figure 11. This

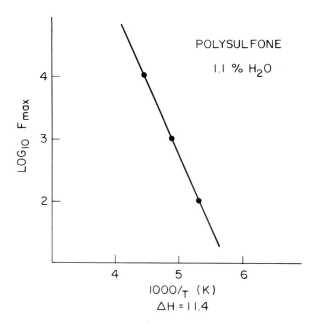

FIGURE 11 — Log dielectric loss maxima vs 1/T for polysulfone's β transition (1.1% bound water).

activation energy is in agreement with Allen's[13] prior determination. The areas under the loss curves of Figure 10 are directly proportional to the amount of bound water present. As with polycarbonate, this dielectric loss increase of the polymer's β mechanism can be explained based on the correlated motion of the water dipoles according to the dynamics of the polymer molecules. A calculation of the added dipolar contribution of the water was made and found to be one quarter the value that would be obtained if the water dipoles participated completely.

When clustered water was present, an additional low temperature dielectric mechanism was inferred in the kHz region. It appeared as a low temperature broadening of the dielectric loss dispersion with 0.04% clustered water and increased with additional clusters (0.16%). This is the same phenomenon observed in the polycarbonate studies as a separate loss peak.

Poly(vinyl acetate)

Compression molded samples of poly(vinyl acetate) also showed a mild temperature dependence in equilibrium absorption. The amount of water went from 4% at 23 C to 6% at 70 C. This polymer was the only one tested that formed clustered water while stored isothermally at room temperature. This clustering was obtained after 17 hours as confirmed by DSC and could be seen visually as a whitening of the polymer.

The DSC cooling curve was obtained for a sample containing 6.3% total water, 2.1% of which was clustered. At a cooling rate of 20 C/minute, the vitrification of the polymer occurred between 20 and 5 C. The next thermal event observed was the onset of freezing of clustered water at -5 C. The crystallization process proceeded sporadically until -35 C. At that temperature, the major portion of the clustered water began to freeze and this process was completed by -38 C. The crystallization at smaller undercooling was believed to be due to freezing of large droplets ($\gg$ 10 microns).

The comparative C_p curves vs temperature of PVAc containing 0.2, 1.8, 4.2, and 4.6 Wt% water are plotted in Figure 12. These C_p curves, in all cases, were run after cooling each sample to -70 C. The lowest curve represents a film that had been vacuum dried overnight at 41 C. The C_p increased linearly with temperature until the glass transition, which was noted as a discontinuity in C_p between 35 and 45 C. T_g, which is defined in this work as the midpoint of the transition, was 43 C. The C_p curve for a PVAc film containing 1.8% water (all unclustered) has the T_g shifted 13 C lower to 30 C. Increasing the level to 4.2% unclustered (bound) water lowered T_g to 19 C. In all of the above cases, the transition width of T_g was the same, 12 C.

The upper curve in Figure 12 represented 4.6% total water, with 0.7% of that in the clustered state. The clustered water melted near 0 C and was followed by a T_g at 19 C as in the 4.2% bound water sample. Thus, clustered water has no plasticizing effect on T_g.

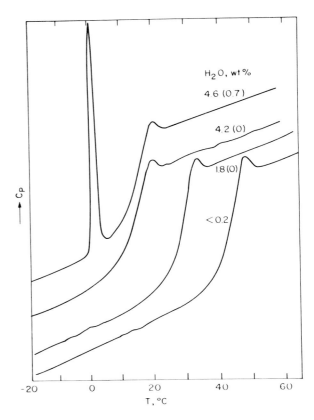

FIGURE 12 — Calorimetric DSC scans of poly(vinyl acetate) as a function of water content.

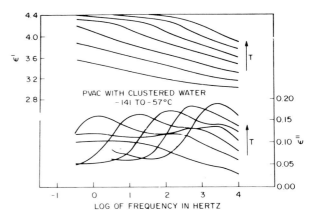

FIGURE 13 — Low temperature dielectric loss curves of poly(vinyl acetate) with clustered water.

The dielectric loss behavior of PVAc was similar to that of the other polymers. An increase in dielectric intensity of the polymer's β mechanism was directly proportional to the amount of unclustered water. In addition when clustered water was present, two separate low temperature peaks occurred as shown in the frequency dependent data of Figure 13. The higher frequency peaks were the result of clustered water. The presence of clustered water is confirmed by the similarity between poly(vinyl acetate) and the clustered water peaks of other polymers as plotted in Figure 14.

With the sample cell, which had low thermal mass, it was possible to measure dielectrically the α transi-

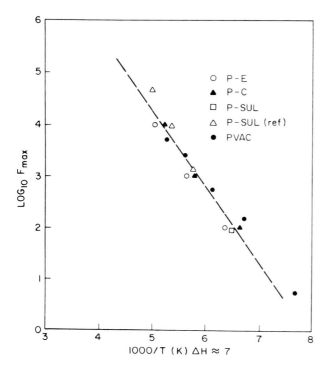

FIGURE 14 — Log dielectric loss maximum vs 1/T for frozen water peak in various host polymers.

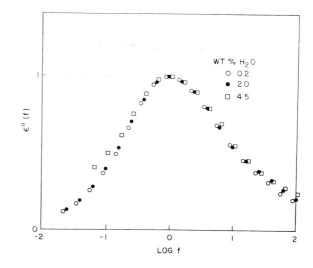

FIGURE 15 — Normalized poly(vinyl acetate) glass (α) transitions.

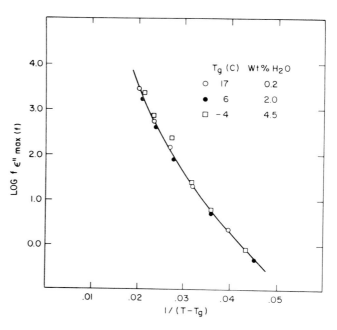

FIGURE 16 — Log dielectric maximum for poly(vinyl acetate) as a function of $1/(T - T_g)$.

tion behavior of poly(vinyl acetate) while retaining water within the sample. Figure 15 shows the normalized loss for the α transition for 0.2, 2.0, and 4.5% unclustered water. The shape of the transition was independent of water content. This behavior was in agreement with the DSC curves on similar samples, which also showed no difference in T_g transition breadth as a function of bound water.

The plot of frequency of ε'' maximum vs reciprocal temperature for the α transition showed typical WLF behavior. All these data were fit to a single WLF equation, with the fractional free volume at T_g equal to 0.0225 and the expansion coefficient of free volume above T_g equal to 5.3×10^{-4}

$$\mathrm{LOG} f_{\varepsilon''\max(f)} = \frac{19.6(T - T_g)}{42 + (T - T_g)} \quad (4)$$

The T_g's of 17, 6, and -4 C for 0.2, 2.0, and 4.5% water in poly(vinyl acetate) were found dielectrically. The reductions of the dry polymer's T_g, 11 and 21 C, were similar to the 13 and 24 C shifts noted by DSC. When all the dielectric maxima were plotted vs $(T - T_g)^{-1}$ in Figure 16, an excellent fit to the above WLF equation was found. The additional fractional free volume obtained when 1% water was absorbed was 0.003. This value is in agreement with earlier polycarbonate data.

Conclusion

Water absorbed in a polymer can exist in a bound state or as a separate phase (cluster). In this investigation the DSC technique of water cluster analysis was used in conjunction with coulometric water content measurements to characterize the water sorption behavior of polymers. All polymers except PVAC had to be saturated at elevated temperature to induce cluster formation. This cluster formation occurs after significant weakening of the polymers. Temperature above T_g is normally a necessary condition with all amorphous polymers, but the plasticizing effect of water lowers the polymer's T_g. The effect of chemical degradation in polycarbonate is an exception.

All amorphous polymers showed an enhancement of their low temperature β-loss transitions in proportion to the amount of bound water present. In addition polyethylene, a polycrystalline polymer, showed similar behavior for its γ transition. Frozen clustered water produced an additional low temperature dielectric loss maximum in PVAc and polysulfone, also common to polyethylene and polycarbonate. Dielectric data obtained on a thin film of water between polyethylene sheets was in quantitative agreement with the clustered water data.

The dielectric data on poly(vinyl acetate)'s α transition showed a good WLF fit, with a shift in T_g occurring with increasing unassociated water content, but no change in the shape of the loss peak. This was in agreement with DSC data on the polymer. Both DSC and dielectric data showed clustered water had no effect on T_g. This plasticization of the polymer by water was also found in polycarbonate. An explanation in terms of added fractional free volume is consistent with the data.

References

1. H.E. Bair and G.E. Johnson, *Analytical Calorimetry*, **4**, R.S. Porter and J.F. Johnson, Ed., Plenum Press, N.Y., p. 219 (1977).
2. G.E. Johnson, H.E. Bair, E.W. Anderson, and J.H. Daane, 1976 Annual Report Conference on Electrical Insulation and Dielectric Phenomena, p. 510 (1976).
3. H.E. Bair, G.E. Johnson, and R. Merriweather, *J. Appl. Phys.* **49** (10), p. 4976 (1978).
4. G. Allen, J. McAinslie, and G.M. Jeffs, *Polymer*, London, **12**, p. 85 (1971).
5. J.B. Johnson, *Polymer*, London, **10**, p. 159 (1969).
6. G.E. Johnson, H.E. Bair, S. Matsuoka, E.W. Anderson, and J.E. Scott.
7. Link, G.L. and G.E. Johnson, Annual Report, Conference on Electric Insulation and Dielectric Phenomena (1972).
8. G.E. Johnson, E.W. Anderson, G.L. Link, and D.W. McCall, *Am. Chem. Soc., Preprints, Div. Org. Coat. Plastics Chem.*, **35** (1), p. 404 (1975).
9. Handbook of Chem. and Phys., Ed. R.I. Weast, 54th Edition (1973-1974) CRC Press B-244.
10. *Progress in Dielectrics*, **7**, edited by J.B. Birks, CRC Press, Cleveland, Ohio (1967).
11. Matsuoka, S., et al., Dielectric Properties of Polymers, edited by F.E. Karasz, New York (1972).
12. Stein, R.S. et al., Anisotropy of the Dielectric Relaxation of a Crystalline Polymer, *J. Polymer Sci.*, Part A-2, p. 1593 (1972).
13. G. Allen, J. McAinslie, and G.M. Jeffs, *Polymer*, London, **12**, p. 85 (1971).

The Inhomogeneous Nature of Polymer Films and Its Effect on Resistance Inhibition

D.J. Mills* and J.E.O. Mayne**

Introduction

In previous studies[1-3] of the ionic resistance characteristics of detached varnish films, two modes of conduction have been observed in 100 mm² areas of film. In I type films, conduction ran counter to that of the external solution, and in D type films conduction followed that of the solution in which they were immersed. This method of characterizing films, although important, meant that the relevance of D and I type films to protection was not immediately apparent. Also, the size of D areas and the cause of their appearance has not been elucidated. The present work shows that D type films always have a resistance which is three or four orders of magnitude less than I type films in solutions of chloride (0.5 - 3M). The relevance of D areas to protection has been established recently.[2] Corrosion of the mild steel substrate coated with a varnish film occurred at D areas in the film.

Experimental

Preparation of Polymer Films

For much of the work described here, the three varnishes studied previously were used. These were a pentaerythritol alkyd, a phenol formaldehyde tung oil, and an epoxy polyamide. Details have been published.[1]

Films were cast by means of a spreader bar on glass plates, dried at room temperature in a glove box with a fan for two days, and then for a similar period in an oven at 65 C. One of two spreader bars were used: the first which had a gap of 100 μm produced films of 30 to 40 μm thickness (thin films); the second with a 250 μm gap produced films 75 to 90 μm thick (thick films). The dried films were then soaked in water, carefully removed from the glass plate and cut into pieces of area slightly greater than 100 mm². Any piece showing a visible defect was discarded. The pieces were mounted in cells shown in Figure 1 which exposed 100 mm² of film to the solution, and the two

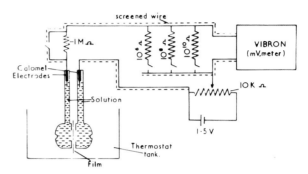

Method used for D.C. resistance measurement of detached Films

FIGURE 1

arms of the cell were filled with potassium chloride solution (0.001M or 3.5M). The resistance of the film was measured by immersing the cell in a water thermostat at 25 C, applying a potential difference of 1 volt across the film in series with a known resistance, and recording the potential drop across the known resistance, (Figure 1).

Resistance Characteristics of Thin and Thick Films

The resistance behavior of films always fell into one of two categories. Either the resistance was high (10^{10} to 10^{12} ohms/100 mm²) and rose slightly when transferred from 0.001M to 3.5M solution (I Type), or the resistance was much lower (10^6 to 10^8 ohms in 3.5M KCl), and fell about two orders of magnitude when transferred from 0.001M to 3.5M KCl solution (D type). Thin films were left half an hour in solution; however, thick films were left 3 to 4 hours before resistance measurements were made to ensure equilibrium had been achieved. Typical results for the effect of solution concentration on resistance are shown in Figure 2. The number of 100 mm² areas that give D type behavior, divided by the total number examined, which was normally 20, gave the figure for the %D type for the film.

Table 1 gives the value of %D Type, Log R_I and Log R_D for thin and thick films formed from the three vehicles. The figures in brackets are the standard

---

*Department of Mechanical and Production Engineering, Trent Polytechnic, Nottingham, England.
**University of Cambridge, Department of Metallurgy and Materials Science, Cambridge, England.

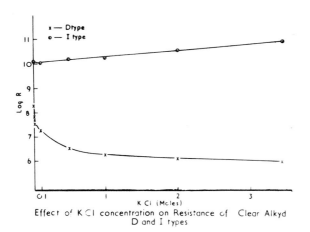

FIGURE 2

Effect of KCl concentration on Resistance of Clear Alkyd D and I types

deviations. Table 2 shows the results for two coat films in which two sets of samples have been taken from a single cast. These were prepared by allowing one thin coat to dry in the normal way, and casting a second coat on top, using the 100 μm spreader bar and allowing that to dry. Two coat films thus had a lower coat of 30 to 40 μm thickness and an upper coat of 25 to 35 μm.

Reproducibility of %D Type

The reproducibility of the %D type measurement has been investigated. Duplicate results of %D type were obtained from a second sample of twenty pieces from the same cast, and also from different casts (Table 3).

Effect of Stoving, Stripping and Substrate on %D Type

Experiments have also been conducted which show that the process of heating the films to 65 C for 2 to 4 days and to 110 C for a short period, in order to melt the picene wax, did not affect the results. Furthermore, the nature of the substrate (that is, glass, mild steel, platinum or tin foil), the methods of stripping the films, (that is, removal from glass after soaking in water or casting on tin foil, which was removed by means of mercury), and the methods of preparation of the films, (that is, casting, dipping or brushing), did not affect the value of the percentage of D type films. Thus, the values of %D type were within 15 to 20% of each other for a particular film at a particular thickness.

Size of D Areas

From Table 1 it can be seen that I type films, in all three systems, exhibit a high resistance in 3.5M KCl.

TABLE 1 — Values of Resistance and %D Type for Different Varnishes Showing Effect of Thickening Film

Varnish	Thickness μm	%D Type	Log R_I (3.5M KCl)	Log R_D (3.5M KCl)
Alkyd	40 to 45	54%	11.3 (±0.15)	6.7 (±0.7)
Tung Oil	35 to 40	72%	12.3 (±0.25)	6.8 (±0.7)
Epoxy	35 to 40	80%	11.3 (±0.2)	6.8 (±0.7)
Alkyd	75	8%	12.0 (±0.2)	—
Tung Oil	85	25%	12.6 (±0.3)	7.5 (±0.5)
Epoxy	75 to 80	50%	11.9 (±0.2)	7.5 (±0.7)

TABLE 2 — Resistance Characteristics of Two Coat Films

Varnish	Thickness (μm)	Log R_I	%D Type Sample 1	%D Type Sample 2
Alkyd	60 to 65	11.5 (±0.2)	5	5
Tung Oil	60 to 65	n.m	5	0
Epoxy	70 to 75	12.0 (±0.2)	0	5

TABLE 3 — Repeat Measurements of %D Type

Varnish	Thickness (μm)	Cast 1 Sample 1	Cast 1 Sample 2	Cast 2 Sample 1	Cast 2 Sample 2
Alkyd	40 to 45	40	50	55	60
Epoxy	35 to 40	80	65	85	77
Alkyd	75	8	15	—	—
Epoxy	75 to 80	50	40	38	45

This value is reproducible for any one system, but depends on the thickness. D type films of 100 mm² have a much lower resistance (10^6 to 10^8 ohms in 3.5M KCl) and the individual values show greater variation.

The effect of thickening the film is to reduce dramatically the number of D type films observed (by a factor of 3 or 4 in the alkyd and tung oil systems, and by a factor of 2 in the epoxy system). If D areas are produced isotropically, then the fact that there is a large reduction in the number observed suggests that D areas are small. The diameter of some of them may be of the order of the thickness of the film (75 μm).

If D areas occupy only a fraction of a D Type film, then a model of D areas as small (75 to 250 μm diameter), isolated and randomly distributed, can be put forward. A statistical equation for use where an event either occurs (probability p), or does not occur (probability q = 1 – p) can give an idea of the accuracy or reproducibility of a %D type measurement. The standard error (S.E.) equals $\sqrt{pq}/n$, where n is the number of samples. A %D type result of 50% will come from a population whose actual value lies somewhere between 35 and 65%. Error bars of ± 11% can be assigned to %D type figures. Thus, two different values of %D Type must differ by 20% or more to be significant.

These comments serve as the basis for expecting some scatter in the repeat measurements in Table 3. However, the relative reproducibility of the %D Type indicates that this parameter is a characteristic of the resin system and is not influenced by external variables. The ± 11% criteria was also used to show that the alteration of the various parameters in the effect of stoving, stripping, and substrate on %D Type section did not significantly change the value of %D Type.

This model also means that if a varnish film has a high %D Type then many of the individual 100 mm² areas will contain more than one D area. In a polymer film, with a %D Type of 50% or less, most of the D type films will contain only one small D area within them. However, in a film with a %D type of 80% or more, many D type films will contain two, three, four or even more D areas within them. The number will follow a Poisson distribution.

The results in Table 2 also support the theory that D areas are small. The chance of one D area overlapping another D area is low; hence, two coat films in all three systems have none, or only one D type film in a sample of twenty. Thus, multi coat films are more effective in improving the resistance character than a single thick film of the same equivalent thickness.

Other evidence for this model of D areas as small and randomly distributed is provided by the experiments where breakdown, that is, corrosion of the mild steel, under a varnish film occurred in small isolated spots.[2] A diagram from that paper is reproduced in Figure 3. The small area (10 mm²) resistance and microhardness measurements by Scantlebury[3] also provide evidence in support of the model. This work

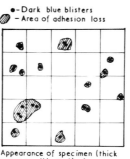

FIGURE 3

led to the suggestion that D areas were regions of reduced crosslinking in the film. Morris and Longhurst[4] investigated the microhardness of an epoxy/melamine film. They observed a bimodal distribution of hardness, and from the line scan reproduced in their paper the areas of reduced hardness appear to be about 250 μm in diameter.

Real Resistance of D Areas

The real resistance of D areas in 3.5M KCl for films 35 to 45 μm in thickness may well be less than the 10^6 to 10^8 ohms/100 mm² observed for D type films. The large scatter in the resistance of D areas presumably relates to the size of the D area (and possibly how open the structure is). If we assume just the former and take two D areas one of 75 μm diameter and one of 250 μm diameter then a comparison of resistance values can be made. This is done in Table 4. The electrolyte is taken to be 3.5M KCl (specific resistance 50 ohm-mm).

The results in Table 4 show that D areas could have an even lower ionic resistance with respect to the bulk of the film (that is, I areas) than was suspected previously. Their "real" resistance is only 1000 to 10,000 ohms/100 mm². Taking a typical value of I type resistance, assumed uniform as 10^{11} ohms/100 mm², then the ratio I area resistance to D area resistance is about 10^7 to 10^8 for a 50 μm film. This helps to explain how locally sufficient chloride penetrates these films in a few days to cause the breakdown shown in Figure 3. One can also calculate that the "real" resistance of D areas is much higher (40,000 X) than if they were pores of the same assumed area filled with 3.5M KCl solution. The "pore" diameter would need to be small (0.4 μm or 1.3 μm) to give the observed D type resistance, that is, 2×10^7 ohms, or 2×10^6 ohms, respectively.[3]

Effect of Temperature on Resistance

These results were all obtained at 25 C. The effect of temperature over the range 0 to 50 C was investigated, and the result for an I Type tung oil film is shown in Figure 4. This figure is in the form suggested by Arrhenius, where Log R_I has been plotted against

TABLE 4 — Comparison of Resistance of D Type Film, and D Area

Assumed Diam. of D area (μm)	Area of D area (mm²)	100 mm² D type film Typical Res. (ohms)	Resistance of D area per 'real' 100 mm² (ohms)
75	0.5×10^{-2}	2×10^7	1000
250	0.5×10^{-1}	2×10^6	1000

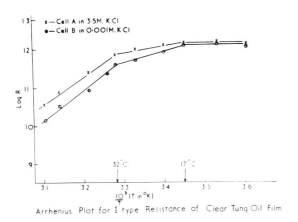

FIGURE 4

Arrhenius Plot for I type Resistance of Clear Tung Oil Film

the inverse of the absolute temperature. This plot is typical of all I types films examined, that is, thin and thick in all three systems. I type conduction is highly activated, at temperatures above about 30 C for tung oil films, Q = 160 kJ/mole, but it is less activated at lower temperatures. Sharp changes in slope occur, and these have been shown to correspond to structural transitions in the polymer.[5]

The effect of temperature on D type films was also examined over the 0 to 50 C range. D type conduction was much less activated at all temperatures, Q = 25 to 40 kJ/mole. This value is only about 1½ times the value of Q for KCl alone. For both D and I type films, the curves obtained in 0.001M and 3.5M KCl were similar. D type conduction was also not significantly affected by the structural transitions occurring in the polymer, that is, there were no sharp changes in slope.

I type films retained a high resistance (>10^8 ohms) up to 50 C. D type films had a low value (<10^8 ohms in 3.5M KCl) over the whole range investigated.

Causes of D Areas

The evidence is against D areas being pores in the polymer. No visible defects can be seen when a dry D type film is examined, using either optical or electron microscopy. If the D areas were acting as porous systems, then the Donnan equilibrium would be obeyed. When gases are diffused through these polymer films, the Donnan equilibrium is not obeyed. The results indicate that the D areas represent areas with a different structure, probably having considerably reduced crosslinking compared with the rest of the film. However, D areas are much smaller than 100 mm². The term "virtual pores" has been used to describe the D type conduction occurring after an I type film has undergone ion exchange.[6] The authors now prefer to use the phrase "conducting polymer phase" to describe them. This idea of a separate phase in certain areas of crosslinked polymer was suggested by Goldring[9] from results on ion exchange resins. It seems reasonable, because the properties are different locally; e.g., high water uptake, low resistance, higher capacitance, etc. Kendig and Leidheiser[10] used a model of a polymer, as heterogeneously penetrated by a conducting phase, to explain their impedance measurements on polybutadiene coated steel immersed in 0.52N NaCl. The fact that D areas are small means that attempts to see differences in bulk properties between D and I films will be difficult.

The following theory as to the cause of D areas in these polymer films is advanced tentatively. It was observed that when resins gelled (that is, became viscous or end of batch) then 100% D type films resulted. It is, thus, possible that D areas are caused by innate material in the resin. All these vehicles have been partially polymerized, and some large (molecular Wt. of 10^6 or more) gel like or "dead" molecules may be present in the resin before casting. During drying, these molecules congregate, form micelles, and due to low functionality, fail to crosslink to anything like the same extent as the rest of the film. These areas may also trap low molecular weight material which would normally escape (solvent, etc.) before the crosslinking stage. Thus, material that started in the can with more crosslinks, ends up with less crosslinks in the dried film. From the observed %D types, it appears that there is rather more of this "dead" material in the epoxy (assumed to be in the epikote) than there is in the tung oil, which has a little more than the alkyd.

To test this theory, some alkyd and some epoxy resin was centrifuged at 25,000 g for several hours, and the %D type from a fraction taken from the top compared with the %D type from a fraction taken from the bottom. The fraction taken from the top had, in both cases, two or three fewer D type films in a sample of twenty, than the fraction from the bottom. Because of the already discussed statistical nature of a %D type measurement, these results cannot be described as highly significant. However, they are encouraging, and suggest that it may be possible to reduce the number of D types by separating out and removing the higher molecular weight material in the resin. Solvent precipitation might also be tried.

TABLE 5 — Resistance Characteristics of other Resin Systems

Varnish	% D type	Log R_I (3.5M KCl)	Log R_D (3.5M KCl)
Urethane Alkyd	57%	>10.0	5.8 (± 1.2)
Polyurethane	78%	>10.0	6.2 (± 0.8)
Chlorinated Rubber	100%	—	5.3 (± 0.8)

All films were thin (35 to 50 µm)

D Areas in Other Systems

Since this work was completed, the resistance characteristics of three other varnish systems has been investigated by Heyes.[7] These were a urethane alkyd (Beckurane 79—60) as a 60% solution in white spirit with 0.05% cobalt in the naphthenate form as driers), a polyurethane (formed by mixing Beckurane 2-253 as a 62% solution in a 1:1 mix in xylene/cellulose acetate with Beckosol 3020 as a 51% solution in xylene in the weight ratio 10 to 22.2) and a chlorinated rubber (based on Alloprene R 20 mixed with Ceredor plastisizer in the ratio 2:1 as a 50% solution in Aromasol H). Detached films were prepared, and resistance measurements were obtained in the same way as the alkyd, tung oil and epoxy films. The results given in Table 5 were obtained for the %D type, Log R_I and Log R_D.

Thus, typical D and I type films were observed in the urethane alkyd and the polyurethane systems. The polyurethane had a similar %D type to an epoxy film. The urethane alkyd was similar to a tung oil film, although the Log R_D was slightly lower. The chlorinated rubber was 100% D type with a low resistance (average about 200,000 ohms/apparent 100 mm²).

Effect of Pigments

Results for the effect of pigments will be given in more detail in a later paper. However, a brief summary will be given here. The effect of incorporating the pigments iron oxide, zinc oxide, and red lead into the three vehicles alkyd, tung oil, and epoxy showed that D and I type films were still observed in all three systems at all PVCs (pigment volume concentrations) up to the critical PVC (about 40%), when pores appeared in the film and the films became 100% D type with low (<10^5 ohms/100 mm²) resistance. Generally, the average resistance of D and I type films was not changed significantly compared with the unpigmented vehicle, except in the case of ZnO in alkyd, when the I type resistance dropped with increased PVC. Also, highly pigmented ZnO in alkyd films were observed to become 100% D type after a day or so in the solution. This change was attributed to soluble zinc soaps forming and drawing more water into the film. These three pigments are virtually insoluble in water themselves.

Similar behavior was observed with another insoluble pigment, zinc phosphate, in the urethane alkyd vehicle.[7] Up to 15% PVC both D and I type films were seen. However, the more soluble pigment, strontium chromate, when incorporated into urethane alkyd or epoxy polymide vehicles, gave 100% D type films at 10 and 15% PVC. The emulsion paint system examined in Reference 2 was also 100% D type with a fairly low (10^6 ohms/100 mm²) resistance with little scatter among individual pieces of film. This paint was based on a styrene butyl acrylate emulsion pigmented with zinc chromate (primer) or TiO_2 (top coat).

The conclusion from this work is that a lightly or noncrosslinked vehicle will lead to paints that are 100% D Type. A crosslinking resin will always produce films with both D and I areas and incorporating insoluble pigments does not change the resistance behavior much. A crosslinking resin incorporating a soluble pigment, or one that leads to soluble soaps, will rapidly become 100% D Type.

The uniformity of the resistance of noncrosslinking systems (e.g., clorinated rubber or the emulsion paints) needs to be investigated. It is possible that the D areas is crosslinking systems have a lower resistance to the penetration of ions than the D areas in a typical noncrosslinked system. This could happen if the resistance measurement on 100 mm² areas of a noncrosslinked system is representative of smaller areas, and not just one weak point. A single coat of chlorinated rubber, of say 40 µm thickness, could give as good, if not better, protection than a single coat of epoxy polyamide at the same thickness. However, crosslinking systems offer the opportunity of getting uniform high resistance by casting multicoat films.

Relation Between Detached and Attached D.C. Resistance

This paper has investigated the inhomogeneous nature of crosslinked polymer films. The importance of this inhomogeneity to the protection afforded by the film, if they are functioning by resistance inhibition, is indicated by the often observed breakdown in spots of protective films, illustrated in Figure 3. The structure contains, to a greater or lesser extent, small weakly crosslinked regions, and these regions represent potential areas where corrosion can occur. Systems which are less crosslinked are 100% D Type.

The work in Reference 2 confirmed that an *in situ* measurement of the D.C. ionic resistance of a paint was a good indication of how effectively the paint was protecting the substrate. The criteria of Bacon Smith and Rugg[8] is that a film is good if the *in situ* resistance is greater than 10^8 ohms/100 mm², and poor if it is less than 10^6 ohms/100 mm². The smallness of D areas, suggested by the work described here, means that the true value of the D.C. resistance measured *in situ*

(Table 2 in Reference 2) was much less than 10^6 ohms/100 mm² and, hence, rapid breakdown ensued. Also, assessment of performance by measurement of the detached film resistance seems to be applicable in the particular case of varnish films on abraded mild steel in 3.5M KCl.

However, one must always keep in mind when making a resistance measurement, on any area greater than about 0.1 mm², that the value obtained may not be representative of the whole area being examined. The resistance may be due to one or more weak points even for what appear to be "perfect" pieces of film.

A practical point arising from the effect of temperature on resistance is that D type films have a low resistance from 0 to 50 C, and thus represent potential corrosion areas at most atmospheric temperatures. I areas have a high ($>10^8$ ohms/100 mm²) resistance over this range, and thus, on the BS and R criteria should resist corrosion up to 50 C, assuming ion exchange does not occur.

It is appreciated that measurement of the D.C. detached film resistance is not always going to be a reliable guide to the performance of a paint. A paint containing soluble inhibitors, either as the salt itself (e.g., zinc chromate), or as a soap formed between the pigment and the vehicle, may well have a low detached D.C. ionic resistance, and will give 100% D type films; yet, the paint protects because the soluble inhibitors passivate the substrate. These low resistance films often will have a sealing coat anyway, as otherwise the inhibitor would be leached out too fast. However, the authors consider that an *in situ* D.C. resistance measurement is still the best way of determining how protective the system will be.

Another case where the detached ionic resistance may not relate too well to the protection afforded by the film, even though the film does not contain inhibitors, is when the adhesion of the film is particularly high (e.g., iron oxide in epoxy). However, again an *in situ* D.C. measurement, if it is high ($>10^8$ ohms/100 mm²), should be a reliable indication that the adhesion is being maintained and the film is protecting.

Conclusions

The following conclusions arise from the work discussed here:

1. Crosslinked polymer films are heterogeneous in structure. They contain small areas which have a vastly lower ionic resistance than the rest of the film.
2. Thickening the film reduces the chance of finding a low resistance area in 100 mm² of film, and a two coat system reduces the chances further.
3. All five crosslinking systems (alkyd, tung oil, epoxy, polyurethane, urethane alkyd) exhibit this bimodal resistance character.
4. Films from noncrosslinking systems yield films of more uniform resistance.
5. Detached D.C. resistance measurements have considerable value in assessing the protective ability of a varnish film.
6. The detached D.C. resistance may not be as relevant to a full paint system; however, an *in situ* D.C. resistance measurement should be high for good protection.

Further work needs to be done to confirm some of the suggestions made here. The resistance homogeneity or otherwise of typical paint systems needs further investigation. The relation between the D.C. *in situ* ionic resistance and protection needs clarifying, particularly in the "grey" area (10^6 to 10^8 ohms/100 mm²). The theoretical basis for the 10^6 ohms/100 mm² *in situ* measurement as an indication of poor protection needs investigation. The behavior of I type films, in the absence of inhibitive pigments, on a corrodable substrate, requires further examination, from the point of view of the onset of corrosion. Finally, ways of modifying resins to produce single coat thin films (<50 μm), but 100% I type, need studying. It is hoped to follow up some of these lines of inquiry in the near future.

Acknowledgments

The authors wish to thank the Procurement Executive, Ministry of Defence, for financial support and encouragement, and Professor R.W.K. Honeycombe for the provision of facilities, without either of which this work could not have been carried out. They also thank Mr. C.A.J. Taylor for preparing the diagrams and the photographs, and Dr. P.J. Heyes for permission to quote some of his results.

References

1. E.M. Kinsella and J.E.O. Mayne. *Br. Poly. J.*, 1969, **1**, p. 173.
2. J.E.O. Mayne and D.J. Mills. *JOCCA*, 1975, **58**, p. 155.
3. J.E.O. Mayne and J.D. Scantlebury. *Br. Poly. J.*, 1970, **2**, p. 240.
4. R.L.J. Morris and E.E. Longhurst. *Fatipec Congress*, 1972, p. 609.
5. J.E.O. Mayne and D.J. Mills. Paper in preparation.
6. B.W. Cherry and J.E.O. Mayne. *1st International Cong. on Metallic Corrosion*, Butterworths, London, 1962, p. 539.
7. P.J. Heyes, Ph.D. Thesis, p. 47, University of Cambridge, 1977.
8. R.C. Bacon, J.J. Smith and F.M. Rugg. *Ind. Eng. Chem.*, 1948, **40**, p. 161.
9. L.S. Goldring. *Ion Exchange: a Series of Advances* (Ed. J.A. Marinsky), **1**, p. 207, Edward Arnold 1966.
10. M. Kendig and H.J. Leidheiser. *Electrochem. Soc.*, 1976, **123**, p. 982.

Polyurethane Films as Electrochemical Membranes: How to Progress in Anti-Corrosion Protection

*Henri Jullien**

Introduction

As has been known for several years, anti-corrosion properties of paint films are related to their electrochemical properties as ion-exchange membranes, and Mayne's scheme must now be considered a classical one. Kittelberger[1] and Mayne[2] proposed three mechanisms by which anti-corrosion protection may occur:

1. Anodic inhibition, suppressing the anodic reaction by means of pigments.
2. Cathodic inhibition, the anti-corrosion coating being a physical barrier preventing water and/or oxygen to come in contact with the metallic surface.
3. Resistance inhibition, the high electrical resistance of the coating preventing the corrosion current to circulate between the metal and its environment.

Without considering the properties and action of included pigments, and any possible interactions between pigments and binders, it can be said that anti-corrosion properties are also related to the electrical conductivity of binders. The corrosion current passes between the electrolytic solution (sea water, or a drop of saline solution, for instance) and the painted metallic substrate, by a circulation of ions (anions, cations or both), through the paint film considered as a membrane. Several parameters appear to be of great importance: the penetration of water into the film; the permeability of the film to water and to oxygen; the diffusibility of ions in the coating; and the nature of the diffusing ions (anions or cations).

When corrosion is initiated under a paint film, diffusion of water and oxygen occurs into and through the coating. It is known that every resin film can be permeated by water and oxygen. It is quite impossible to prevent this diffusion for a long time, and a paint film should protect a metallic surface for many years. Thus, the paint film cannot stop corrosion forever, but it should slow down the corrosion phenomena as efficiently as possible.

Therefore, the aim of our work was to examine how polyurethane films react with water and ions to slow down electrochemical corrosion, and how to improve the properties for commercial application in the future.

*GR n° 35, CNRS, F 94320 THIAIS, France

Experimental

Preparation of Films

The raw materials used were an adduct of toluylene diisocyanate and trimethylolpropane (Bayer Desmodur L 75, a 75% solution in ethyl acetate), and a polyester-polyalcohol (Bayer Desmophen 1100). For a stoichiometric reaction between these two compounds, 100 g of Desmophen correspond to 125 g of Desmodur solution. The resulting product is a film with a nitrogen content of 6.2%.

Films were also made with a nonstoichiometric ratio of the two components:

- Three films poorer in nitrogen (4.3, 4.8 and 5.5% of nitrogen); in these films, prepared with an excess of polyester-polyalcohol, hydroxyl functions were thus greater.
- Three films with a higher content of nitrogen (6.8, 7.35 and 7.7%). In these films, prepared with an excess of triisocyanate, several functional groups appeared: primary amines, disubstituted ureas and biurets, resulting from water action on isocyanate groups, and reactions between these groups and amines.

The films were prepared on a glycol polyterephthalate sheet of a very regular thickness. First, a solution of gelatin was spread over it, in order to obtain a very thin film of dry gelatin (about 5 μm). On this film, the solution of resin was spread with an applicator; this solution was previously diluted with a convenient amount of ethyl acetate, in order to obtain dry polyurethane films of about 60 μm thickness, after hardening in an oven at 120 C. Then gelatin was dissolved by immersion in distilled water, and polyurethane films separated from the plastic substrate. The thickness of each film was determined by means of a microcomparator. The characteristics of the experimental films are given in Table I.

Water vapor permeability. Using a method similar to the Payne's cup method, the water vapor permeability coefficient P was measured. As a function of the nitrogen content in films, the permeability coefficient decreased to a minimum, at the "critical value" of 6.2% corresponding to the stoichiometric

TABLE 1 — Characteristics of Polyurethane Films

	Initial composition		Nitrogen content %	Water content %	Calculated number of functional groups (for 100 g of dried films)			
	Tri-isocyanate (parts)	Polyalcohol (parts)			urethane	hydroxyl	amine	crosslinking
A	100	150	4,3	2,71	0,305	0,133	—	0,051
B	100	125	4,8	2,45	0,343	0,097	—	0,057
C	100	100	5,5	2,53	0,392	0,049	—	0,065
D	125	100	6,2	2,54	0,442	—	—	0,074
E	150	100	6,8	2,87	0,451	—	0,041	0,068
F	180	100	7,35	3,17	0,461	—	0,083	0,063
G	200	100	7,7	3,27	0,466	—	0,108	0,060

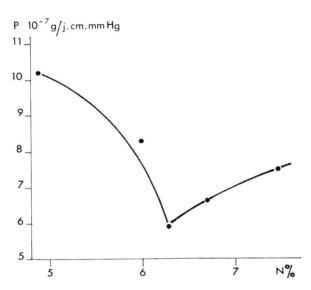

FIGURE 1 — Water vapor permeability as a function of nitrogen content, at 30 C.

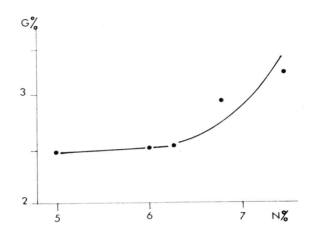

FIGURE 2 — Water uptake variation as a function of nitrogen content (percentage of dry film weight).

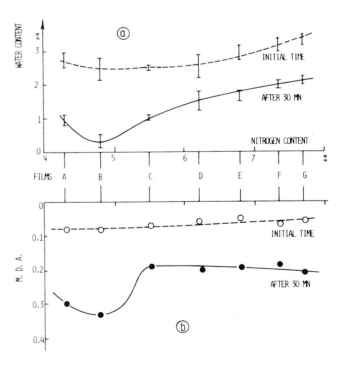

FIGURE 3 — Variations of water content (a) and microwave dielectric absorption (b), at the beginning of dehydration and 30 minutes after, as a function of nitrogen content.

mixture, then slightly increased. Water vapor diffusion was influenced by the crosslinking in the macromolecular network, which was maximum for this critical value (Figure 1).

Water uptake. The films were immersed in distilled water for a week, weighed, dried in a vacuum oven, and weighed again. The water uptake was small, and constant below the critical value of the nitrogen content and increased as the nitrogen content increased and the films became more hydrophilic (Figure 2).

Dehydration properties. In order to investigate the bonds between water and the macromolecule, a method was used to follow the time variation of dielectric losses in hydrated films, when submitted to a microwave electromagnetic field.[3] The frequency of 9.4 GHz was chosen, corresponding to the dielectric relaxation of free water, and very much higher than relaxation frequencies of macromolecules. A qualitative evaluation of bond forces between water and polyurethane is thus possible, especially as a function of applied power, for the dehydration rate was in direct relation with the bonding degree of water.[4]

In Figure 3a, the water weight inside the films is represented as a function of nitrogen content at the initial time and after 30 minutes of natural dehydration in air (films were initially hydrated by immersion in distilled water).

Figure 3b shows the corresponding variation of microwave dielectric absorption (MDA), $\Delta\varepsilon''$; at the "initial" time, the MDA variations are very similar to that

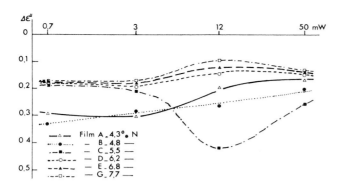

FIGURE 4 — Microwave dielectric absorption (MDA) variations as a function of microwave power after 30 minutes of dehydration for various nitrogen contents.

of water weight. After 30 minutes of dehydration, the MDA variations, below the critical nitrogen content, follow those of water content because the water binding on free hydroxyl groups is very weak. Above the critical content, the MDA variations are in the direction opposite to those of water content because the water molecules are more strongly bound to the macromolecule, probably on the NH sites. It thus appears that MDA variations are related not only to the water content, but also to the nature of the binding sites on the macromolecule.

More interesting is the fact that increasing the microwave power levels allows an acceleration of the film dehydration by microwave heating of films. MDA was measured in the various films after 3, 6, 10, 20, and 30 minutes of dehydration, and at power levels of 0.7, 3, 12 and 50 mW. Figure 4 exemplifies the experimental results after 30 minutes of dehydration.

At the critical content of 6.2%, when the crosslinking of polyurethane is high, a regular dehydration is observed, rather independent of the microwave power level. For a nitrogen content higher than 6.2%, the MDA variations are rather similar: the MDA increases until 12 mW and then decreases with the power level, which decrease can be explained by a heating of the film up to 12 mW and then a forced dehydration above 12 mW which decreases the measured MDA. For nitrogen contents lower than 6.2%, MDA variations are significantly different. For contents as low as 4.3 and 4.8%, the film dehydration is spontaneous, and practically achieved after 30 minutes; the power level increase induces film heating, thus increasing the dielectric losses. At 5.5% nitrogen, losses decrease as the power level increases to 12 mW, and then increases above this level. This behavior can be explained by the fact that microwaves induce a dehydration acceleration up to 12 mW, and then a heating process at high levels when the film is dehydrated.

Curves in Figure 4 thus show clearly that microwave power can induce a forced dehydration process which is closely related to the water binding sites in the polymer film. A microwave power level for an optimum dehydration exists which is mainly determined by the nitrogen content.[5]

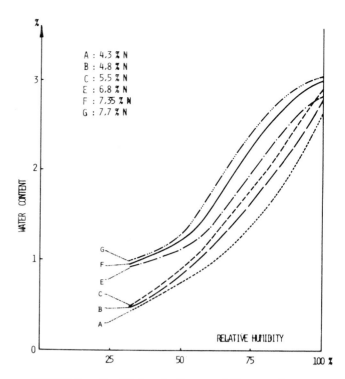

FIGURE 5 — Equilibrium isotherms of films vs nitrogen contents.

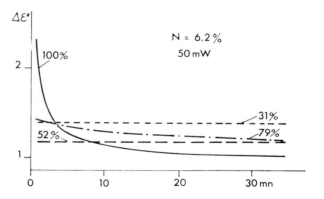

FIGURE 6 — MDA variations of hydrated films as a function of time.

The same method was used to investigate the dehydration of films previously conditioned in the air at different relative humidities (31, 52, 79 and 100%). The water weight in every film, determined by gravimetry, was a function of the air relative humidity (RH) as shown in Figure 5 for the different nitrogen contents. As a typical example, the MDA variations and dehydration time are shown in Figure 6 for a 6.2% nitrogen content film and a 50 mW power level. For 31 and 52%, the dielectric absorption does not vary, even after 30 minutes. In contrast, at 100%, the MDA regularly decreases with time.

From such curves obtained for the various nitrogen contents, it was possible to compute the amounts of free and bound water for films hydrated in air at 100% RH. Free water content showed a minimum value near 6.2%, whereas bound water content continuously increased with the nitrogen content (Figure 7).

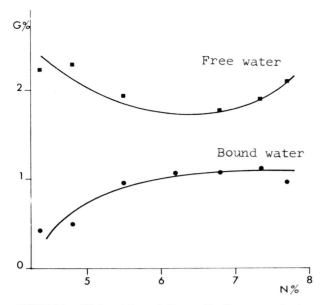

FIGURE 7 — Water state variations with nitrogen content in 100% RH hydrated films.

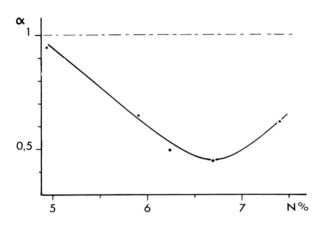

FIGURE 8 — Variations of the efficiency coefficient (measured capacity/theoretical capacity ratio) for the sulfate chloride exchange, as a function of nitrogen content.

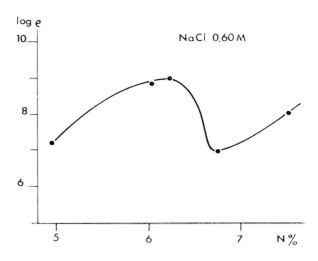

FIGURE 9 — Variations of ionic resistivity, in NaCl 0.6M solution, as a function of nitrogen content.

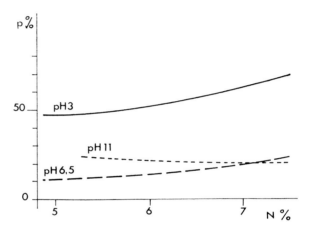

FIGURE 10 — Permselectivity variations as a function of nitrogen content, in acidic, neutral and basic media.

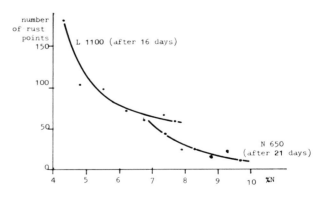

FIGURE 11 — Number of rust spots which appeared at the end of exposure, plotted as a function of nitrogen content.

Anion Exchange Properties

Anion-exchange capacity. Anion exchange capacity was determined after immersion into a 0.1 N sulfuric acid solution for a week. The tested film was then transferred into a 0.1 N HCl solution, the chloride ion activity in which being followed by means of ion specific electrode. Sorbed chloride ions were compensated in the solution by an equivalent amount of a molecular sodium chloride solution, supplied by an automatic buret, to maintain constant the chloride activity. The total supplied amount of sodium chloride gave the anion-exchange capacity.[6]

Data in Figure 8 shows the variations of the efficiency coefficient in ion-exchange (measured capacity/theoretical capacity), for the sulfate/chloride exchange, as a function of the nitrogen content in films. The minimum corresponds to the weakened access of ions to active sites, near the critical value related to a maximum in crosslink density.

Ionic resistivity. Ionic resistivity was obtained by measuring the electrical resistance of a cell in which a tested film was put between two compartments filled with a sodium chloride solution. A typical example is given in Figure 9 of the variations of resistivity in 0.6M

NaCl: below the critical nitrogen content, cross linking and ionizable function density both increase, so that ion diffusion is weakened, and resistivity increases; just above the critical value, the decrease of crosslinking aids diffusion of ions, and thus weakens ionic resistivity, until at higher nitrogen containing functions the ionic resistivity increased again.

Permselectivity. Permselectivity was determined by measuring the variation of the electromotive force of a cell: silver-silver chloride electrode/electrolyte solution/film/electrolyte solution/silver-silver chloride electrode, as a function of the logarithm of the concentration of the electrolyte in one of the compartments. Theoretically, according to the Nernst law, this variation should be linear, with a slope of 0.058. The actual variation was linear, but with a slope less than 0.058. Permselectivity was expressed as a percentage of the theoretical slope, the difference being due to the passage of cations through the film, supposed to be impermeable to these ions and permeable only to anions.

Measurements were made successively in the presence of NaCl, HCl and sodium carbonate solutions to explore 3 pH fields: neutral, acidic, and basic. In Figure 10 we see the very weak permselectivity of polyurethane films in neutral or basic media, and the higher values obtained in acid conditions, moreover largely increasing with nitrogen content in this case.

Salt Spray Tests

According to the preceeding results, it may be supposed that the greater the nitrogen content in films, the better the anti-corrosion protection, for water permeability remains rather low, while DC resistivity notably increases. To confirm this conclusion, two series of steel panels were coated with polyurethane unpigmented varnishes, and submitted to the salt spray test under standard conditions (35 C, 5% NaCl).[7]

The first series, L-1100, was coated with the same Desmodur L polyisocyanate-Desmophen 1100 polyalcohol mixtures as before; nitrogen contents being the same as described (4.3, 4.8, 5.5, 6.2, 6.8, 7.35 and 7.7%). In order to investigate higher values of nitrogen content, the second series (N-650) was made of Desmodur N polyisocyanate and Desmophen 650 polyalcohol, according to the same principle of varying the composition of initial mixtures, so that we obtained the following nitrogen contents in the films: 6.9, 7.4, 7.9, 8.3 (stoichiometric mixture), 8.8, 9.3 and 9.7%.

Panels were coated by means of a centrifugation applicator. The thickness of dry films was approximately 50 μm, controlled by means of a magnetic gauge.

L-1100 coated panels were exposed for 16 days, and the appearance rate of rust spots was inversely proportional to the nitrogen content. N-650 coated panels were exposed for 21 days since corrosion protection was better than for L-1100 films.

In Figure 11, the number of rust spots which appeared at the end of exposure is plotted as a function of nitrogen content: the higher the nitrogen content in resins, the better the anticorrosion protection.

Discussion and Conclusions

The determination of water uptake, and the study of dehydration by means of MDA variation technics, show that polyurethane films absorb water, and a part of the sorbed water molecules are more strongly bound to the macromolecule. As the nitrogen content in polymer is increased the total water uptake and the bound water amount are increased. It is assumed that water is bound to the nitrogen containing functional groups. At the same time, free water shows a minimum amount at the critical nitrogen content, corresponding to the stoichiometric mixture of components and to the maximum in crosslinking density in the polymer. The free water content is related to crosslink and hydroxyl group density. Since bound water is adsorbed by nitrogen containing groups, it becomes apparent that anion fixation is possible, a result confirmed by anion exchange capacity and permselectivity measurements. The accessibility of anions to active sites, however, is sensible to crosslinking. Permselectivity is poor in neutral and basic media; that is to say, polyurethane films are permeated both by anions and cations. However, permselectivity becomes fair in acid medium, and increases with the nitrogen content in films. Electrical resistivity shows a special behavior, being a function of both crosslink density and nitrogen content.

The anion-exchange membrane properties of polyurethane films indicate poor permselectivity. However, it appears that anti-corrosion properties, and especially resistivity are enhanced by high nitrogen contents. This hypothesis was confirmed by salt spray test carried out with polyurethane coatings with a large range of nitrogen contents.

Two essential conclusions now appear to answer the question of how to progress in anti corrosion.

First, polyurethane films containing high nitrogen contents seem to show a better anti-corrosion quality. This quality is assumed to be due to the anion-exchange properties and probably to strong bindings between chloride ions and macromolecule: in ion-exchange terms polyurethane films show a high selectivity for chloride ions, preventing them from diffusing to the metallic substrate.

Second, electrochemical methods allow one to carry out scientific and methodic studies of anti-corrosion problems. By these techniques, we have shown how high nitrogen content in polyurethane films is a good way to increase anti-corrosion qualities. Electrochemical methods do not seem to be quick testing methods for complex paint films. To be successfully used, long time periods are required: and they must be scientifically and carefully applied; that is to say, only one parameter should be varied at a time.

References

1. W.W. Kittelberger, *Ind. Eng. Chem.*, **34,** p. 943 (1942)
2. J.E.O. Mayne, *Off. Dig.,* **24,** p. 127 (1952)
3. A.J. Berteaud, F. Hoffman and J.F. Mayault, *J. Microwave Power,* **10,** p. 309 (1975)
4. A.J. Berteaud, H. Jullien, R. David and C. Her, *J. Microwave Power,* **12,** p. 231 (1977)
5. H. Jullien, A.J. Berteaud and J. Petit, Angew. *Makromol. Chem.,* **62,** p. 241 (1977).
6. H. Jullien, F. Henry and J. Petit, Angew. *Makromol. Chem.,* **52,** p. 179 (1976).
7. H. Jullien, W. Funke and U. Zorll, XVth FATIPEC Congress Book, III, p. 255 (1980).

Permeation Properties of Organic Coatings in the Control of Metallic Corrosion

*M. Yaseen**

Introduction

Small amounts of water, ions, vapors, and oxygen which permeate a protective coating may cause corrosion of the substrate at a rate similar to that of bare metal. Pigments, inhibitors, passivators, and the additives used in coating formulation may protect the substrate by affecting the corrosion reactions and reducing them to a minimum. Therefore, the durability and servicability of coatings depend on their protective properties, which in turn are governed by the properties of the individual ingredients, i.e., binders, pigments, and additives used in the formulation.

Permeation

Protection provided by an organic coating, apart from being related to its physio-chemical properties, also depends on other factors, one of which is its resistance to the permeation of gases, vapors, liquids, and ions.[1] The process of permeation begins when the permeant in contact with the coating is absorbed, moves through the coating, and a state of dynamic absorption equilibrium is attained.[2]

Water

Water in the form of rain, dew, or humidity is always in contact with the coating, since the coating is applied over a metal in order to protect it from corrosion. In the case of exterior coatings exposed to adverse weather conditions, permeation of water is considered important because the water affects the permeation of ions, oxygen, and other corrosive agents. Consequently, the presence of such substances at the coating-metal interface promotes corrosion of the substrate.

Ions

Certain ions, particularly chloride ions, are especially agressive corrosive agents.[3] The rate of movement of ions in coatings is slower than that of water, and for some types of ions it is almost restricted. Coatings in contact with water or electrolytic solution develop charges which restrict the flow of ions through them; sometimes acting as perm-selective membranes.[4-6] During long exposure to sun and

TABLE 1 — Composition of Clear and Pigmented Coatings

Serial No.	Composition of Coating Formulation	Density of Dry Film g/ml
	Para-Phenolic Varnishes	
1.	Tung oil—66.6%	1.166
2.	Tung oil—80.0%	1.106
3.	Linseed oil—66.6%	1.138
4.	Linseed oil—80.0%	1.119
5.	Soya Oil—66.6%	1.131
6.	Soya Oil—80.0%	1.087
7.	DCO—66.6%	1.132
8.	DCO—80.0%	1.109
	Alkyds[1] (Commercial)	
9.	Soya oil 39.6, orthophthalic—39.0	1.202
10.	Soya oil 48.0, orthophthalic—35.0	1.184
11.	Soya oil 56.0, orthophthalic—30.0	1.120
12.	Soya oil 64.0, orthophthalic—25.0	1.108
13.	Soya oil 65.0, isophthalic—25.0	1.10
14.	Soya oil 70.0, isophthalic—20.0	1.09
15.	Soya oil 75.0, isophthalic—18.0	1.06
	Alkyds (Prepared)	
16.	Linseed oil—55.0—orthophthalic—glycerol	
17.	Linseed oil—66.0—orthophthalic—glycerol	
18.	Linseed oil—55.0—orthophthalic—pentaerythritol	
19.	Linseed oil—66.0—orthophthalic—pentaerythritol	
20.	DCO—55.0—orthophthalic—glycerol	
21.	DCO—66.0—orthophthalic—glycerol	
	Clear Commercial Products Pigmented with TiO_2 (rutile)	
22.	Alkydal R35W[1] + Resydrol WM501[1]—clear	1.222
23.	Alkydal R35W[1] + Resydrol WM501[1]—pigmented	1.626
24.	Maprenal WL[1] (Melamine resin) + Resydrol VWY 23[1]—clear	1.152
25.	Maprenal WL[1] (Melamine resin) + Resydrol VWY 23[1]—pigmented	1.532
26.	NC E 510[1] dried (nitrocellulose) + Castor oil + Dioctylphthalate—clear	1.226
27.	NC E 510[1] dried (nitrocellulose) + Castor oil + Dioctylphthalate—pigmented	1.628
28.	Vinylite VY HH[1]—clear	1.232
29.	Vinylite VY HH[1]—pigmented	1.675
30.	Alkydal 251[1] + Maprenal NPX[1]—clear	1.187
31.	Alkydal 251[1] + Maprenal NPX[1]—pigmented	1.620
32.	Electrodeposition paint	1.287

*Regional Research Laboratory, Hyderabad-500009, INDIA.

[1] Registered tradename.

TABLE 2 — Effect of Relative Humidity on Rate of Permeation at 23 C

Coating Formulation No.	Rate of Permeation of Water Vapor — g/m²/h/mil[1] at Different Relative Humidity				
	50	65	80	90	97
Phenolic Coatings					
1.	0.67	0.88	1.07	1.25	1.39
2.	1.12	1.51	1.85	2.20	2.38
3.	0.75	0.96	1.20	1.41	1.60
4.	1.21	1.60	2.02	2.36	2.61
5.	1.11	1.46	1.86	2.19	2.37
6.	1.46	1.91	2.41	2.74	3.05
7.	0.92	1.19	1.50	1.73	1.93
8.	1.20	1.60	2.05	2.45	2.73
Alkyd Coatings					
9.	0.73	1.01	1.57	1.85	2.13
10.	0.86	1.19	1.85	2.21	2.55
11.	1.03	1.41	2.17	2.54	2.97
12.	1.42	2.00	3.11	3.67	4.26
13.	1.30	1.76	2.40	2.88	3.31
14.	1.55	2.12	3.20	3.72	4.17
15.	1.87	2.52	3.75	4.36	4.96

[1] Thickness of coating film.

TABLE 3 — Effect of Temperature on Rate of Permeation at 50% RH

Coating Formulation No.	Rate of Permeation of Water Vapor — g/m²/h/mil[1] at Different Temperatures				
	17.2 C	23 C	29 C	34.4 C	40 C
Phenolic Coatings					
1.	0.41	0.67	1.01	1.59	2.38
2.	0.61	1.12	1.76	2.83	4.36
3.	0.44	0.75	1.15	1.89	2.91
4.	0.68	1.21	1.92	3.04	4.80
5.	0.61	1.11	1.91	3.04	4.64
6.	—	1.46	2.28	3.84	6.27
7.	0.5	0.92	1.46	2.33	3.56
8.	—	1.20	2.10	3.52	5.46
Alkyd Coatings					
	—	23 C	28.3 C	33.3 C	39 C
9.		0.73	1.24	2.09	3.44
10.		0.86	1.45	2.41	3.79
11.		1.03	1.68	2.64	4.35
12.		1.42	2.39	3.63	6.46
13.		1.30	2.12	3.15	5.44
14.		1.55	2.51	3.71	6.27
15.		1.87	2.98	4.38	7.39

[1] Thickness of coating film.

water, the coatings lose resistance to the permeation of ions and, in longer course, may not serve as efficiently in protecting the substrate from corrosive ions.[7,8]

Knowledge about the permeation properties of organic coatings in the control of metallic corrosion is important, and the choice of exterior coating should take this into account. It is sometimes observed that exterior coatings (that might be expected to last for years) develop blisters and start peeling within short periods, thus suffering premature failure. This failure may be related to low permeability properties. During high humidity conditions the exterior coatings absorb sufficient amounts of water which permeate and reach the metal substrate. In the heat of the day, the condensed water at the substrate evaporates and develops high pressures if it does not find easy egress from the coating. Damage to the coating may occur, especially at weak spots.

The data reported here indicate how water absorption and permeation properties of clear and pigmented coatings change with an increase in temperature and relative humidity of the surroundings, and how the performance of exterior coatings in the control of metallic corrosion is effected.

Experimental

The methods of preparation, the compositions, and the sources of the commercial products are described

TABLE 4 — Effect of Temperature on Rate of Permeation Under Dry and Wet Cup Test Conditions

Coating Formulation No.	Rate of Permeation of Water Vapor — g/m²/h/mil[1]					
	Dry Cup Test 50 to 0% RH			Water Cup Test 100 to 50% RH		
	23 C	30 C	40 C	23 C	30 C	40 C
Clear and Pigmented Coatings						
22 (clear)	0.96	1.71	3.60	1.27	2.76	6.04
23 (pigmented)	0.93	1.59	3.41	2.91	5.34	11.62
24 (clear)	4.92	8.26	16.01	6.91	13.49	26.08
25 (pigmented)	3.24	5.48	11.06	6.41	10.96	22.46
26 (clear)	2.40	4.03	8.08	5.47	6.38	14.43
27 (pigmented)	2.48	4.28	8.54	3.97	7.51	14.70
28 (clear)	1.04	1.58	2.92	1.33	1.94	3.87
29 (pigmented)	0.89	1.29	2.30	1.66	2.96	5.27
30 (clear)	1.33	2.48	5.37	2.23	4.30	10.46
31 (pigmented)	1.18	2.17	4.67	1.74	3.80	8.25
32	0.63	1.11	2.46	1.34	2.72	6.87

[1]Thickness of coating film.

elsewhere.[8,9] In Table 1, details about the composition of the coatings used for the study are given. The methods of determination of water absorption, permeation of water vapor and chloride ions, and calculation of results have also been reported earlier.[9-11]

Discussion

Effect of Relative Humidity on Permeation

The rates of permeation of water vapor through phenolic coatings increase linearly with humidity in relatively less humid environments. However, in highly humid environments, the increase in the rates with relative humidity is a little more than linear (Table 2). This increase may be due to water absorption in the coating at high RH. Also, as the coating swells due to the presence of more sorbed water, there is less resistance to the movement of the permeant.

Permeation rates increase more in humid atmosphere through alkyds than through phenolics. The air-drying alkyd coatings contain some unreacted polar groups and develop hydrophilic characteristics in highly humid conditions; consequently this condition facilitates permeation. The relatively greater amount of absorbed water in alkyd coatings has some plasticizing action on the internal structure of the coating with consequent ease in the movement of molecules through the coating.

Effect of Temperature on Permeation

The fairly low rates of permeation of some coatings at lower temperatures indicate that energy is needed to generate the thermal motion of molecules in the coating so that the permeant may more readily pass through the coating. The increase in temperature by 23 C activates some of the phenolic coatings so that the rate of permeation of water vapor is increased seven times (Table 3). The results also show that increase in permeation rates with temperature depends on the type and content of oil used in the formulation.

The temperature dependence of permeation rates of alkyd coatings is greater than that of phenolics. Among alkyds the increase in rates with temperature is greater in the case of orthophthalic than the isophthalic alkyds. The effect of temperature on permeation properties of coatings can be correlated with the degree of crosslinking, the intermolecular spacings, and the total free energy levels in the molecular structures.

In highly crosslinked coating material, randomness due to the movement of segmental chains and formation of spacings is restricted. In such systems, the movement of the penetrant, and consequently the rate of permeation, is impeded. Therefore, the energy of activation required for permeation of water vapor through a coating may be taken as a measure of its degree of crosslinking.

Permeation Tests Under Dry and Wet Cup Conditions

Some pigmented coatings exhibit a 10 to 12 fold increase in the rate of permeation of water vapor when the conditions are changed from 50 to 0% RH at 23 C to 100 to 50% RH at 40 C (Table 4). However, under similar conditions the maximum increase observed in the case of clear coatings is six fold. These observations indicate that the increase in temperature and humidity does not affect permeation through clear coatings as much as it does in the case of pigmented ones. As a result of such permeation properties, some pigmented coatings may serve better, as exterior coatings, than the clear coatings of the binder. This is because in highly humid and hot weather the condensed vapor at the coating-substrate interface will conveniently permeate during the hot part of the day without causing loss of adhesion or other damage at defective spots in the coating.

Permeation of Chloride Ions

The comparison of data on the rates of permeation of chloride ions and water vapor through the films of the same coating at the same temperature shows that the permeation of chloride ions is about 200 times slower than that of water vapor (Table 5). The results support the view that coatings, when immersed in

TABLE 5 — Effect of Temperature Permeation of Water Vapor and Chloride Ions

Coating Formulation No.	Temperature			
	30 C	35 C	40 C	45 C
Permeation of water vapor—g/m²/h/mil[1]				
16	3.96	6.48	10.4	18.24
17	3.06	5.42	9.05	14.46
18	4.32	6.54	9.80	16.31
19	3.60	5.76	9.21	14.24
20	2.52	4.32	6.48	9.81
21	1.98	3.24	5.04	7.72
Permeation of Chloride ions—g/1000g water/m²/mil[1]/h				
16	.021	.0026	.0036	.0045
17	.018	.0024	.0030	.0038
18	.023	.0029	.0037	.0050
19	.019	.0025	.0032	.0040
20	.014	.0019	.0026	.0035
21	.012	.0017	.0023	.0032

[1] Thickness of the film.

water or electrolytic solution, develop certain charges on the surface which restrict the permeation of ions through them. The rate of permeation of ions through coatings is not accelerated by the rise in temperature, as observed in the case of water vapor. Relatively more energy is needed for the movement of ions through coatings than for water. The ionic permeability and water vapor permeability properties of coatings are interdependent.

Knowledge about the changes in permeability properties of coatings with increasing relative humidity and temperature is important, not only because it indicates how the coating may behave in service, but also because it gives an insight into its protective properties in the prevention of corrosion of the metal substrate.

References

1. W. Funke. *Farbe & Lack*, **73** (8), p. 707 (1967).
2. M. Yaseen. *Deut. Farben Zeit.*, **27** (5), p. 213, (1973).
3. R.S. Saxena. *Paintindia*, **8**, p. 8 (1958).
4. K. Sollner. *J. Phys. Chem.*, **49**, pp. 47, 171, 265 (1945).
5. A.G. Nasini, G. Poli and V. Rava. *Proc. Premier Congress Technique International de l' Industrie des Peintures*, p. 299, Paris, 1947.
6. J.E.O. Mayne. *J. Oil Colour Chem. Assoc.*, **32**, p. 481 (1949).
7. A.L. Glass and Smith. *J. Paint Tech.*, **38**, p. 203 (1966); **39**, p. 490 (1967).
8. R. Vittal Rao, M. Yaseen and J.S. Aggarwal. *Farbe & Lack*, **79** (2), p. 95 (1972).
9. M. Yaseen and H.E. Ashton. *J. Oil Colour Chem. Assoc.*, **53**, pp. 977, 1015 (1970).
10. M. Yaseen and H.E. Ashton. DBR Internal Report No. 430, NRC Canada, 1976.
11. M. Yaseen and W. Funke. *J. Oil Colour Chem. Assoc.*, **61**, p. 284 (1978).

Studies on the Subcoating Environment of Coated Iron Using Qualitative Ellipsometric and Electrochemical Techniques

*J.J. Ritter and J. Kruger**

Introduction

The behavior of organic coatings on metals has been examined with a variety of electrical and electrochemical measurement techniques, including DC resistance, AC resistance, capacitance, corrosion potential, etc.[1] The objectives of these studies has been to evaluate coating efficacy, and to probe the processes which lead to coating failure. This work describes results from the application of a new technique of combined ellipsometric and electrochemical measurements to a study of the fundamental corrosion processes on metals protected with organic coatings.

The details of the technique have been described previously.[2] Briefly, painted metal is simulated using a polished iron specimen coated with transparent cellulose nitrate (collodion). The specimen is provided with a miniature probe to measure pH under the coating while visible, elliptically polarized light passes through the coating and is reflected from the metal surface. Computer modeling and experiments with collodion on relatively inert gold surfaces have indicated that for the ellipsometric conditions chosen ($\lambda = 546.1$ nm, $\Phi = 68°$), the ellipsometric parameters Δ and ψ are most sensitive to changes at the metal-metal oxide surface. Thus, for example, variations in the ellipsometric parameter Δ can be related to the growth or dissolution of the oxide film at the surface of the metal and under the organic coating. At the same time, the pH of the environment to which the metal is subjected can be assessed.

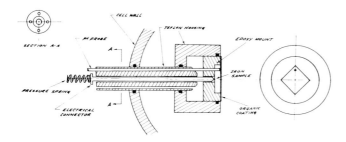

FIGURE 1 — Schematic drawing showing details of the specimen mount.

*Chemical Stability and Corrosion Division National Bureau of Standards, Washington, DC

The technique has been used to study both coated iron and iron with various inhibitors available either in or under the coating material. Due to the complexity of the multilayered system which must be dealt with in these experiments, the ellipsometric results are only useful in a qualitative sense. However, cathodic reduction experiments done on stripped iron specimens in deaerated borate buffer environments permitted some semiquantitative assessment of subcoating events.

Experimental

Specimen preparation details have been published previously.[2] For the present, it suffices to say that the iron specimens were cast in epoxy, polish to a 0.05 μm finish and dipped in a 1:1 collodion-methanol mixture to give a coating of 10 to 30 μm in thickness over the face of both the metal and the epoxy mount. The prepared specimen was inserted in an improved mount (Figure 1) and set in a polytetrafluoroethylene lined cell which was filled with 0.05 N NaCl. Inhibitor materials were either applied directly to the metal before coating or dispersed as a finely ground powder in the collodion methanol mixture. For the case of $SrCrO_4$, the metal specimen was dip coated, finely powdered $SrCrO_4$, applied to the surface of the semicured coating, and a second coat applied to cover the inhibitor.

The phosphate chromate surface treatment was accomplished as follows: Solution I: H_3PO_4 3.0 g, ZnO 1.5 g, $NaNO_3$ 1.0 g, make up with H_2O to 100 ml. Solution II: $K_2Cr_2O_7$ 0.7 g, H_3PO_4 0.5 g, make up with H_2O to 100 ml.

The solutions were used at room temperature. The face of the specimen was rotated in a horizontal plane at 50 rpm in a mixture comprised of 8 ml of Solution I and 2 ml of Solution II for a period of six minutes. This procedure gave a homogeneous surface finish which retained sufficient specular quality to allow ellipsometric measurements.

The ellipsometric measurements on coated iron specimens show changes in time in both the Δ and ψ parameters (Figures 2 and 3). By far, the largest change is seen in the Δ parameter which always shows a decrease. It is known that several factors could contribute to this directional change in Δ;

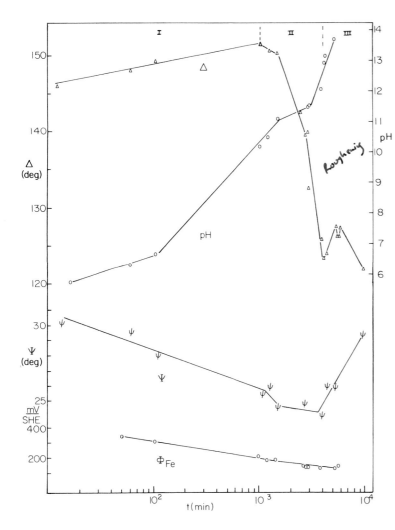

FIGURE 2 — Δ, relative phase retardation, ψ relative amplitude reduction, pH and Φ$_{Fe}$, potential of the iron surface vs standard hydrogen electrode (S.H.E.) vs time for Fe with collodion in 0.05 N NaCl.

dion coated gold specimens in 0.05 N NaCl and in saturated NaOH (pH ~ 15).

Results and Discussion

The examples of experimental results shown in Figures 2 and 3 indicate at least two stages of activity. In the first stage, (generally up to ~1000 minutes) the pH tends to rise gradually while the ellipsometric parameter Δ tends to increase several degrees.[1] As the pH approaches 11, the Δ values undergo a significant decrease, which signifies the beginning of the second stage. Invariably, small anodic regions become visible during the second stage with final overall anodic to cathodic area ratios ranging from 1/30 to 1/600.

In some experiments, particularly in those where no inhibitor is present and the pH remains high, a third stage is observed where Δ fluctuates. This behavior can be seen in Figure 2.

The most significant change occurs in the Δ parameter during the second stage. In cases where no inhibitor is present, this change arises from these three factors along with their estimated contributions to the overall δΔ: (1) metal surface roughening, ~55%; (2) oxide film growth, ~35%, and; (3) changes in coating material in a high pH environment ~10%.

It will be noted that the second stage changes of Δ with time appear to approximate a semilogarithmic relationship, that is, $dΔ = k \log t + c$. This law is observed frequently in the growth of passive films on metals.[4] The rate constants, k, for coated iron with various applications of inhibitors, are summarized in Table 1.

With reference to Table 1, it is readily seen that the rate of ellipsometrically observed subcoating events for coated iron, and for coated iron which had been given a phosphate chromate pretreatment, are comparable. However, when K_2CrO_4 is present at the metal surface, the rate is slower by a factor of ten, whereas $ZnCrO_4$ applied similarly slows the rate by a factor of two. This phenomenon may be partly related to solubility differences between the two inhibitors. Inhibitors dispersed throughout the coating give a similar factor of two reduction in the rate. The mechanism by which chromates inhibit corrosion reportedly involves the incorporation of chromium into an existing oxide film; that is, a chemical and structural modification of the existing oxide film which results in improved protection.[5] The rate constants derived from the present work seem to be related to inhibitor ion mobilities and accessibilities to the substrate surface. Contributions to the Δ parameter due to modifications of the metal oxide (for example, upon the incorporation of chromium) which may change its refractive index, are expected to be small

among them, metal oxide film growth, surface roughening,[3] and possibly coating deterioration. The following describes these effects and their relative contributions to the systems under study were investigated.

The extent of oxide film growth under the coatings was estimated at the conclusion of an experiment by stripping the coating with methanol, and cathodically reducing the metal surface at -1000 mV S.C.E. in deaerated borate buffer solutions (pH 8.5) while observing the surface with the ellipsometer. Oxide film thicknesses were calculated, assuming that the film was essentially Fe_2O_3, and that each degree change in Δ corresponded to 0.6 nm of film.

The presence of surface roughness was qualitatively detected by cycling the specimen potential between -1000 and +200 mV approximately 60 times. A significant increase in the Δ parameter indicated surface smoothing as a result of electrochemical polishing.

Contributions from the coating to changes in Δ were estimated by ellipsometrically monitoring collo-

---
[1] In some experiments, the Δ values tend to decrease several degrees over the first stage. The contributing factors to this portion of the curve are not fully known.

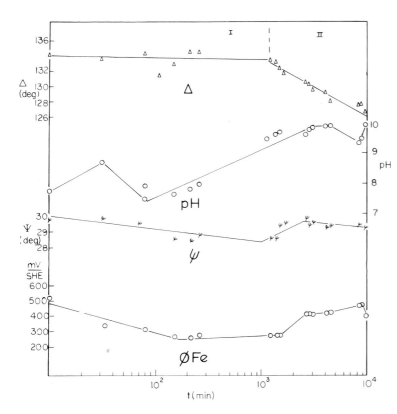

FIGURE 3 — Δ, ψ, pH and Φ_{Fe} vs time for Fe with collodion and K_2CrO_4 islands in 0.05 N NaCl.

for films <5nm in thickness.[6] Thus, the thickness of these modified films can be estimated with reasonable confidence using parameters for Fe_2O_3.

Preliminary results from uncoated iron in a simulated subcoating environment of pH 15 (saturated NaOH), 0.05 M K_2CrO_4 and 0.05 M NaCl show a change in Δ of which approximately 1/3 is due to oxide film growth, and the remainder ascribed to surface roughening. The oxide film growth on uncoated Fe, even in this highly alkaline environment, was only 2.4 nm as compared to the typical 6 nm growth measured for coated iron in the absence of inhibitor. Additional simulated environment experiments with other inhibitors, under a variety of conditions, are currently in progress and will be the subject of a future publication.

During the first stage, oxygen, water, Na^+ and Cl^- permeate the coating. The dilute Cl^- medium generated promotes localized breakdown of the existing air formed oxide film originally present on the metal surface. Active metal sites are exposed, and iron oxidation is promoted in these regions. Less active portions of the surface serve as cathodic regions for the reduction of O_2 to OH^-. Hence, the gradual rise of pH during the first stage.

As the pH reaches high values (>13), other events can proceed depending upon the potential of the metal in its subcoating environment. For example, experiments with uncoated iron in saturated sodium hydroxide solution (pH ~ 15) suggest that this potential may vary over the range from -900 to -150 mV S.H.E. Under these conditions, iron can be brought into solution through the formation of either the $HFeO_2^-$ or FeO_4^{-2} ions and oxide film can grow, the oxidation of $HFeO_2^-$ being one source of Fe_2O_3. The experimental evidence (from cathodic reductions of stripped iron specimens and from deliberate electrochemical metal smoothing) indicates that both surface roughening and oxide film growth phenomena occur under the coating, and are primarily responsible for the large decreases observed in the Δ parameter during the second stage. Thus, the high pH environment and concomitant decreases in the Δ parameter only occur when active corrosion is underway on the surface of the coated metal. The new technique serves to detect the presence of serious subcoating corrosion, and provides a measure of the overall kinetics for some of the subcoating processes.

At this point, the implications of the observed surface roughening and oxide film growth phenomena for coating failure mechanisms remains speculative. For the case of an intact coating, one might envision the initial formation of small, "two dimensional" localized corrosion cells as a result of O_2, water and Cl^- permeation through the coating. The development of these cells may entail local delaminations of the coating. Anodic and cathodic regions develop within the cells, and the anodic/cathodic surface area ratio is small due to a paucity of active sites.[2] Thus, much of the local cell area develops an alkaline environment which is conducive to surface roughening and oxide film growth. Topographical and possible compositional changes of the substrate surface promote more extensive coating delamination, and lead to the coalescence of local cells into larger units. These events probably occur during the first stage noted in the experiments.

The anodic and cathodic reactions continue to promote the enlargement of the delaminated regions by the processes described, and the active area gradually comes within the sensitivity boundary of the ellipsometer. These events are then detected as the second stage. It is believed that by the time the second stage is observed ellipsometrically, the subcoating corrosion processes are widespread over the specimen face.

As corrosion products build up in anodic areas they contribute to the stresses which finally cause coating rupture. With relatively free access to the electrolyte at the rupture site, the anodic activity accelerates, and the electrolyte pH in cathodic regions may exceed 14. Experiments with collodion coated Au in NaOH at pH 15 indicate coating deterioration after prolonged exposure. This deterioration may be partially responsible for the erratic behavior of Δ noted during the third stage.

---

[2] The limitation of active sites may be related to the number of susceptible crystal planes at the surface. For example, the iron [110] plane shows a higher tendency toward passive film breakdown and pitting than do other crystal faces.[7-9]

TABLE 1 — Stage II Rate Constants for Ellipsometrically Measured Subcoating Events on Collodion Coated Iron in 0.05 N NaCl

Summary of Ellipsometric Rate Constants for Stage II Events
Semilogarithmic Behavior Assumed
$d\Delta = k \log t + C$

Specimen	Immersion medium	k(deg)
Uncoated Fe	Saturated NaOH	76
Coated Fe	0.05 M NaCl	83
Coated Fe with deliberate holiday	0.05 M NaCl	73
Coated Fe with phosphate chromate surface pretreatment	0.05 M NaCl	76
Coated Fe with K_2CrO_4 "islands"	0.05 M NaCl	8
Coated Fe with $ZnCrO_4$ "islands"	0.05 M NaCl	35
Coated Fe with ZTOC[1] dispersed in coating	0.05 M NaCl	58
Coated Fe with K_2CrO_4 dispersed in coating	0.05 M NaCl	42
Coated Fe with $SrCrO_4$ located between two layers of collodion	0.05 M NaCl	56

[1]Zinc tetraoxychromate $[Zn(OH)_2]_4 \cdot ZnCrO_4$

The effects noted in the presence of K_2CrO_4 inhibitor are a generally lower subcoating pH, and comparatively less surface roughening and oxide film growth. These results are consistent with the reported behavior of CrO_2^{-2} in supressing anodic dissolution of iron through the formation of an effective passive film. Under these conditions, the cathodic reduction of O_2 must also be supressed, and the subcoating environment becomes only mildly alkaline. Therefore, surface roughening and oxide film growth are also attenuated. Presumably, this minimization of substrate topographical and constitutional changes translates into reduced coating delamination and, hence, a longer coating life.

Summary and Conclusions

1. The combined ellipsometric electrochemical technique provides an effective, nondestructive, *in situ* approach to the study of corrosion processes under transparent coatings.

2. The studies reveal the growth of oxide film and surface roughening on those portions of the substrate under the coating which are subjected to a highly alkaline environment.

3. It is proposed that the initial changes in topography, and possibly in the constitution of the substrate surface, may contribute to coating delamination.

4. The presence of inhibitors such as K_2CrO_4 attenuates active dissolution of iron, and thus suppresses OH^- buildup. The resultant attenuated film growth and surface roughening can be monitored ellipsometrically, thus suggesting that the new technique may be a valuable tool for the evaluation of inhibitor performance.

References

1. Henry Leidheiser, Jr. Progress in Organic Coatings, **7** p. 79, (1979).
2. J.J. Ritter and J. Kruger, Surface Science, 1980, **96**, p. 364, (1980).
3. J. Kruger, *Corrosion,* **22**, p. 88, (1966).
4. F.P. Fehlner and N.F. Mott, Oxidation of Metals, **2**, p. 59, (1970).
5. J.E.O. Mayne and M.J. Pryor, *J. Chem. Soc.,* p. 1831, (1949).
6. J. Kruger, Advances in Electrochemistry and Electrochemical Engineering, R.H. Muller, Ed., **9**, (1973), J. Wiley and Sons, p. 250.
7. J. Kruger, *J. Electrochem Soc.,* **106**, p. 763, (1959).
8. J. Kruger, *J. Electrochem Soc.,* **110**, p. 654, (1963).
9. C.L. Foley, J. Kruger and C.J. Bechtholt, *J. Electrochem Soc.,* **114**, p. 994, (1967).

Electrochemical Values: Their Significance When Applied to a Coated Substrate

*M. Piens, R. Verbist**

Introduction

Electrical and electrochemical methods are often used to study the protective capacity of anticorrosion paints. Several review articles, published recently, have provided a synthesis and a useful critical assessment of these methods.[1-3] Most of the measurements were made under direct current or at a single frequency. Very few studies exist where the coated metal has been investigated over a large range of frequencies.[4-7]

Electrochemical studies of bare metals have, however, shown the complexity of electrode kinetics. The processes which take place at the interphase constituted by the metal and the corrosive medium are numerous and are by nature very different:[8] electrochemical reactions; chemical reactions; solvation; adsorption of the intermediates of reactions; transport of material by migration; diffusion; and natural or forced convection. The presence of an organic coating on the metal in no way lessens this complexity; on the contrary, it introduces additional electrical and electrochemical properties (the dielectric behavior and ionic resistance of the coating, its barrier effect on diffusion of chemical species).

Measurements made under direct current bring into play all these processes, from the slowest to the most rapid. The values obtained this way can therefore be considered as providing an overall indication. When alternating current is used, only those processes which have sufficient time to take place during the alternation of the electric field come into play, and the slower processes no longer occur. When the measurements are made at only one frequency (even a high frequency), one can neither exclude the possibility that the values obtained result from the interplay of several phenomena, nor the possibility that the physical interpretation to be given to the results is affected. Thus in order to determine the precise physical significance of the electrochemical values, the various processes outlined above must be separated. As has been stated, these processes can be separated according to their relaxation time by using different frequencies, provided that the relaxation times are not too close. The measurement of the impedance of an electrode over a wide range of frequencies (from a few mHz to several kHz) thus provides an analytic examination of the various processes brought into play by a direct current in the system under study. The aim of this publication is to illustrate, by means of some selected examples, the varied electrochemical behavior of the metal-paint couple, and to demonstrate the advantages and limitations of the electrochemical method for the study of the protective capacity of paints.

Experimental Method

The electrochemical cell was prepared in the following manner: a circular area was demarcated on the surface of the painted specimen by covering it with an insulating sheet (Scotchrap 50), with this area cut out. A glass cylinder was attached to this sheet with mastic (Gebsicone), and the container thus formed was filled with a 0.5 NaCl solution. The platinum counter electrode had a platinized surface area of 12 cm².

This cell constituted the input resistance of a CA 3140 operational amplifier connected as voltage inverter. Feedback resistors ranging from 10Ω to 10 MΩ were used in order to keep the gain at a level of about 1 for all the impedance spectrum readings.

The experimental set up used was proposed by Scantelbury et al.[7] (Figure 1).

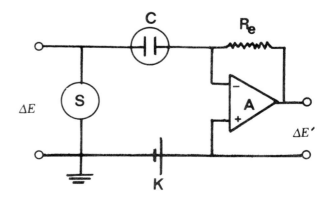

FIGURE 1 — Diagram of the experimental set up, where: S = Solartron 1174 transfer function analyser; K = Keithley 241. d.c. voltage supply; C = electrochemical cell with 2 electrodes; R_e = metal film feedback resistor; A = CA 3140 operational amplifer.

---
*Laboratoire I.V.P., Limelette, Belgium

The sinusoidal output voltage, at an amplitude ΔE of 10 mV, of the generator in S reappears multiplied by the gain at the amplifier's output, i.e., $\Delta E'$.

As the gain is determined by the ratio between the feedback resistor R_e and the impedance Z of the electrochemical cell, the latter is given by:

$$Z = \frac{\Delta E}{\Delta E'} \cdot R_e$$

The impedance spectra were obtained by plotting the measurements made for Z at different frequencies in a complex plane.

The measurements of the polarization resistance were made with a three electrode cell (cell C + calomel electrode), using a "corrovit" type of apparatus (Tacussel).

Discussion

Simple Behavior

The impedance spectra presented in Figure 2 were obtained with a polyurethane varnish of 34 μm

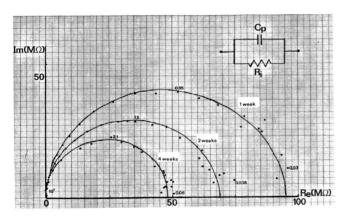

FIGURE 2 — Relation of the impedance spectrum of a polyurethane varnish (34 μm) applied to Armco iron to the period of immersion in 0.5M NaCl; parameter:hertz. (Surface area = 7 cm^2).

thickness applied to polished Armco iron and immersed in 0.5 M NaCl. The shape of each spectrum, a semicircle passing through the origin, indicates[9] that the electrical behavior of the system examined can be represented in each case by an analogous circuit consisting of a capacitor and a resistance in parallel (Figure 2). As the value of the resistance is given by the diameter of the circle, the capacitance can then be calculated from the equation:

$$RC \cdot 2\Pi f_{max} = 1$$

where f_{max} is the frequency associated with the highest point on the semicircle.

The physical significance of the resistance can be shown easily. Electrochemical processes do not obey Ohm's law. The relations between the strength of the current which crosses an electrode and the value of the polarization voltage which produces this current are exponential[10,11] and nonlinear. Thus, in order to measure an electrochemical impedance, the experimenter must use low polarization amplitudes ($\sim$10 mV), so that the exponential relations can be linearized by approximation. But if the diameter of the capacitive semicircles in Figure 2 are measured with increasing polarization voltages, it can be seen that the resistance measured is independent of the polarization amplitude (Figure 3).

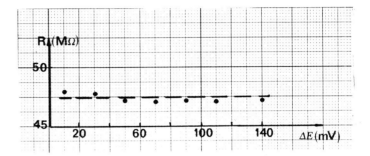

FIGURE 3 — Measurement of R at 0.1 Hz related to the polarization voltage ΔE after 4 weeks of immersion in 0.5 M NaCl. Polyurethane varnish (idem Figure 2).

Since the resistance obeys Ohm's law it cannot be associated with any electrochemical process; therefore it is not a measurement of the corrosion rate. It represents the ionic resistance (R_i) of the coating. Under the effect of the electric field created between the electrodes, the ions are attracted to the anode or the cathode, according to the sign of their charge. The ionic resistance measured is the resistance offered by the coating to this transport by migration. It depends on the concentration of ions within the coating, their valency, their mobility[12] and, of course, on the dimensions and the number of penetration passages available to the electrolyte per unit of electrode surface area. The decrease in R_i during immersion (Table 1) can thus be interpreted as a penetration of the electrolyte into the coating and as an enhancement of its porosity.

TABLE 1.

Immersion time	R_i MΩ.cm^2	C_p pF.cm^{-2}	R_p MΩ.cm^2	$E_{corrosion}$ (SCE)
1 week	665	240	$\sim$700	-234
2 weeks	480	220	546	-414
4 weeks	340	225	364	-361

Table 1 also lists the capacitance values calculated from Equation (1) using the spectra in Figure 2. These values correspond to those which can be calculated for the parallel plate metal/paint/solution capacitor, a relative dielectric constant $\varepsilon = 9$ being ascribed to the coating. This value, which is higher than those obtained for dry resins, is acceptable when one takes into account the fact that the absorption of water by resin leads to an increase in its dielectric constant.[2,14] The capacitance C_p in parallel with the ionic resistance is, therefore, a characteristic of the continuous part of the coating.[14]

On immersion, the C_p value increases, due to the absorption of water;[4,14] then subsequently stabilizes (Table 1).

A behavior which was comparable in all respects to that of the polyurethane varnish (Figure 2) was observed by Scantelbury and Ho with chlorinated rubber.

The anticorrosive properties of paints are often studied under direct current, or at a low frequency, using the polarization resistance technique. Figure 4 presents the polarization curves of the polyurethane varnish recorded immediately after taking readings for the impedance spectrum. The voltage sweep (12.5 mV) was generated at a frequency of 10^{-2} Hz. The above frequency is, in this case (Figure 2), sufficiently low to ensure that the impedance of the capacitor C_p, formed by the coating, is very high in relation to the ionic resistance R_i. The latter short circuits C_p, and therefore one finds no capacitive element in the "forward-return" plot of the polarization curves (Figure 4).

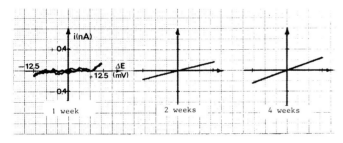

FIGURE 4 — "Forward-return" plot of the polarization curves of polyurethane varnish (idem Figure 2) at different periods of immersion in 0.5 M NaCl.

At a frequency of 10^{-2} Hz, the system under study behaves like a pure resistance. The measurements of the polarization resistance are in agreement with the resistance values obtained from the impedance spectra (Table 1). The polarization resistance technique gives the ionic resistance of the coating; therefore, the physical significance of R_p is precise and unrelated to the corrosion rate of the metal support ($i_{corrosion}$). Since the polarization is completely absorbed by the ohmic drop, the well known equation $R_p = B/i_{corr}$ does not apply.[15-17]

A Complete Description of the Behavior

The preceding electrical model (Figure 2), of a capacitor and a resistance in parallel, is clearly inadequate if one considers the physical significance of R_i and C_p. This model does not allow for corrosion, since the metal-electrolyte interphase at the base of the pores is not represented. However, the corrosion phenomenon takes place at the interface: the transparency of the polyurethane varnish made it possible for us to observe the formation of a spot of rust during the immersion period. The same observation was reported by Scantelbury for chlorinated rubber.[7] One must conclude that the information relating to the electrochemical processes (which should have appeared on the impedance spectra) was, in our case, obscured by the high level of the ionic resistance.

To be a useful tool in electrochemical investigations, a model of the metal-paint couple must be able to describe both the manner in which reality is conceived, and the information obtained by experimental measurements and observation. The model presented in Figure 5 is an attempt along these lines.

The physical significance of the elements C_p and R_i has been described above. The model includes, among other features, the traditional electrochemical representation of the metal-electrolyte interphase: the double layer capacitance C_D in parallel with the faradaic impedance. This was represented in our model by placing the transfer resistance R_t and the diffusion impedance Z_d in series.

The transfer resistance R_t is linked to the activation of the oxidation and reduction processes responsible for corrosion. It is the resistance to the transfer of charge across the interface. It is the only electrochemical value which is closely linked to the corrosion rate of the metallic support[8].

The convection diffusion impedance Z_d was introduced into the model since it is known[19-20] that the transport of matter can have a considerable influence on the electrode kinetics of bare metals. This influence is more likely to be felt when the electrolyte is confined within an organic coating; furthermore, its influence has often been evident in practice. The convective diffusion impedance Z_d[21] given by:

$$Z_d = \frac{\sigma}{\omega^{1/2}} (1-j) \tanh\left[\delta\sqrt{\frac{j\omega}{D^*}}\right]$$

where σ = Warburg coefficient[22]; $\omega = 2\Pi f$; $j^2 = -1$; δ = Nernst diffusion layer thickness; D^* = some average value of the diffusion coefficients of the diffusing species; f = frequency.

appears on the impedance spectrum as a straight line with a unit slope (when $\tanh(\delta\sqrt{\frac{j\omega}{D^*}}) \to 1$) which bends towards the real axis at very low frequencies.

According to the model, a measurement made with direct current, or at a frequency which is sufficiently low for the impedances of C_p and C_D to be high compared to that of the other elements, gives the total resistance $R_i + Z_d + R_t$. The resistances measured on the polyurethane varnish are thus equivalent to the sum $R_i + Z_d + R_t$, but with $R_i \gg Z_d + R_t$. The metal-paint interface is thus incorporated in the representation of the system without altering the physical significance of the measurements.

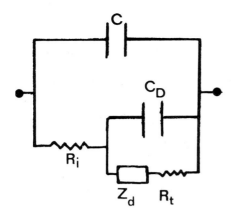

FIGURE 5 — A model of the metal-paint couple, where: C_p = capacitance characterising the continuous part of the coating; R_i = ionic resistance of the coating; C_D = double layer capacitance of the metal-electrolye interface; Z_d = diffusion impedance; R_t = transfer resistance.

The spectrum in Figure 6 is an example in which all the constituent elements of the model can be distinguished. This graph was obtained with chlorinated rubber, pigmented with iron oxide, and brush applied to sand blasted steel. The readings were taken at the corrosion potential (–553 mV/SCE) after 24 hours immersion in 0.5 M NaCl.

The two capacitive semicircles, covering the frequency ranges 10^5 Hz to 280 Hz, and 280 Hz to 0.28 Hz, respectively, correspond to the couples R_i, C_p and R_t, C_D. These values can be calculated, as before, for the polyurethane varnish (Figure 2). The extrapolation of the arcs of the circle to their points of intersection with the real axis, however, introduces a margin of error which could be considered unacceptable. In this case, another way of representing the data which facilitates the determination of the values R, C[9,22] may be used.

The linear section lying between 0.28 Hz and 10^{-2} Hz is characteristic of the diffusion impedance Z_d. Its presence in the spectrum proves that the rate-determining step, in the corrosion kinetics of painted metal, is the transport of matter by diffusion through the coating.

The measurement of the impedance of an electrode over a wide range of frequency constitutes a form of analytical examination of electrochemical behavior. It enables one to separate the various processes which play a role under direct current conditions in the system studied; in our case, the various processes participating in the corrosion phenomenon, and to determine their relative contribution. By contrast the polarization resistance technique and measurements made with direct current always produce overall values.

In Figure 7, the measurement under direct current of the resistance of the sample is given. The 12.5 mV anodic polarization applied initially was cut after 250 seconds; the intensity of the current measured under steady state conditions can be used to calculate the total resistance (R_{DC}), namely 13.3 kΩ.

This value is higher than the greatest value of R found in Figure 6, since the readings for the impedance spectrum were not continued beyond 10^{-2} Hz and we know from Equation 2 that the diffusion impedance must bend over further towards the real axis.

The "forward-return" plot of the polarization curves for the chlorinated rubber was made at 10^{-1} Hz and at 10^{-2} Hz. The R_p values obtained were 8.0 kΩ and 10.6 kΩ, respectively. They agree with the resistance found on the impedance spectrum (Figure 6) for the corresponding frequencies.

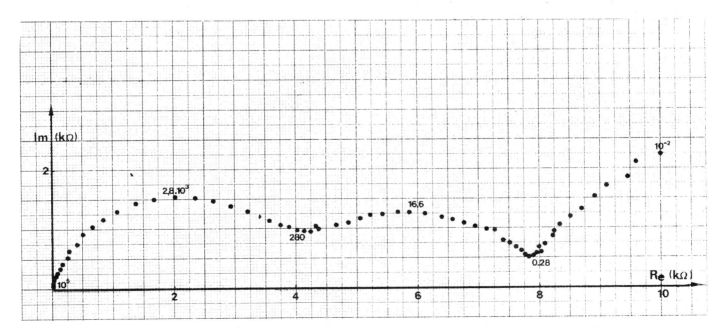

FIGURE 6 — Impedance spectrum of a chlorinated rubber-iron oxide paint (40 μm) applid to sand blasted steel after 24 hours immersion in 0.5% NaCl. Surface area = 28 cm²; parameter: Hz.

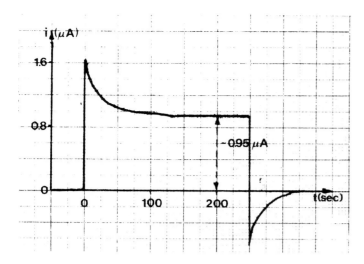

FIGURE 7 — Variation of the current intensity with time after applying an anodic overvoltage of 12.5 mV. Chlorinated rubber (idem Figure 6).

These results demonstrate the difficulties faced in the interpretation of data obtained for painted metals (under direct current or with the polarization resistance technique), as the measured values have no precise physical significance. Thus, in the example (Figure 6), the R_p value at 10^{-1} Hz is by chance close to the $R_i + R_t$ value, but at 10^{-2} Hz its value is the sum of three contributions: the ionic resistance; the charge transfer resistance; and an indeterminate fraction of the diffusion resistance. The resistance measured under direct current (R_{DC}) is (provided that a steady state is reached) the sum of the resistances associated with the processes which describe the behavior of painted metal in a corrosion situation[1], but since the relative contribution of each one remains unknown, identical values of R_{DC} can correspond to different situations. The resistances R_p and R_{DC} only take on a precise meaning when one of the processes predominates; in the case of painted metal, when the polarization is completely absorbed by the ohmic drop (Figure 2). But under these conditions, the use of low polarization amplitudes (~10 mV), as an experimental necessity, is no longer justified (Figure 3), and industrial laboratories will find it advantageous to use simple and inexpensive equipment to measure R_i.

Unfortunately, the relationship between ionic resistance and protective capacity is only qualitative.[2] The charge transfer resistance R_t is the only electrochemical value which is closely linked to the corrosion rate of the metal support. The equivalence between the corrosion rate calculated from R_t, and that measured with the weight loss, was established in a 1 N H_2SO_4 solution for iron protected by an epoxy paint.[23]

---

[1] The relaxation of the adsorption of intermediates, such as $(FeOH)_{ads}$ and $(FeH)_{ads}$ in the case of the corrosion of iron in a sulphuric medium, can give rise to very low frequency inductive semicircles on the impedance spectrum. These semicircles were observed even when an epoxy paint covered the metal.[23] In that case, the measurement of R_{DC} loses the significance described here.

A More Frequently Observed Type of Behavior

Measurement of the corrosion rate of painted metal is, without doubt, the best method of quantitatively determining the protective capacity of a coating. To calculate this rate it is necessary to separate the transfer resistance R_t from the other elements in the impedance spectrum. This separation is possible if an acid medium is used,[23] or if the total resistance of the coating is low (Figure 6); in other words, if the protective coating has a poor level of effectiveness. As long as the coating continues to act essentially as an insulator (Figure 2) the corrosion rate is not measurable. In addition, once a deterioration in its behavior takes place, it is often the contribution of the diffusion impedance which first makes its presence felt on the spectrum, and R_t, unfortunately, remains indeterminable.

To illustrate this evolution, which we have observed on numerous occasions, we chose (Figure 8) the case of the sample of polyurethane identical to that studied previously (Figure 2), but which was naturally aged for 13 months in a rural environment before the measurements were made.

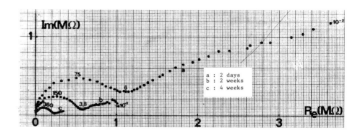

FIGURE 8 — Relation of the impedance spectrum of a polyurethane varnish (30 μm) applied to Armco iron to the period of immersion in 0.5 M NaCl after a natural aging of 13 months; parameter:Hz (Surface area = 7 cm²).

In addition to the semicircle due to R_i, C_p, a contribution was observed at lower frequencies which we ascribed to diffusion, despite the fact that it was not in the form of a straight line with unit slope. Until now, few studies have been made dealing with the behavior of painted metal at low frequencies, and the theoretical framework which would make it possible to exploit fully this part of the spectrum is lacking. However, we have good reason to believe, after some preliminary simulation tests carried out with a computer using Equation 2, that the shape of this part of the spectrum is due to the heterogeneity of the diffusion conditions in the different pores of the coating. This heterogenous aspect of the coatings is well known[23,24] and has been successfully simulated in the description of high frequency behavior.[5]

At present, the difficulty involved in dealing with the diffusion impedance, and in isolating the transfer resistance, limit the use of electrochemical methods for the quantitative determination of the protective capacity of a coating. We emphasize that this limitation is not inherent in the electrochemical method, but is due to the nature of the electrode studied: coated metal.

By contrast with the measurement of the polarization resistance and of the d.c. resistance, an impedance spectrum recorded over a wide range of frequencies has the great advantage that it brings into view the difficulties involved in the interpretation of certain values, and it thus gives a better preception of the real state of things.

Conclusion

In a corrosion situation, the metal-paint sysem is the site of various electrical and electrochemical processes, whose relative contributions evolve during aging. The measurements made under d.c. conditions reflect the totality of these processes. To obtain precise, physically significant values, it is, therefore, necessary to separate the different contributions involved in d.c. behavior. Impedance measurements over a wide range of frequency makes it possible to realize this aim provided that the resistances associated with the different processes are of the same order of magnitude and that the relaxation times are not too close.

Unfortunately, in most cases, the ionic resistance of an intact paint film is high and obscures other components of the behavior of the paint-substrate system. But, when the ionic resistance is sufficiently low, or has become so after ageing of the paint film or immersion in an electrolyte, these other contributions, i.e. that of the diffusion of chemical species through the coating and that of the paint-substrate interface, appear in the impedance spectrum and the paint-substrate system can be characterized completely.

The only value allowing the measurement of the corrosion rate of the coated metal and which, as a consequence, is a quantitative expression of the protective efficiency of the paint is the charge transfer resistance. The other values (ionic resistance, dielectrical capacity, diffusion impedance) are but qualitatively related to the protective efficiency.

Acknowledgment

Acknowledgment is made to Dr. P. Janssen Bennynck, Director of the Laboratoire I.V.P., for the stimulating discussions which we have had. This work was supported by l'Institut pour l'Encouragement de la Recherche Scientifique dans l'Industrie et l'Agriculture (I.R.S.I.A.).

References

1. H. Jullien, Progr. Org. Coatings, **2**, p. 99 (1973).
2. H. Leidheiser Jr., Progr. Org. Coatings, **7**, p. 79 (1979).
3. J. Wolstenholme, Corros. Sci., **13**, p. 521 (1973).
4. H. Leidheiser Jr., M.W. Kendig, Corrosion, **32**, p. 69 (1976).
5. M.W. Kendig, H. Leidheiser Jr., J. Electrochem. Soc., **123**, p.982 (1976).
6. G. Menges, W. Schneider, Kunststofftechnik, **12** (11), p. 316 (1973).
7. J.D. Scantelbury, K.N. Ho, J. Oil Col. Chem. Assoc., **62**, p. 89 (1979).
8. T. Erdey-Gruz, Kinetics of Electrode Processing, Adam Hilger, London, p. 18 (1972).
9. R.D. Armstrong, M.F. Bell, A.A. Metcalfe, J. Electroanal. Chem., **77**, p. 287 (1977).
10. J.O'M. Bockris, A.K.N. Reddy, Modern Electrochemistry, MacDonald, London, p. 883 (1970).
11. P. Janssen Bennynck, M. Piens, Progr. Org. Coatings, **7**, p. 114 (1979).
12. G. Kortum, Treatise on Electrochemistry, Elsevier, Amsterdam, p. 25 (1965).
13. F.W. Billmeyer, Textbook of Polymer Science, J. Wiley, New York, p. 502 (1966).
14. D.M. Brasher, A.H. Kingsbury, J. Appl. Chem., **4**, p. 62 (1954).
15. F. Mansfeld, Advances in Corrosion Science and Technology, **6**, p. 163 (1976).
16. E. Heitz, W. Schwenk, Br. Corros. J., **11** (2), p. 74 (1954).
17. LM. Callow, J.A. Richardson, J.L. Dawson, Br. Corros. J., **11** (3), p. 132 (1976).
18. A. Caprani, I. Epelboin, Ph. Morel, H. Takenouti, 4th Eur. Symp. on Corr. Inhib., Ferrara (1975).
19. J.N. Agar, Discussions Faraday Soc., **1**, p. 26 (1947).
20. T. Erdey-Gruz, Kinetics of Electrode Processes, Adam Hilger, London, p. 97 (1972).
21. D. Schuhmann, C.R. Acad. Sc. Paris, C, p. 624 (1966).
22. J.H. Sluyters, Recueil Trav. Chim., Pays-Bas, **79**, p. 1092 (1960).
23. L. Baunier, I. Epelboin, J.C. Lestrade, H. Takenouti, Surface Technology, **4**, p. 213 (1976).
24. J.E.O. Mayne, D.J. Mills, J. Oil Col. Chem. Assoc., **58**, p. 155 (1975).

Properties and Behavior of Corrosion-Protective Organic Coatings as Determined by Electrical Impedance Measurements

John V. Standish[1] *and Henry Leidheiser, Jr.**

Introduction

Over the past 25 to 30 years, studies have appeared in the literature which describe the use of electrical impedance measurements as a technique for characterizing organic coatings. These measurements have been used, for example, to investigate diffusion of water into coatings,[1-5] corrosion beneath organic coatings,[5-7] and the effect of aging and pigment on the dielectric properties of coatings.[8] Many of these studies have been reviewed by Leidheiser.[9] In addition, impedance measurements have been used to study the curing of thermoset polymers[10] and to identify the presence of condensed water within polymers.[11-13]

In spite of the many applications described in the literature, impedance measurements have yet to find widespread use in the development of new coatings, or in the study of corrosion phenomena involving polymer coated metals. One of the larger factors limiting the use of impedance measurements was, until recently, the availability of experimental equipment that is convenient to use, as well as applicable to a variety of coatings related problems. Impedance measurements have traditionally been conducted by bridge balancing methods, in which the sample is placed in one arm of a bridge and the values of a variable capacitor and resistor in parallel in a second arm are used to balance the bridge to determine the equivalent parallel capacitance (C_p) and resistance (R_p) of the sample.[14,15] Each bridge balancing operation is tedious and time consuming. If measurements are to be made as a function of time or temperature, constant operator attention is required.

Several of the more recent methods used for dielectric measurements have been discussed by Hedvig,[10] and the application of such measurements to problems in electrochemistry has been discussed by a few authors.[16-18]. One of these methods, known as phase sensitive detection, can be conducted with a commercially available piece of equipment known as a lock-in amplifier (LIA). The use of an LIA provides advantages over bridge balancing methods: it can be applied to a wide variety of coatings problems and is relatively inexpensive.

This article provides a summary of studies carried out in our laboratory on the application of dielectric measurements to obtain a better understanding of the corrosion mechanisms related to polymer coated metals. The subject areas that will be discussed include: electrical properties as a function of exposure to electrolytes; frequency dependent electrical properties; temperature dependent electrical properties; and the application of electrical measurements to the determination of condensed water in polymer coatings.

Experimental Procedures

In many dielectric studies of polymers, metal electrodes are evaporated onto polymer sheets or films in order to create a parallel plate capacitor with a polymer dielectric. In cases where polymer coatings are studied on metallic substrates, it is necessary to adopt different methods of study. One such technique is shown schematically in Figure 1. A phenolic cylinder, about 3 cm. in diameter, is attached to the surface of a coated substrate in order to create a vessel to hold electrolyte or mercury. The mercury provides electrical contact to the sample surface, but can be easily removed to expose the sample to an aqueous environment. It is also possible to make electrical contact (by immersing an inert electrode in the electrolyte) through the electrolyte and thereby continuously monitor any changes in electrical response as a function of exposure time to an electrolyte.

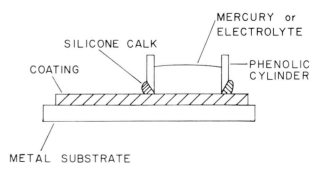

FIGURE 1 — Experimental arrangement used in making dielectric measurements as a function of time of exposure to an electrolyte.

---
[1]Presently, ARMCO Inc., Middletown, Ohio
*Center for Surface and Coatings Research Lehigh University, Bethlehem, Pennsylvania

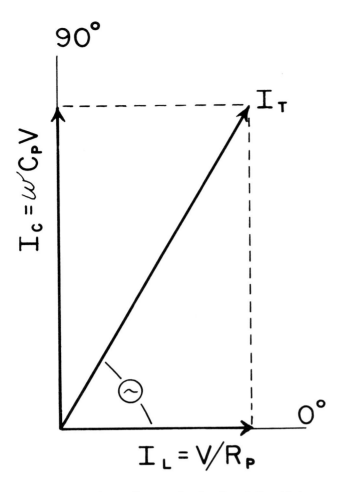

FIGURE 2 — Vector diagram showing the relationship between I_T, I_L, and I_C.

A second method uses a flat metal plate which is clamped against the coating surface under moderate pressure. Tinplate was used as an electrode for the results reported. Small air gaps may exist between the electrode and the polymer surface, but they do not generally introduce appreciable errors, given the accuracy of the LIA and the magnitude of change resulting from the experimental variables.

The application of an alternating excitation voltage to a sample develops a current which is out of phase with the excitation voltage, as shown by the vector diagram in Figure 2. I_T is the total current in the circuit which leads the excitation voltage by a phase angle, Θ. I_T is resolved into its vector components, the loss current, I_L, which is in phase with the excitation voltage, and the charging current, I_C, which is 90 C out of phase with the excitation voltage. The values of I_C and I_L are used to calculate the equivalent parallel capacitance and resistance of a sample through the relationships

$$I_C = \omega C_P V \quad (1)$$

$$I_L = (1/R_P)V \quad (2)$$

where ω is the angular frequency and V is the RMS value of the excitation voltage. The equivalent parallel conductivity, G_P, is equal to $1/R_P$.

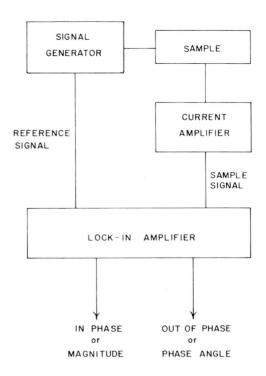

FIGURE 3 — Schematic outline of experimental equipment.

The LIA is capable of providing, automatically and continuously, analog output voltages proportional to each value of I_T, Θ, I_L and I_C. The experimental equipment is shown schematically in Figure 3. A wave generator supplies an alternating voltage to the sample and to the reference channel of the LIA (Princeton Applied Research model 5204). The sample current is input to a current sensitive amplifier (PAR model 181), which both amplifies the current and converts it to a proportional voltage which is input to the LIA. Any phase shift in the signal which does not originate in the sample can be nulled out before an experiment by placing an air capacitor in place of the sample. The value of I_L for an air capacitor is zero, and, therefore the reference signal can be adjusted to compensate for any in-phase component which might exist. References 19 and 20 discuss LIAs and phase relationships in greater detail. The output voltages of the LIA are conveniently read from a digital voltmeter, or recorded on a strip chart or x-y recorder.

The coatings used in this study were primarily epoxy resins (Epon 828 or 1002) cured with a polyamide resin (Versamid 125 or 100). Samples were prepared by mixing stoichiometric amounts of the epoxy and polyamide resins and either casting or spraying films on metal substrates from solvent. The solvent used was usually a blend of xylene, methylisobutylketone, and n-propanol. Samples were baked at 100 C for 20 minutes after coating. It is recognized that epoxy/polyamide coatings are not often baked in practice, but were in this case to eliminate or reduce any effects of residual solvent or sample aging. Free polymer films were made by applying the coating on tinplate and dissolving the tin substrate with mercury.

Experiments were carried out as a function of time of exposure to an electrolyte or frequency using the sample configuration shown in Figure 1. Experiments carried out as a function of temperature were conducted by clamping a sample between two metal plate electrodes. The sample, placed in an insulated container, was cooled in liquid nitrogen. Measurements were made as the sample warmed at a rate of a few degrees per minute.

Results and Discussion

Electrical Properties as a Function of Time of Exposure to an Electrolyte

An example of the effect of exposure of a coated metal to an electrolyte on the electrical properties will be given. Figure 4A shows the change in C_P at 100 Hz

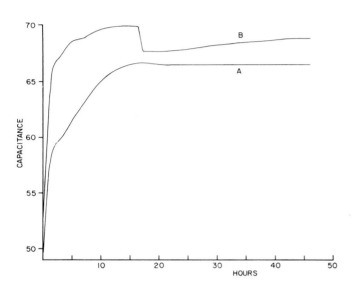

FIGURE 4 — Capacitance changes at 100 Hz during exposure of a coating of Epon 1002/Versamid 125, 61 μm in thickness, to 3% NaCl solution. (A) Initial exposure, and (B) second exposure after drying sample in desiccator for 2 days.

for a sample of Epon 1002/Versamid 125 (61 μm thick) exposed to 3% NaCl for 46 hours. The capacitance of the sample increased for about the first 16 hours of exposure, then leveled off. The increase in capacitance that is observed is generally considered to be caused by the diffusion of water and ions into the coating. After a 46 hour exposure, the sample in Figure 4A was dried in a desiccator for two days and exposed again to 3% NaCl. Figure 4B shows the C_P results obtained during the second exposure. For the second exposure, the value of C_P at $t = 0$ was greater than for the initial exposure. Further, the value of C_P rose more rapidly during the second exposure. These results demonstrate that the effects of exposure of a coating to an electrolyte are not totally reversible. It is our view that the first exposure and drying cycle either caused physical damage within the coating, or ionic species remained trapped within the film which on second exposure to the electrolyte allowed water to diffuse into the coating at a faster rate.

Figure 4B also demonstrates that values of C_P do not always increase with time of exposure; C_P decrease in some cases. The decrease in C_P shown in Figure 4B took place gradually over a period of about 12 minutes in this particular experiment. The reason for the decrease in C_P is not known, but probably reflects some change that took place in the coating. Based on experience with a variety of coatings, it was often observed that changes in the impedance values took place which were difficult to interpret. Impedance values are influenced by a variety of chemical or physical changes in the sample and many of these changes are not readily recognizable without elaborate experimental planning.

Changes in conductivity, G_P, are usually much larger in magnitude than changes in C_P. Figure 5 shows the changes in G_P which took place during exposure of the same sample used to obtain the data in Figure 4. Changes in G_P are often so large that they are best shown on a semilog plot.

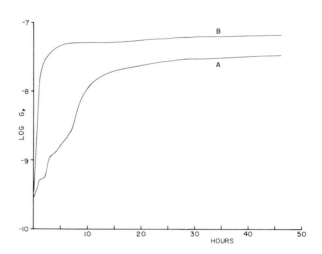

FIGURE 5 — Changes in equivalent parallel conductivity, G_P, for the same sample shown in Figure 4. (A) Initial exposure, and (B) second exposure.

As mentioned above, the values of C_P and G_P leveled off after only about 16 hours exposure time. It is important to keep in mind that the coating used to obtain Figures 4 and 5 was an unpigmented coating without the several additives typically used to improve the performance of commercial coatings. The coating may also be thinner than what would be used for many corrosion resistant applications. In our experience with commercially applied coatings about 250 μm thick, the values of C_P and G_P may increase for several days before they level off to a more or less constant value. Both the composition and the thickness of a coating influence the rate and magnitude of change of the impedance values.

Measurements of capacitance change during exposure to an electrolyte have been used to calculate the water uptake of organic coatings.[2,5] These calculations relied on empirical equations derived to determine the volume fraction of a given component in a two-phase dielectric, where the dielectric constants of each phase and of the composite are known. The use of these equations requires the assumptions that: the water exists as a second phase within the coating; that the dielectric constant of water in the film is the same as that of bulk water; and that water sorption does not appreciably swell the film. The best known empirical equation is one derived by Brasher and Kingsbury[2,9] and another derived by Bottcher has been used by Holtzman.[1,9]

Values of water uptake, as determined by the Brasher equation, have been compared by others to values determined gravimetrically under conditions in which the experimental error is high. The method consists of immersing the sample in the electrolyte for a period of time, removing it from electrolyte, drying the coating superficially and weighing. In some cases fair agreement was found between capacitance and gravimetric values for small amounts of water uptake,[5] and in other cases agreement was poor for large amounts of water uptake.[3]

A film of Epon 1002/Versamid 100 was used to evaluate these empirical equations for this system by employing a more accurate gravimetric means than has been used before. The water sorption of the film was determined by exposing the film to various relative humidities in a quartz spring microbalance as described in the literature.[21] A value for 24-hour immersion in deionized water was also obtained with a five-place analytical balance. The sample used for the capacitance measurements was exposed over various salt solutions to obtain the desired relative humidities. A sample was exposed to each relative humidity for 24 hours before the measurement was taken. Results are shown in Table 1.

First, it is of interest to note that the film contained about 2% water at 84% relative humidity, and the water content of the film did not increase greatly after water immersion. Second, it is apparent that the empirical equations do not predict water content at high relative humidities for the epoxy-polyamide coating.

TABLE 1 — Comparision between Water Sorption by Epoxy-Polyamide Coating as Determined by Capacitance and Gravimetric Measure

Relative Humidity to Which Coating Exposed for 24 Hours	Weight Percent Water Uptake		
	Gravimetric	Calculated from Brasher Equation	Calculated from Bottcher Equation
20%	0.47	0.33	0.50
43%	0.95	1.19	1.83
65%	1.54	4.67	7.33
84%	2.02	6.05	9.16
Water Immersion	2.18	7.15	10.99

Differences between gravimetric and capacitance results have been used to draw conclusions concerning the distribution of water in the coating.[4,9] It is more likely, however, that the equations are not adequate for predicting the water content of all types of polymeric films. Better relationships for determining the water content of polymer films and coatings are required. Other authors[22,23] have assumed that water diffuses into a coating as a layer parallel to the surface, and have used capacitance measurements to determine the thickness of the coating penetrated by water as a function of time. The assumption that water diffuses as a layer through a polymer is not consistent with generally accepted theories of diffusion through polymers.[24] The theoretical approaches and conclusions described in References 22 and 23 are, therefore, of uncertain value.

In spite of the uncertainty in using capacitance values for calculating water content of coatings, Holtzman[1] has successfully used capacitance measurements to determine the permeability of coatings. The Holtzman method utilizes the logarithm of the rate of change with time of the water content of the film for exposure times to an electrolyte of less than 5.5 hours. Since the deviation between gravimetric values of water content and those determined by the empirical equations discussed above is less at lower water contents (shorter exposure time), and since the logarithm of the rate of change is the basis of determining permeability values, errors in the absolute value of water content can apparently be tolerated.

Frequency-Dependent Electrical Properties

Leidheiser and Kendig[5] have shown that the frequency response of a coating depends on whether or not: the coating has been exposed to an electrolyte; and whether or not the coating has been penetrated by the electrolyte to create conductive pathways. Similar behavior has been noted by Bellobono, et al.[25] Table 2 shows C_P, G_P and Θ values obtained for Epon 1002/Versamid 125 coating, 28 μm thick over cold rolled steel. Sample A is an area of the coating overlaid with mercury, after exposure to the laboratory atmosphere, but before exposure to an electrolyte. Sample B is an area of the coating exposed to 3% NaCl for about 5 days. Sample C is an area of the coating showing one small corrosion spot, after about a 5 day exposure to 3% NaCl. Sample D is an area of the coating purposely damaged with a pin and exposed to 3% NaCl for about an hour.

The results for A are characteristic of dry polymer films not penetrated by electrolyte: C_P varies little with frequency; G_P increases rapidly with frequency; and Θ is about 89° and little influenced by frequency. The results for area B indicate that exposure to an electrolyte can influence the frequency behavior dramatically. Assuming that there is a conductive pathway through the coating at the corrosion spot in C, and that it is probably smaller than the pinhole placed in D,

TABLE 2 — A.C. Electrical Parameters of an Epoxy-Polyamide Coating as a Function of Frequency and Different Conditions of the Coating

Frequency	C_P in picofarads/cm.^{2}	G_P in mhos/cm.^{2}	Θ in degrees
Before Exposure to Electrolyte			
3	77.1	1.17×10^{-11}	89.5
10	79.3	3.12×10^{-11}	89.4
100	77.5	3.58×10^{-10}	89.3
1000	79.6	4.68×10^{-9}	89.2
10,000	80.0	1.50×10^{-7}	88.0
After Exposure to 3% NaCl for 5 Days (no corrosion visible)			
3	183	1.24×10^{-8}	13.9
10	145	1.32×10^{-8}	34.2
100	128	1.44×10^{-8}	79.0
1000	129	2.18×10^{-8}	88.0
10,000	128	2.22×10^{-7}	87.0
After Exposure to 3% NaCl for 5 Days (corrosion spot visible)			
3	580	1.00×10^{-7}	5.6
10	217	9.67×10^{-8}	7.4
100	138	1.04×10^{-7}	38.5
1000	134	1.11×10^{-7}	81.7
10,000	133	4.68×10^{-7}	86.2
After Introduction of Intentional Defect			
3	36,700	1.50×10^{-5}	3.0
10	19,800	1.56×10^{-5}	4.0
100	8,550	1.40×10^{-5}	22.0
1000	2,730	3.56×10^{-5}	25.0
10,000	343	6.83×10^{-5}	16.9

it is possible that the magnitude of the impedance values is related to the area of substrate exposed to electrolyte. These results also show that the effects of exposure are greater and more easily observed at the lower frequencies.

Mayne and Mills[26] and Kinsella and Mayne[27] have described so-called "direct" films where the D.C. conductivity of the film increases as the electrolyte concentration is increased, and so-called "indirect" films where, due to osmotic pressure,[3] the conductivity of the film decreases as the electrolyte concentration increases. "Direct" and "indirect" areas can occur on the same sample. Buller, Mayne, and Mills[28] measured the impedance of polymer films using a bridge at 1592 Hz and found that while DC resistance values could clearly distinguish between "direct" and "indirect" areas, impedance values could not. These authors concluded that only DC methods should be used to assess coating performance. The results in Table 2 show, however, that AC measurements can be used to distinguish between coatings having different physical characteristics. Table 2 also suggests that impedance measurements at about 10 Hz would be more sensitive to "direct" and "indirect" areas than are measurements above 1000 Hz.

Sample C in Table 2 shows relatively high values of G_P soon after introduction of electrolyte but only after several days of exposure did evidence of corrosion (one small black spot) appear. Experience has shown that samples exhibiting high G_P values are likely to show evidence of corrosion after a period of exposure to electrolyte. It is not always true, however, that samples showing corrosion have high values of G_P. The impedance measurements described here reflect the properties of the polymer film and, only to a small and probably insignificant amount, the polymer/metal interface. During long-term studies of commercial coatings, one case was observed in which localized corrosion occurred with no concurrent change in the impedance values. The corrosion was traced to calcium chloride residue that was present on the metal surface prior to coating.

Impedance measurements are useful for determining the presence of defects in a coating, but in some cases it is valuable to know the distribution of defects, or the exact location of the defect. Spatial resolution is especially important in those cases where it is desired to cross section the coating and determine the nature of the defect. A method has been developed in our laboratory which involves passing a probe over the surface of a coated sample and using an LIA to detect variations in impedance on a highly localized scale. This method, which is described in detail in another publication,[29] is based on a method first described by Isaacs and Kendig,[30] and has also been improved upon by Hughes and Parks.[31]

Temperature Dependent Electrical Properties

Measurement of the dielectric properties of materials as a function of temperature is known as dielectric spectroscopy. Such measurements have long been conducted by bridge methods in order to determine glass and secondary transition temperatures, as well as to obtain information concerning the effects of different exposures or additives on polymer properties.[10,14] This type of measurement is greatly facilitated with the LIA.

Figure 6 shows results obtained at 1000 Hz for a sample of Epon 1002/Versamid 100 after storage in a desiccator, after exposure at 65% relative humidity for one day, and after immersion in deionized water for one day. These exposures result in water contents of about 0, 1.5, and 2.2 Wt. % of water, respectively, as determined gravimetrically. Exposure to humidity eliminates the maximum in I_L observed at 75 C for the dry sample which corresponds to the glass transition temperature T_g. The increase in I_c associated with T_g, however, is seen clearly to move to lower temperatures as a result of exposure to water vapor and liquid water. Exposure to 65% relative humidity lowered T_g about 14 C, while water immersion for one day reduced T_g about 23 C. These results are significant in terms of corrosion protective coatings for two reasons. First, the mechanical properties of films are dependent upon water content of the film and may influence tests which evaluate properties such as impact resistance, abrasion resistance, or tensile strength. Second, as a general rule, the permeability of a polymer to water, oxygen, or ionic species is expected to increase as T_g decreases.

The information obtained by dielectric spectroscopy is qualitatively similar to that obtained from

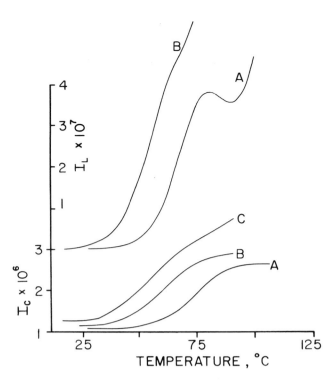

FIGURE 6 — Dielectric results obtained at 1000 Hz for a coating of Epon 1002/Versamid 100 after (A) Storage in a desiccator, (B) after one day at 65% relative humidity, and (C) after one day immersion in water. The I_L data for sample C are similar to those of B and are not shown.

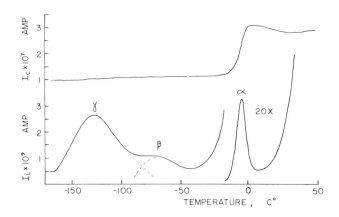

FIGURE 7 — Dielectric results obtained at 1000 Hz for Epon 828 resin.

dynamic mechanical spectroscopy of polymers. The great advantage of dielectric methods is that experiments can be conducted on coatings without removing them from the metal substrate. Dynamic tests, on the other hand, are difficult to conduct with thin films. Torsional braid analysis has recently been applied for investigating the physical properties of coatings, but the method has the shortcoming that impregnation of a glass braid with the coating is required. Roller and Gillham[32] have discussed the application of dynamic mechanical testing to coatings with special emphasis on the use of the torsional braid.

Figure 7 shows results obtained at 1000 Hz for a sample of Epon 828 which is a liquid at room temperature. It is likely that dielectric spectroscopy could be a valuable tool for quality control of coating resins as well as cured films. Molecular weight distribution and purity of a sample, for example, are reflected in dielectric results.

Application of Electrical Measurements to the Determination of Aggregated Water

Bair, Johnson, et al.[11-13] have reported using dielectric spectroscopy to identify the presence of condensed water within polymers such as polysulphone and polycarbonate. Dielectric results were correlated with differential scanning calorimetry and electron microscopy, which showed the volume of condensed water detectable by dielectric means in their experiments was as little as 0.37 Wt. %[11] and the water was located in domains about 2 μm in diameter.[13] When condensed water was present in a sample, a dielectric loss peak was observed at about -90 C at 1000 Hz. This loss peak is thought to reflect a bulk property of ice.[33] Ice is known to exhibit several dielectric relaxations, but the physical origin of these relaxations is not known.[34]

It would be of great value to know if condensed water existed either within a paint film (perhaps within voids, cracks or polymer/pigment interfaces), or at the paint/substrate interface from the standpoint of corrosion. A series of experiments was conducted to investigate the applicability of dielectric spectroscopy for locating condensed water within paint films, or at paint/substrate interfaces.

The rationale for the experiments described below is the following. Small amounts of water are placed between a polymer film and a metal plate electrode, and the assembly is compressed in order to model condensed water at the coating/metal interface. In a similar way, water is compressed between two polymer films attached to metal electrodes in order to model condensed water within a polymer coating. It was observed that the water spread rather uniformly over the interface, although it is not certain that the water film was continuous. The average thickness of the water film at the interface was determined by knowledge of the volume of water applied to the interface and the area over which the water spread. The experiments summarized here were carried out on Epon 1002/Versamid 100 films which had been soaked in 3% NaCl for four days prior to the experiment.

Figure 8a shows results obtained at 1000 Hz for a sample about 150 μm thick after four days in 3% NaCl, but with no electrolyte placed at the polymer/electrode interface. Figures 8b and 8c are results obtained where 3 μL (a layer about 1 μm thick) and 1 μL 3% NaCl, respectively, was placed at the polymer/electrode interface. In 8b relaxations at -105 C and -25 C are seen, as is a discontinuity at -5 C. In 8c the relaxation at -105 C is decreased in intensity, the -25 C relaxation is still clear, and there is no longer a discontinuity at -5 C. When deionized water or .01M NaOH was placed at the polymer/metal interface, or if any solution was

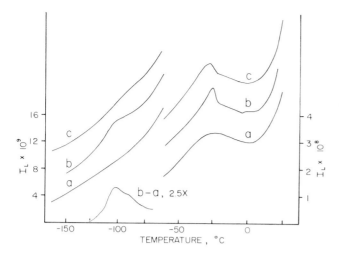

FIGURE 8 — Dielectric results obtained at 1000 Hz for a coating of Epon 1002/Versamid 100 (A) After four days in 3% NaCl, (B) with 3 µL of 3% NaCl at the film/electrode interface, and (C) with 1 µL of 3% NaCl at the film/electrode interface.

placed between two films, the lower temperature relaxation was observed, but the relaxation at -25 C was not evident. Further, when deionized water was used, the lower temperature relaxation occurred at -97 C and, if a discontinuity appeared, it was at about 0 C.

The lower temperature relaxation is thought to be the relaxation observed by Bair, Johnson, et al. This relaxation is a bulk property of water, and is dependent upon the purity and the amount of water in the sample. Whether or not a discontinuity appears at -5 to 0 C is also dependent upon the volume of water, since this continuity results from the melting of ice or electrolyte. The relaxation at -25 C is, perhaps, due to an interfacial phenomenon which takes place with the 3% NaCl and the electrode, and is independent of volume.

The water layers used in these experiments are only a fraction of a percent of the total sample thickness. However, it is recognized that the volume of water in those layers may be large, with respect to the amount necessary to initiate corrosion, or to contribute to destruction of a polymer film. One approach to increasing the sensitivity of the method is by a differential technique such as shown in the lower part of Figure 8. Figure 8a was graphically subtracted from 8b to reveal the -105 C relaxation more clearly. This subtraction was done manually, but could be done quickly and accurately with computer assistance to reveal small differences between various samples.

Summary and Conclusions

Impedance measurements, involving phase sensitive detection, are useful in studies of protective organic coatings. Impedance values are influenced by exposure to an electrolyte or other environment, and provide information concerning the integrity of a coating. The impedance values reflect the properties of the coating; therefore, a coating having a high conductivity is corrosion prone. However, a coating with low conductivity will not necessarily always be providing corrosion protection.

The physical properties of polymer films can be evaluated conveniently by impedance methods using the metal substrate as one electrode. Impedance measurements show promise in identifying the presence of condensed water at the polymer/electrode interface or in the bulk of polymer coatings.

References

1. K.A. Holtzman. *J. Paint Technol.*, **43**, No. 554, p. 47 (1971).
2. D.M. Brasher and A.H. Kingsbury. *J. Appl. Chem.*, **4**, p. 62 (1954).
3. D.M. Brasher and T.J. Nurse. *J. Appl. Chem.*, **9**, (1959).
4. J.K. Gentles. *J. Oil Col. Chem. Assoc.*, **46**, p. 850 (1963).
5. H. Leidheiser, Jr. and M.W. Kendig. *Corrosion*, **32**, p. 69 (1976).
6. L. Beaunier, I. Epelboin, J.C. Lestrade, and H. Takenouti. *Surface Technol.*, **4**, p. 237 (1976).
7. J.D. Scantlebury and K.N. Ho. *J. Oil Col. Chem. Assoc.*, **62**, p. 89 (1979).
8. J. Devay, L. Meszaros, and F. Janaszik. *Ind. Eng. Chem. Prod. Res. Dev.*, **28**, p. 13 (1979).
9. H. Leidheiser, Jr., *Prog. Org. Coatings*, **7**, p. 79 (1979).
10. P. Hedvig. Dielectric Spectroscopy of Polymers, John Wiley and Sons, New York City (1977).
11. H.E. Bair, G.E. Johnson, and R. Merriweather. *J. Appl. Phys.*, **49**, p. 4976 (1978).
12. G.E. Johnson, H.E. Bair, and J.E. Scott. Conf. on Elec. Insulation and Dielectric Phenomena, Natl. Acad. Sciences, Washington, D.C. (1979).
13. G.E. Johnson, H.E. Bair, and S. Matsuoka. Am. Chem. Soc. Symp. on Water in Polymers, Washington, D.C., September 10-13, 1979.
14. N.G. McCrum, B.E. Read, and G. Williams. Analastic and Dielectric Effects in Polymeric Solids, John Wiley and Sons, New York City (1967).
15. S. Negami. Treatise on Coatings, Vol. 2, R.R. Meyers and J.S. Long, Editors, Marcel Dekker, New York City (1976).
16. D.D. MacDonald. Transient Techniques in Electrochemistry, Plenum Press, New York City (1977).
17. G. Gabrielle and M. Keddam. *Electrochim. Acta*, **19**, p. 355 (1974).
18. M. Sluyters-Rehback and J.H. Sluyters. Electroanalytical Chemistry, **4**, A.J. Bard, Editor, Marcel Dekker (1970).
19. S.G. Letzer. *Electronic Design*, October 11, 1974.
20. H.V. Malmstadt, C.G. Enke, and E.C. Toren. Electronics for Scientists, W.A. Benjamin, Inc., New York City (1963).
21. Pittsburgh Soc. for Paint Techno. *J. Paint Technol.*, **42**, p. 730 (1970).
22. R.N. Miller. *Materials Protection*, **7**, No. 11, p. 35 (1968).
23. H.C. O'Brien. *Ind. Eng. Chem.*, **58**, No. 6, p. 45 (1966).
24. J. Crank and G.S. Park. Diffusion in Polymers, Academic Press, New York City (1968).
25. R. Bellobono, B. Marcandalli, R. Massara, and L. Guaschino. Atti Accad. Naz. Lincei, Cl. Sci. Fis., Mat. Nat., Rend. **64**, p. 293 (1978).
26. J.E.O. Mayne and D.J. Mills. *J. Oil Col. Chem. Assoc.*, **58**, p. 155 (1975).
27. E.M. Kinsella and J.E.O. Mayne. *Br. Polym. J.*, **1**, p. 173 (1969).
28. M. Buller, J.E.O. Mayne, and D. J. Mills. *J. Oil Col. Chem. Assoc.*, **58**, p. 155 (1975).
29. J.V. Standish and H. Leidheiser, Jr. *Corrosion*, **36**, p. 390 (1980).
30. H. Isaacs and M.W. Kendig, *Corrosion*, **36**, p. 269 (1980).
31. M. Hughes and J. Parks. Paper presented at this conference.
32. M.B. Roller and J. K. Gillham. *Am. Chem. Soc. Org. Coatings and Plastic Preprints*, **37**, No. 2, p. 135 (1977).
33. G.E. Johnson. Private communication.
34. G.P. Johari and S.J. Jones. *J. Chem. Phys.*, **62**, p. 4213 (1975).

An AC Impedance Probe as an Indicator of Corrosion And Defects in Polymer/Metal Substrate Systems

*Michael C. Hughes and Jeffrey M. Parks**

Introduction

Over the years a number of electrochemical techniques have been used to study surface properties of metals, including corrosion and dissolution. The subject has been recently reviewed.[1] Direct current methods have been by far the most widely employed, but in recent years there has been a growing number of reports on the application of the more powerful AC impedance techniques.[1] Almost all applications of AC impedance techniques to surface studies of metals have involved "bare" or anodically coated metal surfaces, but a few workers have reported results from polymer coating/metal substrate systems.[2-6]

Recently Isaacs and Kendig[7] have reported an alternating current probe system based on a square wave excitation function which was used to map metal surfaces. A number of types of corrosion could be identified by the AC response exhibited by the probe. An AC probe based on a different electrical circuit and using a sinusoidal excitation function has been reported by Standish and Leidheiser[8,9] and has been applied to measurements of polymer-metal-electrolyte systems. We report here results obtained with an AC probe, which differs from the probe of Standish and Leidheiser primarily in that it is designed to be used with a conventional three-electrode potentiostat. The probe and the associated electronics have been designed with the expectation that its reponse will be interpretable in terms of the standard theory of the AC current response of various electrode processes.[10,11]

The alternating current response of an electrochemical system to a sinusoidal excitation function is normally treated by using a phase/frequency sensitive detector to separate the current response into two vector components, the "in phase" (IP) current and the "quadrature" (Q) current. Once this has been done, the resulting data are amenable to treatment by two approaches: vector analysis[11] and complex admittance or impedance plane analysis.[1] The latter method is a far more sophisticated and powerful technique for the interpretation of the AC current response, but at the basic level, vector analysis, if one keeps in mind its limitations, is often easier to apply.

The results of this work will be primarily phrased within the terminology of vector analysis, since at the present stage of the work we do not understand the nature of the probe response well enough to apply rigorous complex plane analysis.

In this paper we will be presenting data primarily having to do with induced defects in amide cured epoxy coatings on several types of steels. The sensitivity and varying response of the probe to different conditions of corrosion and coating delamination will be noted, and sensitivity to defects underneath an intact coating will be demonstrated. The probe response can also be used to follow the progress of corrosion and/or coating delamination.

Experimental

Figure 1 shows a schematic of the electronic instrumentation used. The potentiostat is a Princeton Applied Research Corporation (PARC) Model 173, using either a PARC 176 or 179 current amplifier plug-in. The Lock-In Amplifier (LIA) is a PARC 5204/90 dual phase LIA with the vector analysis option. A Wavetech Model 185 signal generator provides both the reference and excitation signal. Output is recorded on either a Hewlett-Packard 7047A X-Y recorder, or a Hewlett-Packard/Moseley 7001A X-Y recorder. Voltage measurements are verified using Hewlett-Packard digital multimeters, and frequency is

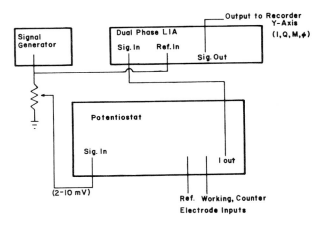

FIGURE 1 — Block diagram of probe electronic circuit.

*Department of Chemistry and Center for Surface and Coatings Research, Lehigh University, Bethlehem, Pennsylvania

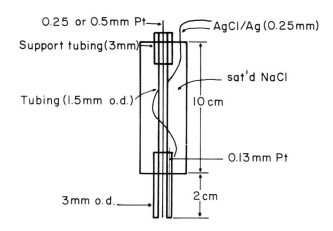

FIGURE 2 — Detail drawing of probe construction.

measured with a Hewlett-Packard 5381A electronic counter. An applied AC potential of 10mV (RMS) was used for all experiments. Varying DC bias potentials were applied to the samples to keep the steel substrate in its passive region. The system was configured such that the probe wire was the counter electrode and the metal substrate was the working electrode in the three-electrode system.

Figure 2 shows the design of the probe. It is a variation on the probe reported by Isaacs and Kendig[7] specifically designed so that a true reference electrode (Ag/AgCl saturated NaCl) could be used instead of a pseudo-reference electrode. Platinum and silver wire were purchased from Alfa Inorganics. The silver wire was anodized in 0.1M NaCl for a few seconds to coat it with AgCl. The tip of the probe was polished to a smooth flat surface. In operation it was normally set so that the tip of the probe rested gently in contact with the sample surface.

Metal and coated metal samples were normally 5 × 7.5 cm or 7.5 × 10 cm. An insulated ring 3.4 cm in diameter and about 2.5 cm high was glued to the sample to provide a container for the electrolyte. The electrolyte used for all experiments reported here was 3.5% NaCl. The sample was held on a machinist's table which could be moved in the X and Y directions by hand wheels calibrated in thousands of an inch (0.0025 cm). A ten-turn potentiometer powered by flashlight batteries was connected to the X axis drive of the stage and this signal drove the X axis of the X-Y recorder. Movement of the stage in both axes was made manually.

Steel samples were conventional cold rolled mild steel and galvanized steel. Cold rolled steel was phosphated by immersion in a solution of 50 ml Granodine phosphating agent per liter at a temperature of 70-80 C, followed by rinsing the steel in chromic acid (1.5 g/L).

Two types of polymer coatings were used. For the first coating, ten grams of Epon 828™ were mixed with 8.28g Versamid polyamide in toluene solvent. The second coating was 5.00g Epon 1010F™ mixed with 2.36g Emerez 1511™ polyamide in toluene solvent. In both cases, the coating was applied with a draw bar apparatus to give a coating thickness in the 50 to 75 μm range. The coated samples were baked for 20 minutes at 100 C after application.

Results and Discussion

Figures 3a-c show the types of response which may be obtained from the probe. The sample was a phosphated cold rolled steel coated with the Epon 828/Versamid epoxy. A DC bias potential of −0.56V (vs Ag/AgCl, saturated NaCl) was applied to the probe and the applied AC potential was 10mV at 6.49 kHz. The DC bias potential was determined by monitoring the DC current flowing at the substrate as the probe potential was made increasingly negative (biasing the substrate — the working electrode — positive) until the DC current was less than 1 μA. This procedure was used to prevent rapid damage to the sample, which usually occurred when the DC potential was not in the passive region. The X axis and Y axis resolution, in Figures 3a-c are typical of the probe resolution, although much finer resolution may be obtained (vide infra).

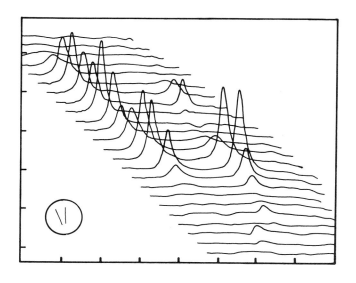

FIGURE 3a — I component response of a defect in Epon 828/Versamid epoxy coating on phosphated cold rolled steel. DC bias potential = −0.56V. Applied AC potential = 10mV. Frequency = 6.49 kHz. X-axis scale: 1 division = 0.254 cm (0.100 inch). Y-axis scale: 0.0635 cm (0.025 inch) between scans. Current sensitivity = 5 μA/division. Inset shows the shape of the defect.

The defect was a deliberate one made by cutting the polymer surface gently with a razor blade. The shape of the defect is shown in the inset in Figure 3a. Figures 3a and 3b, taken immediately after the defect was produced, and the 3.5% NaCl electrolyte added, show that the I and Q responses are essentially parallel in both distribution and magnitude. When contrasting this behavior with that reported for some other systems, it should be kept in mind that the defect was fresh, showing no visible signs of corro-

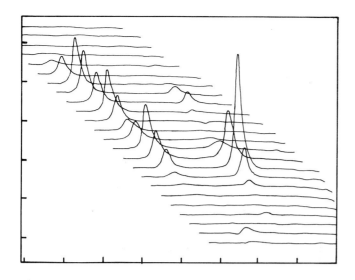

FIGURE 3b — Q component response for the defect shown in Figure 3a. All parameters are the same for Figure 3a.

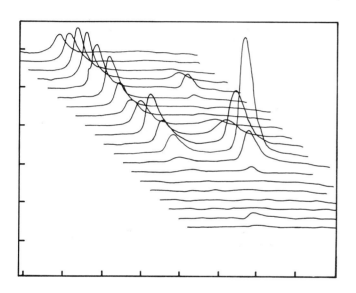

FIGURE 3c — Relative phase angle response for the defect shown in Figure 3a. Initial value of $\phi = 76°$ for all scans, with ϕ decreasing with increasing response on the vertical axis. All other parameters the same as for Figure 3a.

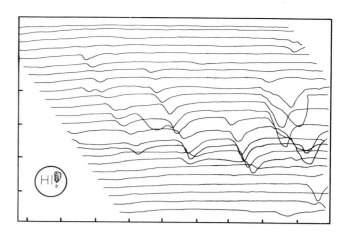

FIGURE 4a — Q component response of a defect on galvanized steel underneath Epon 1001/Emerez 1511 epoxy. DC bias potential = 0.0V. Applied AC potential = 10mV, frequency = 10.9 kHZ. X-axis scale: 1 division = 0.254 cm (0.100 inch). Y-axis scale: 0.0635 cm (0.025 inch) between scans. Current sensitivity = 125 μA/division. Inset shows the shape of the defect and the delamination area (stippled region).

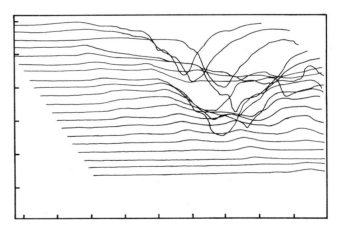

FIGURE 4b — I component response for the defect shown in Figure 4a. Current sensitivity = 10 μA/division. All other parameters the same as in Figure 4a.

sion or delamination. Figure 3c shows the relative phase angle response for the same defect. The phase angle measured is the angle between the applied alternating potential and the total current vector, measured directly by the lock-in amplifier; it does not correspond to the phase angle of the faradaic current usually encountered in AC electrochemistry. Also, the measured values are not corrected for the complex response function of the probe itself. Nonetheless, Figure 3c does show the phase angle decreases as the probe passes over the defect region. which is the expected behavior if the complex AC impedance of the system becomes more resistive, and less capacitive, in the region where the polymer coating is cut.

Figures 4a and 4b illustrate the sensitivity of the probe, and the utility of the phase sensitive detection. The sample is Epon 1001/Emerez 1511 epoxy coated on a galvanized steel substrate. The scans were run at a DC bias potential of 0 V (potentiostated, not open circuit) vs Ag/AgCl (saturated NaCl), with an applied AC potential of 10 mV at 10.9 kHz. The defect consists of scratches made on the galvanized steel surface immediately *before* the application of the coating. The inset in Figure 4a shows the shape of the defect, and the location of the regions of corrosion and delamination, which developed after the sample remained in contact with the 3.5 NaCl solution overnight. The coating was completely intact when the measurements were made. Figure 4a is the Q component response. The pattern of the scratches underneath the coating is quite obvious, even in the areas where delamination has occured. In contrast, the I component response (Figure 4b) shows essentially no signal except in the delamination region, where a massive response is obtained.

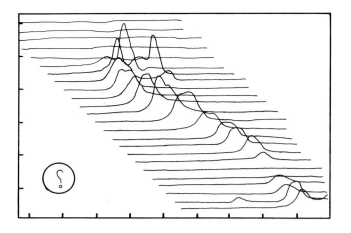

FIGURE 5a — I component response of a defect in Epon 1001/Emerez 1511 epoxy coating on cold rolled steel. DC bias potential = -0.39V. Applied AC potential = 10mV, frequency = 6.48 kHz. X-axis scale: 1 division = 0.254 cm (0.100 inch). Y-axis scale: 0.0635 cm (0.025 inch) between scans. Current sensitivity = 10 µA/division. Inset shows the shape of the defect.

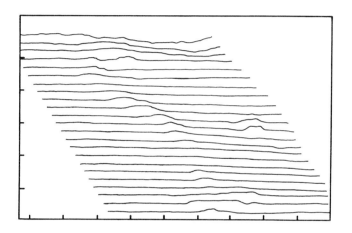

FIGURE 5b — I component response of defect shown in Figure 5a after 48 hours exposure to 3.5% NaCl. All parameters the same as in Figure 5a.

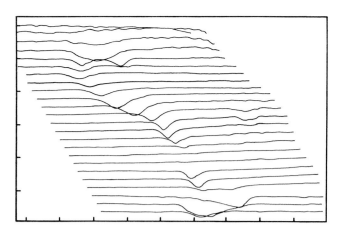

FIGURE 5c — Q component response of defect shown in Figure 5b. All parameters the same as Figure 5b.

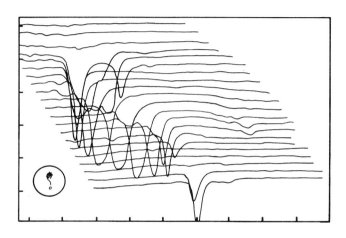

FIGURE 5d — Q component response of defect shown in Figures 5b and 5c after disturbance (see text for details). All parameters the same as for Figure 5a. Inset shows the location of delamination (stippled area).

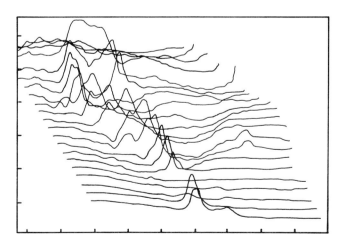

FIGURE 5e — I component of defect shown in Figure 5d. All parameters the same as for Figure 5a.

Figures 5a through 5e further illustrate the use of the phase sensitive detection. The sample was the Epon 1001/Emerez 1511 epoxy coated on a cold rolled steel substrate. The defect was a scratch made in the coating with a razor blade. The shape is shown in the inset in Figure 5a. A DC bias potential of -0.39V vs Ag/AgCl (saturated NaCl) was applied to the probe, and the applied AC potential was 10mV at 6.48 kHz. Figure 5a is the I component response run approximately 1 hour after immersion of the defect region in 3.5% NaCl. The shape of the defect is clearly delineated.

After 48 hours of immersion in the 3.5% NaCl, the entire defect became covered with brown iron oxide corrosion. Figure 5b and 5c are the I and Q component responses respectively, run at this point. The defect area appears to be passivated and virtually no I component signal is observed. The Q component still shows the defect clearly enough that its general shape can be readily discerned.

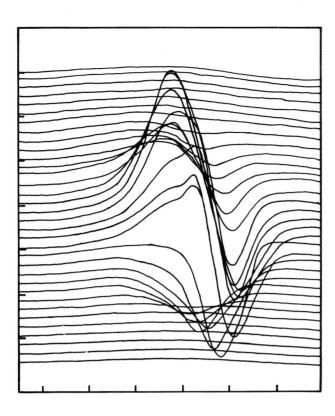

FIGURE 6 — I component response for a point defect in Epon 828/Versamid epoxy coating on phosphorated cold rolled steel. DC bias potential = −0.57V. Applied AC potential = 10mV, frequency = 22.2 kHz. X-axis scale: 1 division = 0.082 cm (0.0323 inch). Y-axis scale: 0.013 cm (0.0005 inch) between scans. Current sensitivity = 1 µA/division.

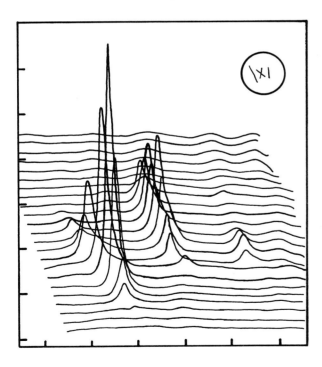

FIGURE 7a — I component response of a defect in Epon 1001/Emerez 1511 epoxy coated on cold rolled steel. DC bias potential = −0.54V. Applied AC potential = 10mV, frequency = 2.07 kHz. X-axis scale: 1 division = 0.410 cm (0.061 inch). Y-axis scale: 0.051 cm (0.020 inch) between scans. Current sensitivity = 10 µA/division. Inset shows the shape of the defect.

After the scans in Figures 5b and 5c were run, the corrosion coating covering the defect was deliberately broken up, the entire system stirred and allowed to sit for about one hour. At the end of this time, massive corrosion was present on the defect and distributed in the electrolyte. Delamination was visible in the region indicated in the inset in Figure 5d, as well as along both edges of the defect. Figure 5d shows the Q response. The signal intensity greatly increased with the onset of active corrosion, and the outline of the defect became clearly visible again. The cause of the change in polarity of the Q component relative to the earlier measurements is not understood at this time. This behavior was observed for a number of other experimental runs, and may be an artifact of the response function of the probe itself. Figure 5e shows the I response. As in the example cited previously (Figure 4b), an especially strong I response clearly outlines the region of delamination. The remainder of the defect shows a complex response of multiple peaks in the I component compared to the Q component. This apparently reflects the delamination which was beginning to occur visibly all along the defect.

The magnification of a defect which is possible with the probe is illustrated in Figure 6. The sample was phosphated cold rolled steel coated with Epon 828/Versamid epoxy. The DC bias to the probe was −0.57V vs Ag/AgCl (saturated NaCl), and the applied AC potential was 10mV at 22.2 kHz. The defect was a spot defect produced by piercing the coating with the tip of a needle. The sodium chloride electrolyte was added and the defect region was scanned immediately. Figure 6 shows the I component response; the Q component was essentially identical. Resolution is better than 250 µm (0.01") on both the X and Y axes.

Figures 7a and 7b illustrate how the variable magnification effect of the probe can be used to follow the progress of corrosion and delamination in a somewhat complex defect, the outline of which is shown in the inset in Figure 7a. The system was Epon 1001/Emerez 1511 epoxy on cold rolled steel. The DC bias potential on the probe was −0.45V vs Ag/AgCl (saturated NaCl), and the applied AC potential was 10mV at 2.07 kHz. Figure 7a shows the I component of the entire defect region immediately after the electrolyte was added. After 48 hours of exposure to the electrolyte, delamination had begun in a portion of the defect region (see inset in Figure 7b). Figure 7b shows I component scans run at a higher resolution than in Figure 7a, and covering only the delamination region, so that the pattern in that region can be seen more clearly. The shape of the defect is well defined, and, in agreement with other results shown, the magnitude of the I component response increased considerably, relative to the fresh defect (Figure 7a).

The results reported indicate that the AC potential probe is capable of providing a great deal of information regarding the location of defects in or under a polymer coating on a metal surface. Further, definite qualitative distinctions between some types of defects can be seen, and the progress of corrosion

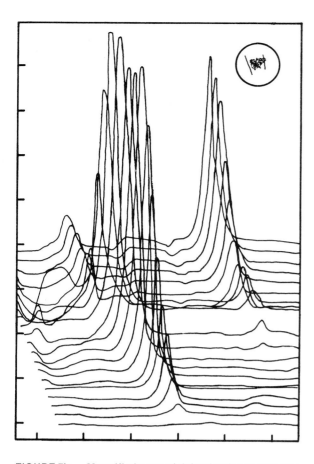

FIGURE 7b — Magnified scan of delamination region (see inset) in the defect shown in Figure 7a after 48 hours of exposure to NaCl electrolyte. X-axis scale: 1 division = 0.254 cm (0.100 inch). Y-axis scale: 0.025 cm (0.010 inch) between scans. Current sensitivity = 20 μA/division.

and delamination can be followed. The resolution of the probe is great enough so that detailed "impedance maps" of a defect region can be made. Thus, the probe can and is being used at the present time to locate and follow the changes in small defects in polymer coated steels. At the same time, we are working to define the complex impedance plane AC response function of the probe in its normal measurement configuration so that quantitative data, amenable to treatment by complex impedance plane analysis, may be obtained from the probe. A consideration of the literature on AC voltammetry[1] indicates that such data could be used to provide mechanistic information regarding defect formation in the polymer film, and the growth of corrosion and delamination, *in situ.* Further, it is expected that more accurate determinations of defect type and status can be made.

Acknowledgments

This work was supported by the Office of Naval Research, to whom grateful appreciation is expressed.

References

1. D. MacDonald. Transient Techniques in Electrochemistry, Plenum Press, New York (1977).
2. G. Reinhard, K. Hahn, and R. Kaltofen. *Plaste Kautsch.*, **22**, p. 522 (1975).
3. A. Farse and M. Paul. *Z. Phys. Chem.*, **252**, p. 198 (1973).
4. N. Gaynes. *Met. Finish.*, **74**, p. 52 (1976).
5. W. Scheider. *J. Phys. Chem.*, **79**, p. 127 (1975).
6. H. Potente, and F. Stoll. *Farbe Lack,* **81**, p. 701 (1975).
7. H. Isaacs, and M. Kendig. *Corrosion,* in press.
8. J. Standish, and H. Leidheiser, Jr. *Corrosion,* **36**, p. 390 (1980)
9. J. Standish and H. Leidheiser, Jr. This volume.
10. M. Sluyters-Reyback and J. Sluyters. Electroanalytical Chemistry, **4,** ed. by A. Bard, Marcel Dekker, New York (1970).
11. D. Smith. Electroanalytical Chemistry, **1,** ed. by A. Bard, Marcel Dekker, New York (1966).

Impedance Techniques for the Study of Organic Coatings

J. David Scantlebury and Graham A. M. Sussex*

Introduction

This paper outlines and discusses impedance responses which have been obtained on immersed coated metals. The technique has been more fully described elsewhere.[1,2] Briefly, a small sinusoidal voltage is applied to the specimen, and the current response of the specimen at that frequency is measured by a transfer function analyser (Solartron 1172). The output data are analysed by computer to produce plots of impedance response over a range of frequencies. The first set of experimental results is for intact paint films. The characteristic shapes and changes of these impedance plots are interpreted in terms of the usual Randles equivalent circuit, with a diffusion term included if necessary. For comparison, and to assist in separating the effect of various processes on the impedance response, experimental results are presented for bare mild steel, a pinholed film, a contaminated thin epoxy, and a contaminated artificial blister under a thick epoxy paint. These results are analysed similarly, and the electrical components are related to corrosion processes and paint film parameters.

Intact Paint Film on Uncorroded Coated Mild Steel

Using both coal tar epoxy and plasticized chlorinated rubber films on mild steel, the impedance response (during the first few days of immersion in 3% NaCl) was that of a series capacitor, resistor network with the coating acting as a dielectric barrier. The plots, up to 3 or 4 days immersion, in Figures 1 and 2 are typical for both coating systems examined. Figure 3 shows the variation in capacitance (on immersion) of a chlorinated rubber film vs the reciprocal of film thickness.[3] A linear plot is obtained, and it may be concluded that, at least on immersion, the paint film behaves similarly to the dielectric of a parallel plate capacitor. The slope of this graph gives an estimate of 5 for the relative permittivity of chlorinated rubber. ICI manufacturers' data for a similar chlorinated rubber give 3.9 at 1 kHz.

One experimental point was also obtained using the same chlorinated rubber mixture on a copper sub-

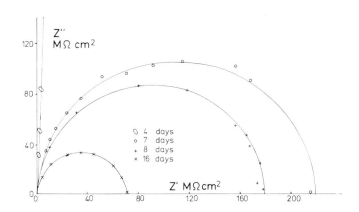

FIGURE 1 — Chlorinated rubber on mild steel after various immersion time in aerated 3% NaCl.

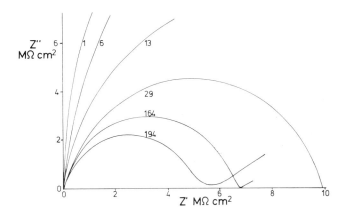

FIGURE 2 — Coal tar epoxy on mild steel in aerated 3% NaCl. Number are days of immersion.

strate. This point lies on the straight line produced by plotting of the mild steel data. This result is predictable, as even if copper and mild steel bond differently to the chlorinated rubber coating, any interfacial effects only affect a small percentage of these relatively thick films.

Similar plots were made of data extracted from Touhsaent and Leidheiser's[4] single frequency work on thin polybutadiene films on mild steel. The plots

---

*Corrosion and Protection Centre, University of Manchester Institute of Science and Technology, Manchester, England

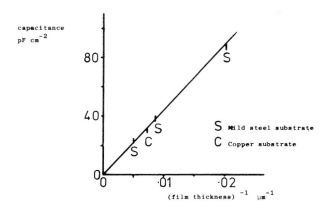

FIGURE 3 — Chlorinated rubber films in KCl solutions.

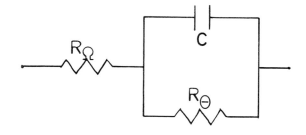

FIGURE 5A — Simple Randles circuit.

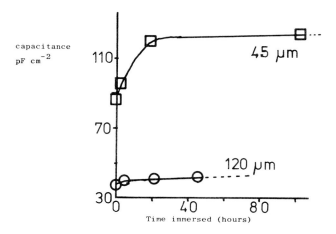

FIGURE 4 — Capacitance change for chlorinated rubber films in distilled water.

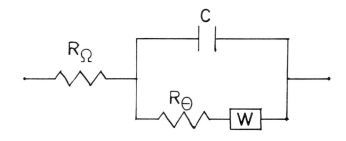

FIGURE 5B — Warburg diffusion term added.

were fairly linear for films in the range of 10 to 20 μm. The slope, and hence the relative permittivity, rose by a third after 10 days immersion, and by 30 days the slope was twice its immersion value.

The simple parallel plate capacitor model is accurate for measurements made on immersion; however, after some days the plots no longer pass through the origin, and a more sophisticated model (perhaps allowing for swelling) may be required. Hence, the values obtained for relative permittivity after long immersion may not be accurate.

Figure 4 shows the results of another experiment in which the variation in capacitance with time was studied for two chlorinated rubber coatings in distilled water.[3] The thinner coating shows a greater change with time. Assuming that the change in film capacitance is proportional to the water uptake of the film,[5] then it seems quite reasonable that the thin 45 μm single coat will take up proportionately more water because it contains more "D," or more open structure areas, than the 120 μm thick double coat.

Figures 1 and 2 show that after several days immersion both coal tar epoxy and chlorinated rubber coatings depart from purely capacitive behavior and the impedance response becomes approximately semicircular. These impedance response curves are for up to 16 days immersion, in the case of the chlorinated rubber, and up to 194 days in the case of the coal tar epoxy. A simple RC circuit has been used to model this type of behavior (Figure 5A). Ho[6] found that the values of R_Ω remained relatively constant with time in both systems at about 10 kohm cm².

Further work[7] was carried out using a higher starting frequency (200 kHz), to obtain a more precise high frequency intercept on the real axis. It was found that, for a range of KCl solutions from, 0.1 mM to 0.1M, R_Ω decreased uniformly as the solution conductance increased. This behavior is shown in Figure 6. The actual values of R_Ω agree well with the resistance calculated for the solutions used.

Ho found that the capacitance values showed an increase with immersion time: for the chlorinated rubber, from 195 pF cm⁻² to 215 pF cm⁻²; for the coal tar epoxy, from 90 pF cm⁻² to 202 pF cm⁻². Assuming that this change is due to water uptake, application of the Brasher and Kingsbury method[5] gives about 3% uptake in the chlorinated rubber (in 16 days), and 23% uptake in the coal tar epoxy (in 194 days). The calculated uptake for the coal tar epoxy is surprisingly large. It may be a real effect due to inadequate curing or possible errors in the calculation caused by the presence of uncured coating.[5] The value of R_Θ, the diameter of the semicircle, decreases markedly with time, from 1000 Mohm cm² to 0.74 Mohm cm² for the chlorinated rubber and from 10 Mohm cm² to 5 Mohm cm² for the coal tar epoxy. The values of capacitance are similar to those obtained for a polymer film about 50 μm thick, and so these R_Θ values must be due to the resistance of the paint film. The size of the R_Θ values shows that the paint film is substantially intact.

After long immersion, the low frequency response of the coal tar epoxy coating shows a linear region of positive slope. This behavior has been modelled by inserting an extra impedance (Warburg impedance) into the equivalent circuit, as in Figure 5B. An expression for this Warburg impedance may be derived from Fick's laws of diffusion, which indicate that processes which cause this linear low frequency response are diffusion controlled. However, further work is required to relate this response to the reactions which are proceeding.

Uncoated Mild Steel and Porous Films

This work is part of a continuing investigation of electrochemical phenomena under coatings, encompassing defects in films, porous films, blistered coatings, and cathodic protection. The coatings used were a coal tar epoxy and a solvent based epoxy. Figure 7 shows the frequency response for uncoated mild steel in aerated 3% NaCl solution. The curve may be explained as a charge transfer semicircle, followed by a slow diffusion response at lower frequencies. The calculated value for the capacitance was of the order of 10 μF cm^{-2}, a reasonable value for the double layer of a freely corroding metal. The low value for R_Θ (144 ohm cm^2) reflects the rapid rusting which occurs.

In order to simulate the behavior of a tank or ship coating when it is mechanically damaged, a pinhole was made in a 60 μm thick coal tar epoxy coating. Curve (a) of Figure 8 shows the impedance response of the pinholed coating 20 minutes after the damage was inflicted, and curve (b) the response after 18 hours. In curve (a), the high frequency semicircle is attributed to corrosion at the base of the pinhole; the loop is due either to potential drift during the measurement, or to some sort of adsorption process. The low frequency arc is attributed to diffusion.

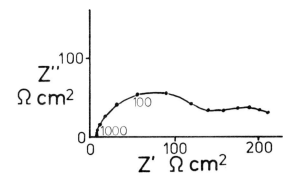

FIGURE 7 — Mild steel in sodium chloride solution (frequency in Hz).

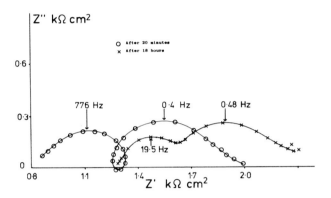

FIGURE 8 — Pinhole in coal tar epoxy on mild steel.

Curve (b) of Figure 8 shows that after 18 hours the low frequency arc is virtually unchanged; that is, the same time constant and diameter. Because rusting was observed with consequent changes in the response for any localized faradaic process, it seems likely that the diffusion explanation for this arc is correct. The inductive loop of curve (a) vanished. If this loop was due to adsorption, then it would not be surprising if active adsorption sites had been blocked after 18 hours immersion. However, this argument is not conclusive, as, if potential drift has caused the loop, then 18 hours corrosion at the pinhole site seems long enough for a stable potential to develop.

After 18 hours the high frequency semicircle is approximately the same diameter, but its time constant is longer, and its high frequency intercept is increased. In terms of the simple models of Figure 5, this suggests an increased R_Ω. However, the pinhole means that it is not clear what area should be used in the calculations; therefore absolute values of R_Ω, R_Θ, and C cannot be obtained. The increase in time constant may be due to corrosion products either clogging up the pinhole, or simply covering the bare steel and possibly slowing the corrosion process. Whatever the mechanism, this increase in time constant is the reverse of the change observed for an intact coal tar epoxy film (Figure 2).

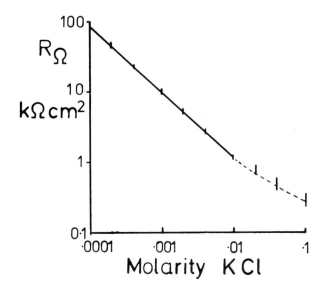

FIGURE 6 — High frequency real impedance intercept for chlorinated rubber on immersion in KCl solutions.

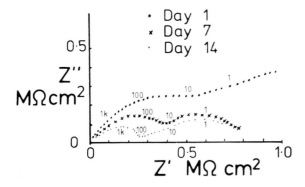

FIGURE 9 — Impedance spectrum for innoculated and coated mild steel (frequency in Hz).

Figure 9 was obtained from a sample with a mild steel substrate, which was contaminated by 100 μl cm^{-2} NaCl before it was coated with a single coat of 40 μm thick epoxy paint. The double arc response, obtained after 7 days, is similar to curve (b) of Figure 8. The time constant for the second arc is of the same order for the two experiments and Figure 9 shows that it does not change between 7 and 14 days immersion. However, the high frequency arc shows the same behavior as the chlorinated rubber and coal tar epoxy films in Figures 1 and 2; that is, both the diameter of the semicircle and its time constant decrease.

Although no pores were observed in the coating, the most likely explanation for this behavior is based on a porous film model in which corrosion commenced almost immediately over the contaminated area. The high frequency semicircle is probably due to corrosion sites at the base of the micropores in the epoxy film. This semicircle is rather depressed. This depression has been modelled by a distribution of time constants;[8] in this case, a distribution of R_Θ and C values. It is likely that there is a spread of pore diameters and depths, and this spread offers the variation of R_Θ and C which could explain the depression of the semicircle.

Since the pore sizes are not known, precise rate values cannot be extracted from these data. However, the reduction in the diameter of the high frequency semicircle suggests that either the corroding area is increasing, or the corrosion rate per unit area at the pore bases is increasing. The former seems more likely.

The immersion time required before the second semicircle developed, its invariance once it had developed, and its similarity to the diffusion arc in the pinhole case, make it fairly certain that it is associated with a diffusion limited reaction, possibly related to the micropores.

Blisters

Figure 10 was obtained from a 200 μm thick coating which was prepared in a manner similar to the 40 μm thick coating which gave the response shown in

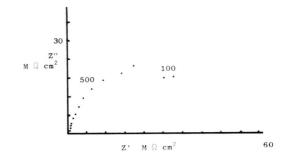

FIGURE 10 — Epoxy coating one electrode under blister, one in solution (frequency in Hz).

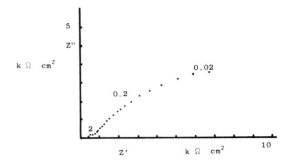

FIGURE 11 — Epoxy coating mild steel electrodes under same blister (frequency in Hz).

Figure 9. As in all the previous cases, the steel substrate was one electrode, and the other was platinum in the solution. One slight modification is that the steel electrode was split to enable beneath film experiments. After immersion for several days, a blister formed under the coating, enclosing green corrosion product. Values for the semicircle diameter and capacitance, 50 Mohm cm², 10.6 pF cm^{-2}, indicate that this impedance response relates solely to the film, and not to the corrosion process at all. Hence, the response is similar to that obtained in the earlier curves of Figures 1 and 2. The green corrosion product shows that the blister environment has a low oxygen concentration. This low oxygen concentration is consistent with the evidence from the impedance response that the film is intact.

Figure 11 was obtained, using the same specimen as the previous figure, but now utilizing the corroding split steel substrate as two electrodes under the same blister.[9] A response dependent mainly on the corrosion reactions in the blister is to be expected. The impedance response is similar to that obtained for bare steel in sodium chloride solution: a high frequency semicircle associated with the corrosion process, and a low frequency diffusion response. However, since the diffusion processes occur across an intact paint film, the contribution of diffusion to the impedance response is overwhelming.

A series of experiments was also carried out to observe the effect of cathodic protection on the impedance response of the split electrode specimen. It was found that the response shown in Figure 11 did not change at all when the substrate was polarized to −850mV and −1050mV (SCE). Apparently, the IR drop across the high resistance intact paint film means that if a local corrosion cell is established under an intact paint film, then it is effectively screened from cathodic protection. Thus, to prevent corrosion, it seems that either the substrate must be thoroughly cleaned prior to application of the coating, or a porous coating should be used and cathodic protection applied. The former is not possible in the field.

Summary

The AC impedance technique is a useful tool for studying immersed coatings. It can be used to calculate film thickness and water uptake. Charge transfer and diffusion processes give different impedance responses; therefore, it is often possible to identify what type of reaction is dominant. It is possible to use AC impedance measurements to determine when corrosion has commenced and when a film has broken down. Some comparison of corrosion rates may be made, although interpretation is not easy because the reaction area usually changes. It appears that with an intact film, AC impedance techniques cannot be used to observe corrosion under that film.

Acknowledgments

The authors wish to thank G.C. Wood for provision of laboratory facilities. One of us (GAMS) wishes to thank the Procurement Executive, Ministry of Defence, for financial support.

References

1. J.D. Scantlebury and K.N. Ho. *J.O.C.C.A.* **62,** p. 89 (1979).
2. J.D. Scantlebury, K.N. Ho and D.A. Eden. Progress in Electrochemical Corrosion Testing. ASTM 1980.
3. J.P. Lomas. M.Sc. Dissertation, The Victoria University of Manchester, 1979.
4. R. Touhsaent and H. Leidheiser, Jr. *Corrosion,* **28,** p. 435 (1972).
5. D.M. Brasher and A.H. Kingsbury. *J. Appl. Chem.* **4,** p. 62 (1954).
6. K.N. Ho. Ph.D. Thesis, The Victoria University of Manchester, 1978.
7. G.A.M. Sussex. Unpublished work.
8. K.S. Cole and R.H. Cole. *J. Chem.* Physics **9,** p. 341 (1941).
9. J.D. Scantlebury, D.A. Eden and M.J. Schofield. Proc. XV FATIPEC Congress, Amsterdam, June 1980.

Dielectric Study of Paint and Lacquer Coatings

J. Devay,* L. Meszaros* and F. Janaszik**

Introduction

The three main types of polarization are termed: electronic, atomic, and orientation polarization. The latter is the most relevant from the point of view of the study of organic coatings.

Orientation polarization is observed only in polar molecules. The electric field tends to turn the permanent dipole moments in its own direction. This effect is hindered by motion due to the thermal energy of the molecules. As a consequence of the latter phenomenon, orientation polarization decreases when the temperature is increased, whereas deformation polarization is independent of temperature. It is to be noted, however, that orientation polarization is not observed when the thermal motion of the molecules is hindered, since rigidly bound molecules or polar groups are unable to move under the effect of the electrical field. This situation is encountered at low temperatures. Orientation polarization becomes possible when a certain temperature is attained during the heating of the material. By increasing the temperature, the thermal energy of various polar molecules and groups is successively increased, permitting orientation. Thus polarization is found to increase in a relatively narrow range of temperature (for example, in the neighborhood of the softening point, or in that of the transition between two vitreous states of polymers), then at higher temperatures, the more intense thermal motion favors the disorientation of the dipoles, and polarization decreases again.

The variation of the field is followed by polarization with a certain delay. This delay is a phenomenon common to all three types of polarization, and it is termed "relaxation effect," characterized by the relaxation time, i.e., the time interval during which polarization decays to the e-th fraction of its initial value. The relaxation times of the various types of polarization differ considerably. The decline of electronic polarization is the most rapid because of the low inertia of electrons, while that of atomic polarization is slower. In fact, the relaxation time of the latter is in the range of the period of infrared electromagnetic radiation, while that of the former is in the range of the period of visible light.

The relaxation time of orientation polarization depends on the size, surroundings, hindrance of motion, etc., of the polar molecules (gas, liquid or solid state crystal or amorphous structure) and it ranges from the period of the radiowaves to several weeks.

Dielectric polarization can also be characterized by the relaxation frequency which is defined as the reciprocal of the relaxation time.

$$\omega_r = 2\pi f_r = \frac{1}{\tau_r} \quad (1)$$

that is, the largest frequency of the electrical field variation which can be followed by orientation in accordance with the definition of relaxation time.

Both relaxation time and relaxation frequency are independent of temperature in the case of deformation polarization. However, in the case of orientation polarization, relaxation time decreases and relaxation frequency increases when the temperature is increased. The dependence of the former quantities on temperature can be given by an Arrhenius type equation

$$\frac{1}{\tau_r} = \omega_r = Ae^{-\frac{E_A}{RT}} \quad (2)$$

where E_A is the activation energy of the orientation, A is a factor corresponding to the relaxation frequency extrapolated to infinite temperature, and R and T have the usual meaning.

Dielectric polarization of a certain material is quantitatively defined by permittivity ε which denotes the ratio of capacity C of a condenser when the dielectric material is placed between its plates and the capacity of C_o of the same condenser in vacuo

$$\varepsilon = \frac{C}{C_o} \quad (3)$$

Polarization depends on the polarizing voltage and the temperature of the dielectric material as shown in Figure 1. The frequency ranges corresponding to

---
*Hungarian Academy of Sciences, Research Laboratory for Inorganic Chemistry, Budapest, Hungary.
**Department of Physical Chemistry, University of Veszprém, Veszprém, Hungary.

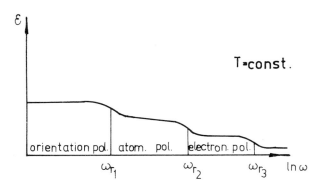

FIGURE 1 — Frequency dependence of permittivity.

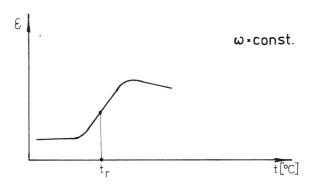

FIGURE 2 — Temperature dependence of permittivity in the case of orientation polarization.

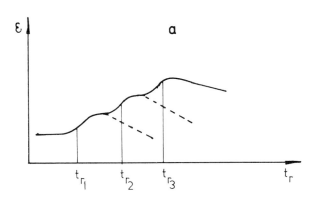

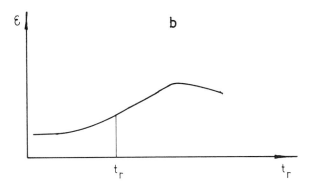

FIGURE 3 — Temperature dependence of permittivity in the case of system exhibiting a. several discrete relaxation times b. continuous distribution of relaxation times.

various types of polarization and the corresponding relaxation frequencies are also shown in Figure 1. For the sake of simplicity, the effects of the resonance of quasi-elastic particles, such as atoms and electrons, have been omitted from Figure 1. The temperature dependence of orientation polarization at constant frequency is shown in Figure 2. This curve is termed a "thermodielectric spectrum." Permittivity is constant at low temperatures since the polar groups are frozen and only atomic and electronic polarizations are observed. An increasing number of polar groups can respond to the orienting field with increasing temperature. Finally, the disorienting effect of thermal motion is manifested in the decrease of ε at high temperatures. The relaxation temperature t_r is at the inflexion point of the curve.

The relaxation effects are superimposed when the dielectric contains various polar groups or molecules having different relaxation temperatures (Figure 3a). The thermodielectric spectrum is flattened when the relaxation temperatures of polar groups exhibit a continuous distribution. The dispersion range is widened in such cases (Figure 3b), and the inflexion point can be considered as an average relaxation temperature.

Polarization loss is another important characteristic of dielectrics. It is due to the work of frictional forces acting during the periodic polarization of the dielectrics. Dielectric loss can be represented by an equivalent circuit consisting of a resistance in parallel with an ideal condenser. The loss resistance also depends on frequency and temperature. Dielectric loss is negligibly small as compared to permittivity in the case of organic coatings, and, consequently will not be considered.

Dielectric Properties of Organic Coatings

The permittivity observed during dielectric polarization of organic coatings depends considerably on frequency and temperature. Thus, orientation polarization is exhibited in addition to electronic and atomic polarization. This phenomenon can be understood if one considers that the molecules of macromolecular compounds are not tightly packed because of the entanglement of the molecular chains; consequently, side-groups and end-groups as well as various segments of molecular chains can move rather freely. The mobility of polar groups is relatively large at sufficiently high temperatures, thus the relaxation time of their orientation is short. This phenomenon can conveniently be observed in the range of audio frequencies (20 Hz to 20 kHz) in the temperature range from 20 C to 150 C.

Lacquer and paint coatings consist of several components, in addition to the organic binder, such as softener, solvent residue, pigments, and other coloring agents. The latter play an important role in the formation of the structure of the coating, and considerably influence both the mechanical and dielectric properties of the paint layer.

The freedom of orientation of polar groups is determined by the interaction of the latter with their surroundings, namely by the extent to which the various groups are bound to the network. Loosely bound polar groups, in a network composed of a relatively low number of crosslinks, exhibit relatively high mobility; consequently, the thermal energy necessary for orientation is acquired at relatively lower temperature. Thus the relaxation temperature observed in such cases is lower than that of a network composed of a large number of crosslinks. Other components of the coating modify these effects.

The structure of polymers and the surroundings of the polar group also determine the activation energy of orientation. The activation energy depends on the height of the energy barrier between two equilibrium positions of the polar group. This energy barrier is affected by the mobility of the polar group built into the network, by the hindrance of the surroundings, and by the free volume available for the motion, as well as by electrostatic and other interactions.

The structure of the films continuously changes during the lifetime of the coating, from the time of its application until the complete degradation of the film. The solvent is evaporated, and the network is built at a relatively high rate in the period of formation of the film. The formation of the network, as well as oxidation and degradation, continue at a low rate during environmental exposure to radiation, humidity, heat, and various gases and vapors. The above processes involve the formation or the splitting of polar groups and the evaporation of components of lower molecular mass. These processes tend to attain a stationary condition; thus, a relatively stable structure is formed after some months, depending on the type of the coating. Finally, degradation tends to prevail and the coating deteriorates.

These processes are also reflected in the dielectric properties of the coating, and can be followed by the determination of the relaxation temperature and activation energy of dipole orientation. Both characteristics exhibit large and rapid variations in the period of formation of the network and attain steady values when the coating has stabilized. A change in both the relaxation temperature and activation energy indicate the start of degradation. The direction of these changes depends on the character of the process. The softening of the coating is caused by the formation of a large amount of components of lower molecular mass, resulting in the decrease of cohesion forces; consequently, both the relaxation temperature and activation energy are decreased. On the other hand, both characteristics tend to increase in the case of embrittlement of the coating.

Literature data reveal that a straight parallelism exists between mechanical and dielectric properties. In fact, mechanical effects and dielectric measurements involve the relaxation of identical polar groups.

On the basis of theoretical considerations and earlier experimental works, the following working hypotheses were formulated:

1. Relaxation temperature can be related to the hardness of the coating: higher relaxation temperatures correspond to higher degrees of hardness. This phenomenon can be explained by the fact that a larger amount of thermal energy is required for the orientation of the polar groups if the latter are bound in a network containing a large number of crosslinks.

2. Activation energy can be related to the elasticity of the coating. High activation energy indicates that movement of polar groups is hindered to a large extent, and in such cases the coating is insufficiently elastic. If the movement of the polar groups requires smaller activation energy, a larger deformation is possible within the limits of elasticity; consequently, the film is more elastic. Thus, the coating is hard and brittle when the relaxation temperature is high and the activation energy is large. On the other hand, low relaxation temperature and small activation energy correspond to a soft and plastic film. The quality of the coating is optimum from a mechancial point of view if it is hard and elastic, i.e, its relaxation temperature is relatively high while its activation energy is relatively small.

The Role of the Pigment in Dielectric Studies

A pigmented coating can be considered a heterogeneous dielectric where the binder forms a continuous matrix containing uniformly distributed inclusions having approximately spherical shape. The permittivity of such a heterogeneous dielectric is given by Wiener's formula

$$\varepsilon = \varepsilon_1 \left(1 + \frac{\Theta_1}{\frac{1-\Theta_1}{3} + \frac{\Theta_1}{\varepsilon_2 - \varepsilon_1}} \right) \quad (4)$$

where ε_1 and ε_2 are the permittivity of the matrix and that of the inclusions, respectively, and Θ is the volume fraction of the inclusions. In the case of pigmented paint coatings the permittivity of the polymer matrix has an order of manitude equal to 10^0 while that of the pigment is considerably larger, e.g., it is 10^2 to 10^3 for TiO_2, hence

$$\frac{\Theta_1}{\varepsilon_2 - \varepsilon_1} \approx 0$$

and

$$\varepsilon = \varepsilon_1 \left(1 + \frac{3\Theta_1}{1-\Theta_1} \right) \quad (5)$$

It is apparent that the permittivity of the polymer matrix is altered by a pigment having a large permittivity. However, the frequency and temperature dependence of the permittivity of the matrix is not affected, since the coefficient of ε_1 depends only on the volume fraction of the pigment. In other words, the relaxation temperature and the activation energy obtained from the thermodielectric spectra are charac-

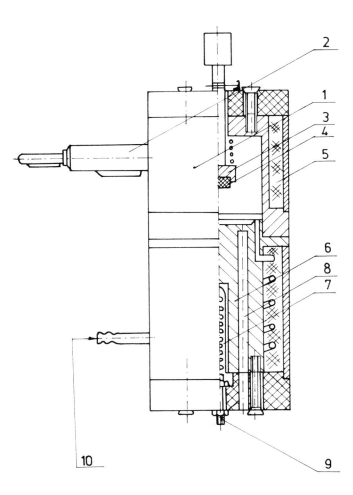

FIGURE 4 — Measuring cell. 1) Lid 2) Connector 3) Spring mechanism 4) Silicon rubber disk 5) Heat insulation 6) Metal block 7) Heating filament 8) Bore for thermoelement 9) Connector for heating element 10) Inert gas inlet.

teristic of the properties of the binder. By virtue of these facts, the interaction between the pigment and the binder can also be studied on the basis of the thermodielectric spectra, since the character of the latter is independent of the dielectric properties of the pigment, as reported earlier. It is noteworthy that this statement applies to the dielectric measuring technique only, since the properties of the coating (such as durability) are greatly influenced by the pigment.

Experimental

Our experimental work was primarily aimed at the study of the relation between mechanical and dielectric properties. Some results will be shown to illustrate the foregoing considerations.

The coatings under study were applied to a stainless steel disk, 30 mm in diameter. The coating constituted the dielectric of a plate condenser one plate of which was the steel disk while the other was an aluminum foil pressed on the film by the spring mechanism in the measuring cell. (Figure 4).

The measuring cell consisted of a cylindrical metal block equipped with a lid. The disk shaped sample was accommodated on the metal block. A silicon rubber disk covered with the aluminum foil was pressed onto the sample by a spring mechanism, built in the lid. The diameter of the rubber disk defined the area of the plate condenser. The plates of the condenser were connected to the measuring unit by a coaxial lead. A heating element enclosed in a silica tube was inserted in a bore of the cylinder. Another bore accommodated an iron-constantan thermoelement. The metal block was surrounded by a coil used to introduce an inert, dry, and preheated gas into the measuring cell to prevent the oxidation of the coating during the measurement. The thermal insulation of the measuring cell was asbestos.

Thermodielectric spectra were obtained by recording the capacity of the plate condenser, assembled in the above manner, as a function of temperature, since the capacity of the condenser is proportional to the permittivity of the dielectric and the proportionality factor is a constant depending on the geometry of the condenser and the permittivity of the vacuum.

The principle of the measurement is based on Ohm's law. The block diagram of the instrumental unit is shown in Figure 5. The voltage of the sinusoidal signal generators of frequencies $f_1 = 2000$ Hz, $f_2 = 1000$ Hz, $f_3 = 500$ Hz and $f_4 = 200$ Hz, respectively, were connected to the plate condenser by a revolving switch driven by a synchronous motor. The speed of the switch was 1 rpm. The current flowing through the condenser under the effect of the imposed voltage was recorded on the Y axis of the X-Y recorder. The heating rate of the sample was adjusted to 2

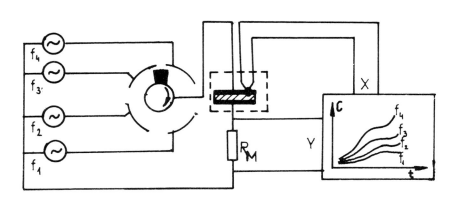

FIGURE 5 — Block diagram of the measuring system.

C/minute. The temperatures of the sample was measured by an iron-constantan thermocouple controlling the X axis of the X-Y recorder. The temperature of the measurement ranged from 20 to 150 C.

Thus, the thermodielectric spectra were simultaneously recorded on one sample at four different frequencies. This arrangement permitted a more accurate determination of the activation energies, since an uncertainty would have arisen in the measurement if the spectra were recorded on different samples.

Determination of the Activation Energy and Relaxation Temperature

The relaxation temperature pertaining to a given frequency can be defined as the inflexion point of the thermodielectric spectra (Figure 2), while the activation energy can be evaluated on the basis of the logarithmic form of the Arrhenius equation (Equation 2).

$$\ln \omega_r = \ln A - \frac{E_A}{RT} \quad (6)$$

and

$$\frac{d \ln \omega_r}{d\left(\frac{1}{T}\right)} = -\frac{E_A}{R} \quad (7)$$

Materials

Various types of paints produced by various manufacturers were studied in order to verify the relationship between the dielectric and mechanical properties (mainly hardness and elasticity) of the coatings. The following paints were examined. The properties of the products are described below on the basis of the information obtained from the manufacturers.

A. Astralin; synthetic enamel for automobile; yellow. The binder consists of a synthetic alkyd resin; curing by oxidation at room temperature.

B. Neolux; synthetic lacquer, yellow. The binder consists of an oil-extended alkyd resin; quick-drying by oxidation.

C. Rezokril Super; synthetic enamel; yellow. The binder consists of a polyacrylate synthetic resin. The film is formed by addition of an aliphatic isocyanate.

D. Resistan Super; enamel; yellow. The binder is a synthetic alkyd type resin, curing by a polyfunctional aliphatic isocyanate.

E. Supren H; enamel; white. Hardening by heat treatment. The binder is a combination of alkyd and amine resin. Diluted with an aqueous solution of an amine such as triethylamine or ammonium ion.

F. Tiszalux; enamel; white. The binder consists of an alkyd resin, curing by oxidation.

G. Trinát; enamel; white and ochre respectively. The binder is a quick-drying alkyd resin, curing by oxidation.

Preparation of the Sample

The coatings were applied to discs of 30 mm in diameter, made of stainless steel, cold-rolled steel

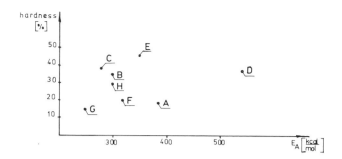

FIGURE 6 — Hardness vs. activation energy diagram.

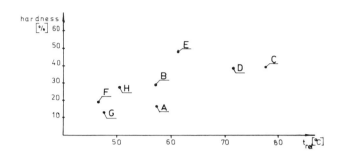

FIGURE 7 — Hardness vs. relaxation time diagram.

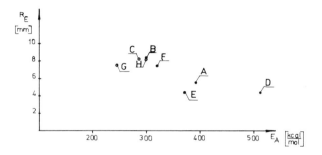

FIGURE 8 — Elasticity vs. activation energy diagram. (Letters refer to the coatings mentioned in the text.)

and glass plates for the dielectric studies, elasticity measurements and hardness determinations, respectively. The coatings were applied by spraying in two layers; the total thickness was 50 to 60 μm. The samples were stored in the laboratory and protected from light for two weeks before exposure. The exposure consisted of placing the samples on a rack tilted at 45° facing southward and aging in normal weather conditions. Measurements were performed after 1, 3, 7, 14, 30, and 60 days.

Parameters of the Coating

Various parameters, necessary for the evaluation of the coatings, were determined both before exposure and after a given time of aging, according to the following methods.

Dielectric Measurements. Dielectric measurements were performed with the aforesaid method. The variation of the activation energy values, determined from the dielectric spectra and that of the relaxation temperature related to $f = 1000$ Hz, were detected during aging.

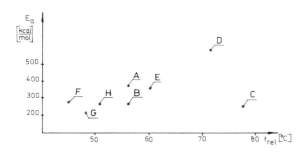

FIGURE 9 — Activation energy vs. relaxation temperature diagram. (Letters refer to the coatings mentioned in the text.)

Hardness Measurements. The hardness of the coating was determined by the pendulum hardness meter according to Konig (manufactured by Erichsen Co.). The results were expressed as the percentage of the hardness of the standard glass plate.

Elasticity Measurements. The elasticity of the coating was determined by the Erichsen cup test. The impression (Erichsen elasticity R_E) was determined in mm.

Discussion

The results of the measurements are summarized in Figures 6 through 9. The data refer to the samples measured after 60 days of aging.

The hardness of the coating of various samples is plotted as a function of the respective activation energy in Figure 6. No correlation is observed between those two parameters. However, the relationship between the hardness and the relaxation temperature of the coating, according to the hypothesis stated above, is manifested as a definite tendency in Figure 7. The same conclusion can be drawn regarding the relationship between elasticity and activation energy shown in Figure 8.

The activation energy of the coatings under investigation as a function of the relaxation temperatures is shown in Figure 9. It is apparent in the figure that a lower activation energy was observed with coatings having a lower relaxation temperature, i.e., a lower hardness value of the coatings can be correlated with higher elasticity. On the other hand, a higher relaxation temperature, which corresponds to a larger hardness value of the coating, may be related to either high or low activation energy indicating either a brittle or an elastic character, respectively.

It can be concluded that the above results support our hypotheses in the case of the coatings under investigation. It is apparent that the measurement of the dielectric properties of paints and lacquers can be useful in the study of aging, and in that of other properties of organic coatings.

The Transport Properties of Polyurethane Paint

R.T. Ruggeri and T.R. Beck*

Introduction

Paint has been used to inhibit corrosion for a great many years, but a complete quantitative model of paint properties and behavior has not been presented. In an effort to facilitate a more quantitative understanding of how paint protects metals from corrosion, a mathematical model has been developed.[1] The model is based on ideas presented by several authors. Mayne,[2] for example, has pointed out that unpigmented paints retard corrosion primarily by acting as a diffusion barrier. Several authors[2-6] have recognized the importance of the diffusion of ions, and have studied the membrane potential of paint films. These studies indicate that paint behaves as a low capacity ion exchange membrane. Ulfvarson[7-9] and others,[10,11] have measured the ion exchange capacity of a variety of paints in an attempt to relate it to the protective performance of the paints. The diffusion of water in polymers has been considered by Crank and Park.[12] They discuss the two major schools of thought concerning water sorbtion and diffusion: adsorption of water by a porous polymer, or dissolution of water into a solid organic polymer phase. The Brunauer, Emmett and Teller[13] (BET) theory of adsorption has been applied to the porous-media model of paint. Solution theories like those presented by Flory[14] have been used to describe polymers as single-phase solutions.

The concepts presented by these authors form the basis for the mathematical description of the transport properties of paint films. The model describes paint as a protective layer limiting the mass transfer of reactants toward, and products away from, the corroding metal. The paint possesses ion exchange capacity and is treated as an organic solution rather than a porous solid. In this manner, the model attempts for the first time to describe quantitatively the movement of mobile species through the paint, and further, to relate the various fluxes stoichiometrically to the corrosion process occurring at the paint metal interface. In order to describe the diffusion of mobile species through paint, a set of equations has been used which are similar to those proposed by Newman[15] for diffusion in liquids. Newman's equations are, likewise, similar to the Stefan-Maxwell equations describing diffusion in gases.[16] These equations are very general in nature, and are consequently widely applicable. A program was written to solve the equations using a digital computer. When the correct set of parameters (diffusion coefficients, transference numbers, etc.)[1] are used, the program calculates all the fluxes and concentrations within the paint as functions of the external conditions (solution concentrations, pressure, etc.) The generality of the diffusion equations represents one advantage of the current model over what has been presented previously. A disadvantage of this model is that the equations used here are not as well known as other descriptions of diffusion, and they have only recently been applied to polymer systems.[17] In order to test the application of the equations to paint systems and evaluate the parameters required for the model, an experimental investigation of the diffusion of water and ions (sodium and chloride) through polyurethane was undertaken. The experiments have confirmed many of the basic concepts upon which the model was constructed, and have led to a new postulate concerning the passage of current through paint membranes.

Experimental

Hittorf Experiments

The classic Hittorf method[18] has been used in an attempt to determine ionic transference numbers and diffusivities in polyurethane films. Hittorf experiments involve recording the current and electrolyte concentrations as functions of time after the application of a fixed potential across the paint films. The polyurethane (#14-9-3-900, Desoto, Inc., Berkeley, California) is used on commercial airplanes, and was obtained from The Boeing Company, Seattle, Washington. The paint meets Military Specification C-83286B, USAF. Free films of the polyurethane were prepared by spray painting sheets of decal paper and subsequent stripping. The paint was dried in laboratory air at room temperature and ambient relative humidity. After the paint had dried, it was aged for a period not less than thirty days at ambient temperature and humidity. The paint was stripped from the decal paper, just prior to use, by soaking in

---

*Electrochemical Technology Corporation, Seattle, Washington

water for a few minutes. The free paint film was rinsed until it no longer felt slippery to the touch, and then given a final rinse in distilled water. The paint film was placed in a Plexiglas Hittorf cell in such a way that it formed the separating barrier between two compartments containing electrolyte solutions of identical concentration. After the paint was placed in the cell, the two compartments were filled and the entire apparatus enclosed in an aluminum box acting as a Faraday cage. A potential was applied across the membrane with two identical silver-silver chloride electrodes. The electrodes were parallel to, the same diameter as, and equidistant from, the paint film. The aluminum box was opened periodically for sampling of the aqueous solutions in the two compartments. The solutions were made from reagent grade laboratory chemicals and distilled water. In some cases, radioactive tracers of sodium ($^{22}Na^+$) and chloride ($^{36}Cl^-$) were used. These radionuclides were purchased as soluble salts from commercial sources, and were certified to be 99% chemically pure. No attempt was made to control the carbon dioxide or oxygen content of the solutions. The experiments were run at ambient room temperature ($\approx$23 C).

Transference-Number Experiments

Transference numbers were obtained by measuring the membrane potential[6] in the Hittorf-cell apparatus. The basic measurement involved placing a membrane between two solutions of different composition and recording, with an electrometer, the open-circuit potential between the silver-silver chloride electrodes. Radioactive sodium and chloride were used to trace the diffusion of ionic species, and tritium labled water was used to measure the water flux. Small (10 μl) samples were periodically drawn off each aqueous solution, and the radioactive tracer contents analyzed by liquid scintillation counting the beta emissions.

Time-Transient Current

The Hittorf-cell apparatus has also been used to obtain measurements of the current transient after application of a potential step function. The current was measured on a Keithley model 616 digital electrometer which produced a voltage output proportional to the current. For experiments shorter than about 1000 seconds, the output voltage was recorded on a Hewlett-Packard model 7046A XY recorder. For longer experiments (t > 1000 seconds) a Hewlett-Packard model 7132A strip chart recorder was used. The time constant of the Keithley electrometer was about 0.1 second while that of the XY recorder was about 0.5 second.

Ionic Capacity

The measurement of the ionic capacity of polyurethane has been attempted employing two different methods: specific ion titrations and radiotracer counting. The latter technique has been the most successful.

Van der Heyden[10] and Khullar and Ulfvarson[9] have indicated that the ion exchange capacity of paint varies between about 0.01 m mole/g and 0.5 m mole/g, or from about 1 to 50% of the capacity of commercial ion exchange materials. For materials with this high an exchange capacity, aqueous titration methods should be appropriate. Such techniques have been applied to polyurethane with little success; the ion exchange capacity was measured to be 0.00 ± 0.01 m mole/g.

For the second method, a solution of known sodium chloride composition was prepared using sodium-22 and chloride-36 radiotracers. A preweighed paint sample was allowed to stand in contact with this solution for two weeks. Small samples of the solution were withdrawn periodically for liquid scintillation analysis. After two weeks, the paint sample was removed, rinsed three times (for five seconds each) in distilled water, and dried. The paint was then analyzed for sodium-22 and chloride-36.

The energy of the sodium-22 beta decay is similar to that of the chloride-36. Liquid scintillation techniques cannot, therefore, be used to distinguish between these two elements; however, sodium-22 emits a positron (β^+) with a decay efficiency of 89.8%.[19] Each positron produces two 0.511 Mev gamma rays when it is annihilated in the surrounding medium. This fact allows one to coincidence-count the gamma radiation from the sodium-22. The technique involves two gamma ray detectors and a timing mechanism. The clock is started when one of the detectors is activated by a gamma ray. If the second detector is excited within 2×10^{-6} seconds, a simultaneous event is assumed and a decay is recorded; otherwise the event is ignored. With such apparatus, the counting efficiency is low ($\approx$7.5%), but the chloride-36 interference is reduced nearly to zero.

Using the coincidence counting technique, the sodium-22 content of paint films has been determined. Liquid scintillation methods were then used to establish the sum of the chloride-36 and sodium-22 present in the paint. The chloride content was calculated by difference. Some difficulties have been experienced using this method. They stem primarily from the low salt content (low radioactive activity) of the paint and the low counting efficiency of the coincidence technique; nevertheless, the order of magnitude of the salt concentration has been determined. Further refinements of the experimental and analytical techniques will reduce the uncertainty.

Results and Discussion

The flux (J) of a neutral species through a film can be described as follows:

$$J = - \frac{P \Delta C}{\ell} \qquad (1)$$

where P = the permeability coefficient, ΔC = the external solution concentration difference, and ℓ = the film thickness (Reference 12, p. 5). Application of the concepts illustrated in Equation 1 to the flux equation for electrolytes in dilute solutions (Reference 15, Equation 82-3) yields

$$J = -P \frac{\Delta C}{\ell} - \frac{zF}{RT} PC \frac{\Delta \phi}{\ell} \quad (2)$$

where z = the ionic charge, F = the Faraday constant, R = the gas constant, T = the absolute temperature, and ϕ = the electric potential. Equation 2 can be thought of as a definition of the permeability coefficent.

Hittorf experiments have been performed with polyurethane in aqueous sodium chloride solution. These experiments were designed to establish the fluxes of all mobile species under varying conditions of applied potential and electrolyte concentration. The sodium and chloride fluxes were to be determined by adding known quantities of radioactive tracers.

$$(10^3 < a < 10^5 \ \mu Curi/mole)$$

to the solution in one of the Hittorf-cell compartments. The results showed that the ionic species fluxes were below the detectable limit; i.e., the concentration change of radiactive tracers was too small to be determined. In contrast, the transport of tritium labled water was easily determined. The results of several experiments are summarized in Table 1. The maximum permeability coefficients for ionic species are indicated. These coefficients were calculated from the maximum fluxes, which in turn were calculated from the detection limits of the appropriate isotopes. It is apparent from Table 1 that the permeability coefficients of all the mobile species are small.

TABLE 1 — Permeability Coefficents in Polyurethane Free Films at ≈22 C

Mobile Species	P (cm²/s)	External Conditions	
		Salt Concentration	Potential Difference
Cl⁻	< 6.0 E-14	0.1 N	1.5 V
Cl⁻	< 6.0 E-14	0.018 N	1.5 V
Cl⁻	< 3.8 E-11	0.018 N	0.0 V
Na⁺	< 3.0 E-14	0.1 N	1.5 V
Na⁺	< 3.3 E-11	0.09 N	0.0 V
H₂O	1.23 E-10	2.5 N	0.14 V
H₂O	3.28 E-10	0.2 N	0.1 V
H₂O	4.06 E-10	0.1 N	1.5 V
H₂O	3.84 E-10	0.09 N	0.0 V
H₂O	4.56 E-10	0.02 N	1.5 V
H₂O	4.56 E-10	0.02 N	0.0 V

The permeability coefficient can be written in terms of the diffusivity and the dimensionless solubility

$$P = DS \quad (3)$$

where D = the diffusivity, and S = the solubility (Reference 12, p. 5). The small magnitude of the permeability coefficent could, therefore, indicate either a low solubility or a low diffusivity.

The solubility of water in polyurethane has been measured in absorption isotherm experiments and has been found to be 1.5% by weight at about 20 C.[20] This information has been used to calculate the diffusivity of water in polyurethane. An average value of

$$D_{H_2O} = \frac{P}{S} = 3.2 \times 10^{-8} \pm 35\% \ cm^2/sec \quad (4)$$

was obtained. This average includes permeability data obtained from diffusion cup experiments (ASTM E96-66, procedure BW) as well as that listed in Table 1.

Radioactive sodium and chloride have been used to measure ion concentrations in the paint. The procedure was the same as that discussed for measuring the ionic capacitance. The paint was allowed to stand in contact with a solution of known compositon for a period of time. Samples of the liquid were withdrawn periodically, and after two weeks the paint was removed from the solution, rinsed in distilled water, and analyzed for radioactive tracers. Conservation of mass requires that the total quantity of radioactive materials in both the paint and the solution remain constant; however, some measurement problems may result because the ion concentrations inside the paint are low. Measurements made on the paint film may suffer from the problems of incomplete phase separation which occur when the paint phase is removed from the solution phase prior to counting the radiotracers.[21,22] Any solution left clinging to the paint after rinsing is counted as though it were absorbed inside the paint. On the other hand, measurements of the changes in solution activity suffer from the classic difficulty of measuring a small difference between large numbers. Despite these problems, the order of magnitude of the ion concentrations have been determined in two different radiotracer experiments, both at about 0.1 N NaCl. The ion concentrations from the first and second experiments were, respectively

$$1.6 \times 10^{-8} < C_i < 1.2 \times 10^{-6} \ mole/g\text{-paint} \quad (5)$$

$$4.6 \times 10^{-8} < C_i < 1.9 \times 10^{-7} \ mole/g\text{-paint} \quad (6)$$

The first experiment was considered less reliable than the second. If a value of $C_i \cong 10^{-7}$ mole/g-paint is taken as an average, the solubility is calculated:

$$S \cong \frac{10^{-7}}{10^{-4}} = 10^{-3} \quad (7)$$

The diffusivity of ions in the paint can then be calculated from the permeability data (Table 1).

$$D_i \cong 6 \times 10^{-11} \ cm^2/sec \quad (8)$$

The ionic capacity of several types of paint has been measured previously.[7-11] In these studies fairly high values of ionic capacity were obtained by conventional titration techniques. Measurement of the ionic capacity for this type polyurethane has been attempted using these same methods. The technique has a sensitivity of about 10^{-2} m mole/g-paint, but no detectable ion exchange capacity was observed. This result is consistent with the results of Khullar and Ulfvarson[9] who reported the ion exchange capacity of another polyurethane as zero at pH values below seven. At a pH of eight they observed a moderate ion exchange capacity (0.35 m mole/g-paint). Higher capacity at high pH was not observed in this investigation. Although the ion exchange capacity has not been determined, an upper bound of about 10^{-2} m mole/g-paint can be established from these results. This upper bounding value is also consistent with the ion content of the paint determined by radiotracer experiments

$$C_i \leq 1.9 \times 10^{-4} \text{ m mole/g-paint } (0.19 \text{ mole/m}^3) \quad (9)$$

Current transients have been recorded after the application of a potential step function across a polyurethane film. These data were analyzed to obtain two pieces of information: the time constant for current decay and the apparent steady state conductivity. The conductivity of the paint film was expected to be qualitatively similar to that for an ion exchange material. Figure 1 shows the steady state conductivity for a polyurethane free film in sodium chloride solution. Activity coefficients were calculated from the data of Robinson and Stokes.[23] The shape of the curve is as predicted, with a conductivity plateau at low solution concentrations and a slope of one at high concentrations. The solid curve represents a mathematical model best fit of the conductivity data. The conductivity becomes constant at low concentrations when the ion content of the paint is limited by the membrane ionic capacity. Donnan exclusion is operating in this concentration region. At higher salt concentrations the ion concentration in the membrane is determined primarily by the external solution concentration and the ionic distribution coefficients.

The second body of information available from the transient-current experiments centers around the determination of the time constant for decay of the transient. Interpretation of these data is complicated by the fact that the results are highly erratic. Similar observations have been reported previously for measurements of other dynamic properties of polymers.[12] The measured properties appear to be related to the history of the polymer. Thus, attempts were made to achieve a uniform condition of the paint prior to testing, but the erratic behavior was unaffected. One of these attempts involved conditioning the paint in the test solution for various lengths of time up to three days (72 hours). Other attempts included conditioning in various solutions (distilled water, for example) at temperatures up to about 55 C, and conditioning in methanol (a water soluble, nonaqueous solvent) then test solution. Despite the failure of these conditioning procedures to eliminate the experimental noise, enough experiments have been conducted to allow qualitative, and even semiquantitative evaluation of the current transient results.

The analysis of the transient current passing through a free film of polyurethane shows that several phenomena are taking place. Figure 2 qualitatively illustrates the current as a function of time after application of a potential step function. The time scale can be divided into four regions representing four types of transient behavior. In region 1 ($t < \approx 4$ seconds) the paint film-solution-electrode circuit can be analyzed as a parallel resistor-capacitor network in series with a second resistor. This equivalent circuit indicates that the current is expected to be high im-

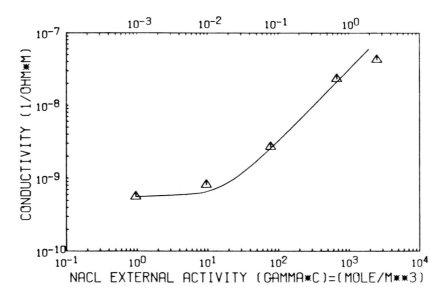

FIGURE 1 — Polyurethane conductivity in aqueous NaCl at 22 C. Top scale units are mole/ℓ. Symbols represent experimental data. Solid curve is the computer calculated "best fit."

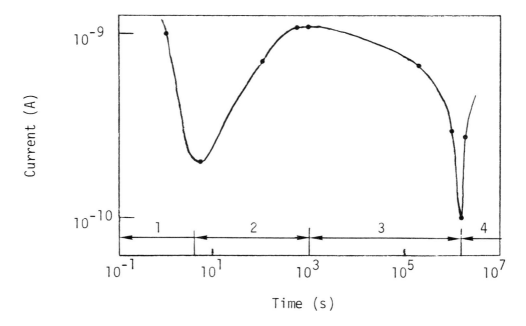

FIGURE 2 — Qualitative transient current through polyurethane immersed in NaCl solution following the application of a 1 V step function.

mediately after the step potential and decay rapidly to a low level. The increasing current in region 2 (4 seconds < t < 1000 seconds) cannot be explained by the R-C network unless the resistance values change. In this region, the conductivity of the paint is increasing and results in an increased flow of current which eventually reaches an apparent steady-state at about one thousand seconds. The reasons for the increasing conductivity of the paint in region 2 are not entirely clear, but must be related to the uptake of either salt or water from the external solution. In either case, the rate of change of concentration inside the paint is expected to be diffusion limited. The time constant for diffusion is thus related to the observed changes of current in region 2. An explanation of the decreasing current in region 3 (1000 seconds < t < ≈ 10^6 seconds) is not yet available, but in this time range molecular rearrangements within the polymer are possible. Thus, such phenomena as stress relaxation or creep may be important. This is also the region which corresonds with the time constant for ionic diffusion ($\tau = \ell^2/D_i = 4.2 \times 10^5$ seconds). In region 4 (t > 10^6 seconds), physical and chemical changes within the polymer are possible. Tests of the quasi-equilibrium properties of paints conducted at this laboratory have shown plasticizer concentrations change in this time region. Also, the diffusivity of water obtained from diffusion cup experiments changes on this time scale. Our understanding of processes occuring throughout the time range studied is incomplete, but clearly more variables enter the analysis at longer times. In an effort to minimize the unknown factors it is desirable to conduct studies over short time intervals. The current transients in region 1 were largely lost because of the relative slow response of the recording equipment. Further discussions will thus be limited to region 2

where the current is expected to be limited by the diffusion rates of mobile ions.

Crank[24] has described the diffusion of neutral species across a membrane under the influence of a concentration gradient. He shows that for this type of diffusion the fluxes can be described by an infinite series involving a single dimensionless parameter

$$\frac{Dt}{\ell^2} \qquad (10)$$

where D = the diffusion coefficient, t = time, and ℓ = the membrane thickness. In concentrated solutions, when Donnan exclusion is not operative, the diffusion fluxes depend on the same parameter (Equation 10) for the case of ions diffusing through paint under the influence of a potential gradient. Crank has also presented a mathematical description of the average concentration of diffusing species in the membrane as a function of the same dimensionless parameter. Thus, one can expect the quasi-steady state current (region 2) to be related to the diffusivity of mobile ions through an equation similar to that presented by Crank. This being the case, an analysis of the transient current can be expected to produce information about the diffusivities of ions in the paint phase.

Time constants for the decay of the quasi-steady state current have been measured using the formula

$$\frac{i_\infty - i}{i_\infty - i_o} = e^{-\frac{t}{\tau}} \qquad (11)$$

where i_∞ = the current at "infinite" time ($10^3 < t < 10^4$ seconds), i_o = the current at time zero, i = the current at time = t, and τ = the time constant. Current decay time constants for polyurethane in sodium chloride solutions have been found to be about 1000 seconds for 5×10^{-3} cm thick paint films. By equating the time constants of diffusion and current decay, an ionic diffusivity can be calculated

$$D = \frac{\ell^2}{\tau} \qquad (12)$$

where ℓ = the paint thickness. For a 5×10^{-3} cm thick paint film, and τ = 1000 seconds, a diffusivity

$$D = 2.5 \times 10^{-8} \text{ cm}^2/\text{sec} \qquad (13)$$

is calculated. It is interesting to note that this is three orders of magnitude higher than the ionic diffusivity estimated previously (Equation 8), but it is nearly identical to that measured for water (Equation 4).

The fact that the current transient time constant (τ = 1000 seconds) is small compared to that estimated previously ($\ell^2/D_i = 4.2 \times 10^5$ seconds) is not entirely surprising. The ionic diffusivity (D_i, Equation 8) was estimated from the maximum ionic flux and a measured value of the ionic concentration. The concentration measurement was subject to a possible error resulting from incomplete phase separation.[21,22] The error would tend to make the ionic concentration appear too high, and this in turn would produce a low diffusivity (Equation 3). It is not likely that these experiments are subject to a three orders of a magnitude error, as would be required to resolve the discrepancy between the two estimates of the ionic diffusivity. The measurement of the ionic concentration in polyurethane has been repeated three times using a single paint sample, and the results were reproduced within ±10%. This consistency is unlikely when incomplete phase separation is a problem. The data indicate, then, that although the apparent ionic diffusivity may be slightly low, a large discrepancy is unlikely, and the ionic diffusivity is much too low to explain the observed time constants for current decay.

A second explanation of the difference between the two time constants should be explored; the presence of additional mobile ions. Only reagent grade (99.8% NaCl) sodium chloride solutions have been investigated in this study. It appears unlikely that a salt impurity could be present and possess a selectivity coefficient high enough to appear in significant concentrations within the paint film. Another possibility exists; hydrogen and/or hydroxyl ions are carrying the current. These ions are fundamentally different from those of a completely dissociated salt like sodium chloride. They are the dissociation products of a neutral molecule, water, which is known to be present in the paint. The fact that the electrical resistance of some paint films increases with the activity of water has been observed previously,[25,26] and is consistent with this possibility. Dissolved carbon dioxide is another neutral molecule which exists in equilibrium with ionic species. Thus, ambiguity exists over which compound, carbon dioxide or water, is providing the ions at low salt concentrations. Experiments have not been performed in a carbon dioxide free atmosphere, making elucidation of the ambiguity impossible at this time. Further research will clarify the point, but for the purposes of this discussion only the water equilibrium will be considered.

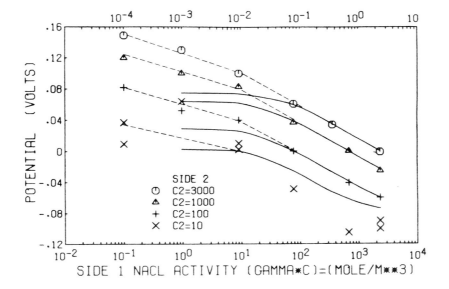

FIGURE 3 — Transference number data for polyurethane in NaCl solution. Top scale units are mole/ℓ. Symbols are experimental data. Solid lines are the computer simulation of an ion exchange membrane. Dashed straight line segments are compatable with the three-transference-number mechanism.

This view of the role water plays in affecting current conduction in paints is further supported by the transference-number data illustrated in Figure 3. The cell potential is shown as a function of sodium chloride activity for experiments in which a polyurethane film separates two solutions containing different concentrations of sodium chloride. In these experiments, the concentration on side two was held constant while the side one concentration was varied. The experimental data are symbol plotted, and the solid curves represent the computer solution of the mathematical model for the case when the paint behaves as a conventional ion-exchange membrane. All the computer generated curves illustrated in both Figures 2 and 3 were calculated using one set of relevant parameter values. The curve through the conductivity data was constructed first, by adjusting the parameters to obtain a good fit. Once the conductivity data had been successfully modeled, only one parameter, the ionic diffusivity ratio (D+/D−) for sodium chloride, was varied to obtain the four "computer curves" shown in Figure 3. Figure 3 shows that when the paint is modeled as an ion exchange membrane, the slope approaches zero at low concentrations. Alternatively, when the conductivity results from water dissociation, one can expect the potential to depend on three transference numbers (Reference 15, Section 17). The exact nature of this dependence cannot be predicted without solving the diffusion equations for this new model, but the equations indicate that, just as in the ion exchange case, two concentration regions can be identified. At high salt concentrations the membrane will behave as before, with the potential falling as the concentration on side one increases. The slope of the potential vs concentration curve in this region is determined primarily by the sodium ion tranference number. In the low concentration region, the slope of the curve depends most on the hydroxyl ion tranference number. Thus, the slope in the dilute region does not go to zero, as for the ion exchange membrane; additionally, if the concentration dependence of the tranference number can be ignored, straight lines can be expected. The dashed lines drawn through the data in Figure 3 illustrate this point. Parallel lines are to be expected in both the low and high concentration ranges if the three transference number mechanism is operative, but are more difficult to explain if the slope changes because the ionic diffusivities are concentration dependent. With the exception of the experimental series having the lowest side two concentration ($C_2 = 10$ mole/m^3), the data appear to be well represented by two straight lines. When the side two concentration was dilute ($C_2 = 10$ mole/m^3), the data were erratic, and subject to considerable experimental error. Despite this difficulty, these dilute side two data can also be approximated by two straight line segments. Thus, the majority of these data qualitatively support the conclusion that the conductivity of polyurethane films exposed to dilute salt solutions is determined primarily by ions resulting from the dissociation of neutral species within the paint. A quantitative check of this new mechanism is planned employing an updated version of the original mathematical model.

The hypothesis that two ionically dissociating species (NaCl and H_2O) are present within the polyurethane membrane, is supported by the results of two different types of experiments: time transient current, and transference number experiments. The time transient current experiments yield low time constants, and hence ionic diffusivities which are three orders of magnitude higher than the best estimates based on NaCl ionic flux and solubility data. These experiments represent dynamic measurements and suffer from the classic difficulties associated with the effects of polymer history. The transference number experiments, on the other hand, represent steady-state measurements and appear to be much less affected by polymer history. The tranference number experiments indicate that two ionically dissociating species (at least three ionic species) are present within the paint, and that the primary current carrying ions change with concentration. The presence of a signficant concentration of a second salt is unlikely. One is thus forced to conclude that water itself represents the most likely second ionically dissociating species. This conclusion is supported qualitatively by Mayne[25] and by the data of Kittelberger and Elm.[27] The hypothesis is also consistent with the results of the Hittorf, ionic capacity, and conductivity data obtained in this laboratory. It should be noted, however, that a limited number of experiments have been performed at this time, and although the results are encouragingly consistent, further tests will be required to verify the hypothesis.

Conclusions

The following conclusions are based on mass transport and supporting experiments carried out on a commercial, unpigmented, polyurethane paint meeting Military Specification C-83286B USAF:

1. The diffusivity of water in polyurethane is about 3.2×10^{-8} cm^2/second.

2. The concentration of mobile ions in polyurethane exposed to 0.1 N NaCl solution at 22 C is

$$C_i < 1.2 \times 10^{-6} \text{ mole/g-paint}$$

and is is probably close to

$$C_i \cong 10^{-7} \text{ mole/g-paint}$$

3. The permeability of polyurethane to sodium and chloride ions is low.

$$P \leq 6 \times 10^{-14} \text{ cm}^2/\text{second}$$

under the exposure conditions investigated

$$C = 0.1 \text{ N NaCl}$$

$$\Delta\phi = 1.5 \text{ V}$$

4. The mechanism of ionic conduction in polyurethane changes as the external electrolyte concentration changes. At high concentrations the conductivity increases as the electrolyte concentration increases, indicating conduction by soluble ionic salts. At low concentrations, the membrane conductivity appears to be determined by the concentration of ions resulting from dissociation of neutral species, such as water.

Acknowledgment

This work was supported by Air Force Office of Scientific Research Contract No. F49620-76-C-0029, and by Naval Ocean Research and Development Agency Contract No. N00014-79-C-0021. Thanks also to William P. Miller, Assistant Director of Reactor Operations, Nuclear Reactor Laboratory, University of Washington, for his assistance with the coincidence counting experiments.

References

1. R.T. Ruggeri and T.R. Beck. "A Model for Mass Transport in Paint Films," Corrosion Control by Coatings, Henry Leidheiser, Ed., Science Press, Princeton, 1979, p. 455.
2. J.E.O. Mayne. J. Oil and Colour Chem. Assoc., 32, p. 481 (1949).
3. E.M. Kinsella and J.E.O. Mayne. Proc. Third Int. Congress on Metallic Corrosion, 3 p. 117, Moscow, 1966 (1969).
4. W.W. Kittelberger and A.C. Elm. Ind. Eng. Chem., 44 (2), p. 326 (1952).
5. C.A. Kumins. Official Digest, 34, p. 843 (1962).
6. C.A. Kumins and A. London. J. Polymer Sci., 46, p. 395 (1960).
7. U. Ulfvarson and M. Khullar. J. Oil Co. Chem. Assoc., 54, p. 604 (1971).
8. U. Ulvarsson, J.L. Khullar, and E. Wahlin. J. Oil Col. Chem. Assoc., 50, p. 254 (1967).
9. M.L. Khullar and U. Ulfvarson. IX FATIPEC Congress, 1968, Section 3, p. 165.
10. L.A. van der Heyden. XI FATIPEC Congress, 1972, p. 475.
11. W.U. Malik and L. Aggarwal. J. Oil Col. Chem. Assoc., 57, p. 131 (1974).
12. Diffusion in Polymers, J. Crank and G.S. Park, Eds., Academic Press, New York, 1968, p.5.
13. Catalysis, P.H. Emmett, Ed., Reinhold, New York, 1954, p. 31.
14. P.J. Flory, Principles of Polymer Chemistry, Cornell University Press, New York, 1953, p. 495.
15. J.S. Newman. Electrochemical Systems, Prentice-Hall, New Jersey, 1973.
16 R.B. Bird, W.E. Stewart and E.N. Lightfoot. Transport Phenomena, John Wiley, New York, 1960, p. 570.
17. W.B. Sunu and D.N. Bennion. Ind. Eng. Chem. Fundam., 16 (2), p. 283 (1977).
18. D.A. MacInnes. The Principles of Electrochemistry, Dover, New York, 1961, p. 65.
19. L. Slack and K. Way. Radiations from Radioactive Atoms in Frequent Use, U.S. Atomic Energy Commission, Washington, DC, 1959.
20. T.R. Beck and R.T. Ruggeri. A Study of Transport Processes and Initiation of Corrosion Under Paint Films, Final Report for Air Force Office of Scientific Research done under Contract No. F49620-76-C-0029, Washington, DC, 1979.
21. G.E. Boyd and K. Bunzl. J. Am. Chem. Soc., 89, p. 1776 (1967).
22. E. Glueckauf and R.E. Watts. Proc. Royal Soc. (London), A268, p. 339 (1962).
23. R.A. Robinson and R.H. Stokes. Electrolyte Solutions, Butterworths, London, 1959.
24. J. Crank. The Mathematics of Diffusion, 2nd ed., Clarendon Press, Oxford, 1975.
25. J.E.O. Mayne. J. Oil Col. Chem. Assoc., 40, p. 183 (1957).
26. C.C. Maitland and J.E.O. Mayne. Official Digest, 34, p. 972 (1962).
27. W.W. Kittelberger and A.C. Elm. Ind. Eng. Chem., 39 (7), p. 876 (1947).

Rate Controlling Steps in the Cathodic Delamination of 10 to 40 μm Thick Polybutadiene and Epoxy-Polyamide Coatings from Metallic Substrates

H. Leidheiser, Jr. and W. Wang*

Introduction

The major function of an organic coating in providing corrosion protection to a metallic substrate is to serve as a barrier to reactants in the environment such as water, oxygen and other gases, and ions. Since all organic coatings are permeable to these species in some degree, the important consideration is whether or not the corrosion reaction occurs when the species are in the vicinity of the metal surface. If the reaction does occur, localized delamination of the organic coating from the metallic substrate is a natural consequence. Alternatively, the presence of a defect in the coating permits electrolyte to reach the metal surface, and the protective value of the coating is determined to a major extent by the ability to resist delamination laterally from the defect. In both the absence and presence of a defect, the precursor to the loss of corrosion protection is delamination of the coating with consequent exposure of the metal to the corrosive environment. Studies of the delamination process are, therefore, relevant to an understanding of corrosion protection by an organic coating.

Delamination of an organic coating by a corrosion reaction is generally considered to result because of the separation of the cathodic half reaction, $H_2O + 1/2O_2 + 2e^- = 2OH^-$, which occurs under the coating, from the anodic half reaction, $Fe - 2e^- = Fe^{++}$, which occurs at a defect. The mechanism of the delamination process itself is poorly understood, but it is thought to result from attack of the coating or the interfacial bond by the alkaline conditions generated at the leading edge of the delaminating region.[1] Acceptance of these concepts implies that delamination should be accelerated under conditions where the metal is made the cathode in an aqueous medium. Cathodic treatment does indeed accelerate delamination and cathodic delamination is widely used as an accelerated test method for determining the quality of an organic coating/metal substrate system.

The study reported herein represents the second stage in efforts to better understand the corrosion phenomena involved in the delamination of organic coatings from metallic substrates where the metal is made the cathode while immersed in an electrolyte. The previous study[2] identified the following variables as important in the delamination phenomenon: oxygen in electrolyte; film thickness; substrate metal; pretreatment of the substrate; type of electrolyte; concentration of electrolyte; and temperature.

The purpose of this presentation is to utilize data obtained in the prior investigation of polybutadiene coatings and new information of epoxy-polyamide coatings in an effort to explain the cathodic delamination process in a more quantitative manner.

Experimental Procedure

Three different substrate materials were used with the epoxy-polyamide coatings: (1) a commercial cold rolled steel, 0.09 cm. in thickness, obtained in the form of square panels, 30 × 30 cm.; (2) commercial galvanized steels in the form of panels, 10 × 30 cm. and 0.05 cm. in thickness; (3) aluminum 1100 panels, 30 × 30 cm. and 0.035 cm. in thickness, obtained from a commercial source.

Several surfacing procedures were used. Aluminum samples were used after degreasing in trichloroethylene. The term "polished" refers to samples that were degreased in trichloroethylene followed by abrasive polishing with metallographic paper. "Acid cleaned" samples were immersed for 5 minutes at 70 to 90 C in a 1:1 solution of concentrated HCl and water, rinsed in distilled water, and then immersed in a 1:1 solution of concentrated sulfuric and nitric acids at room temperature for 10 seconds. The panels were rinsed in distilled water, immersed for 5 to 10 minutes in 5% NaCN solution at room temperature, rinsed in distilled water, dried in acetone and stored in acetone until ready for application of the coating. "Alkaline cleaned" panels were immersed for 5 to 10 minutes at 80 C in a 200 cm³ solution containing 6 g NaOH, 4.8 g Na_2SiO_3, 0.6 g EDTA, 0.48 g Na_2CO_3, and small amounts of the wetting agent, hexadecyl trimethyl ammonium bromide. The panels were thoroughly rinsed in distilled water, dried in acetone, and stored in acetone until ready for the application of the coating.

Information on the polybutadiene coatings has been given previously.[2] Coatings were prepared from

---
*Center for Surface and Coatings Research, Lehigh University, Bethlehem, Pennsylvania

Epon 1001 obtained from Shell Development Company and Polyamine Emerez 1511 obtained from Emery Industries Inc. Both components were dissolved separately in a solvent consisting of 65 Wt.% butyl cellusolve and 35 Wt.% xylene. The coating formulation was a 2:1 mixture of the epoxy and polyamide solutions. The mixture was stirred for one hour before applying to the metal surface using a standard applicator technique. Three different curing techniques were used. The one used in the majority of the experiments consisted simply of curing in air at room temperature for one week. A second procedure involved heating in air at 100 C for one hour before curing at room temperature for a week, and the third procedure involved curing in air for one week followed by heating in air at 100 C for one hour. Unless otherwise stated, the delamination process was studied on samples that had been cured at room temperature.

Coated samples, 7.5 × 10 cm, were cut into pieces, approximately 2 × 6 cm, and the edges and backside were protected from the electrolyte with a thick coating of a commercial pigmented epoxy-polyamine applied with a brush. Just before the experiment was initiated, the center of the sample was damaged by pressing a pointed instrument into the surface. The exposed metal was approximately 0.001 cm² in area.

The experimental procedure has been described previously.[2] The cathode potential was sensed with a capillary tube filled with the electrolyte which was connected by means of a "U" tube to a vessel holding the reference electrode, a saturated calomel electrode (SCE). All potentials are given with respect to SCE. The cathode potential was held constant during the experiment by means of a potentiostat, and the current flowing through the cathode/electrolyte interface was monitored continuously during the experiment.

The cathode potential used with each metal was selected, early in the study, on the basis that a current of approximately 10 μAmp. flowed shortly after the experiment was initiated. Unless otherwise noted in the text, the following cathode potentials were used:

Aluminum	−1.39 v.
Steel	−0.95 v.
Zinc (galvanized steel)	−1.35 v.

It will be noted that these potentials are slightly different than those used in the previous study with polybutadiene coatings. The measured cathode potential was not critically sensitive to the location of the capillary tip so long as the tip was within a mm of the defect and did not cover the defect.

The delaminated area was determined, after the completion of the experiment, by removing the delaminated coating with adhesive tape and measuring by microscopic examination the area that was removed. No coating was removed if the adhesive tape was applied to the defect before exposure to the electrolyte.

The term "delamination parameter" used in the text is defined as the area (in cm²) delaminated divided by the number of coulombs passed through the interface. The term "delamination rate" is defined as the area delaminated divided by the time (in minutes).

TABLE 1 — The Delamination Parameters of 25μm Thick Epoxy-Polyamide Coatings on Steel Surfaced in Different Ways

Pretreatment	Electrolyte	Concentration	Delamination Parameter
Polished	LiCl	0.125M	1.8 cm.²/coulomb
Acid Cleaned	LiCl	0.125M	1.3
Polished	LiCl	0.5M	1.4
Acid Cleaned	LiCl	0.5M	1.0
Polished	NaCl	0.125M	4.4
Acid Cleaned	NaCl	0.125M	3.5
Polished	NaCl	0.5M	3.1
Alkaline Cleaned	NaCl	0.5M	2.5
Acid Cleaned	NaCl	0.5M	2.0
Polished	NaCl	1.0M	1.6
Polished	KCl	0.125M	7.0
Acid Cleaned	KCl	0.125M	5.0
Polished	KCl	0.5M	5.0
Alkaline Cleaned	KCl	0.5M	4.5
Acid Cleaned	KCl	0.5M	3.5
Polished	CsCl	0.125M	14
Acid Cleaned	CsCl	0.125M	6.2
Polished	CsCl	0.5M	9.6
Acid Cleaned	CsCl	0.5M	4.5
Polished	NH_4Cl	0.5M	2.0
Acid Cleaned	NH_4Cl	0.5M	0.45
Polished	KF	0.5M	5.8
Alkaline Cleaned	KF	0.5M	4.8
Acid Cleaned	KF	0.5M	3.9

Experimental Results

The area delaminated as a function of the number of coulombs passing through the interface was reproducible for a standard set of conditions, and was essentially linear over the range used in the experiments. Representative data for a series of halides are given in Table I. Delamination rates for polybutadiene and epoxy-polyamide coatings on a polished steel substrate immersed in CsCl, KCl, NaCl, and LiCl electrolytes are given in Figure 1. For both systems it will be noted that the rate of delamination decreased in the order of CsCl > KCl > NaCl > LiCl. Thus, the rate of cathodic delamination, as well as the delamination parameter, is strongly a function of the electrolyte in which the substrate is immersed. A similar order of delamination efficiency was observed when the substrate was galvanized steel. Delamination was negligible, relative to steel and galvanized steel, in all electrolytes when aluminum was used as the substrate.

Although the delamination parameter was a function of the method of surfacing the substrate before

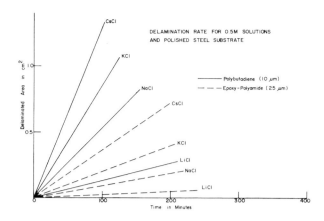

FIGURE 1 — The effect of nature of electrolyte on cathodic delamination of coatings from a polished steel substrate.

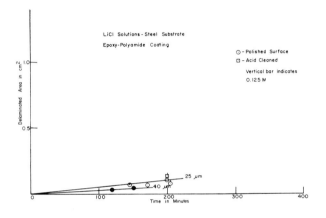

FIGURE 2 — The cathodic delamination of epoxy-polyamide coatings from steel substrates immersed in 0.5 M LiCl.

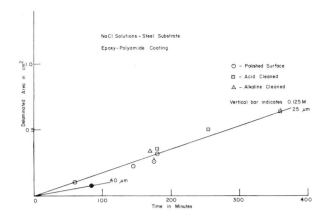

FIGURE 3 — The cathodic delamination of epoxy-polyamide coatings from steel substrates immersed in 0.5 NaCl.

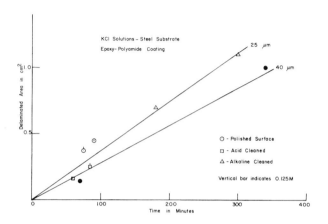

FIGURE 4 — The cathodic delamination of epoxy-polyamide coatings from steel substrates immersed in 0.5 KCl.

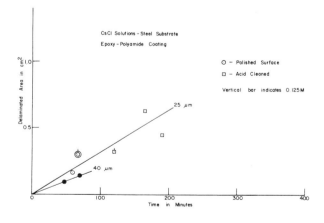

FIGURE 5 — The cathodic delamination of epoxy-polyamide coatings from steel substrates immersed in 0.5 M CsCl.

application of the coating, the delamination rate for the epoxy-polyamide coating was not obviously a function of surface preparation, or even of electrolyte concentration, as shown by the data in Figures 2 through 5. This lack of sensitivity to surface preparation method contrasts sharply with the behavior of polybutadiene coatings, in which the rate of delamination was strongly a function of surface preparation method as shown in Figure 6.

No extensive investigation was made of the method of cure and its influence on the tendency of the epoxy-polyamide coatings to delaminate during cathodic treatment. However, three different cure methods were used, as described earlier. Data are given in Figure 7 for the delamination as a function of coulombs passing through the interface and in Figure 8 in terms of delamination rate. It will be noted that the samples cured at room temperature had the largest delamination parameter and the lowest rate of delamination.

The effect of cathode potential on the delamination parameter of epoxy-polyamide coatings on a polished steel substrate in 0.5M NaCl is shown in Figure 9 where it will be noted that the delamination parameter increases as the cathode potential is made less

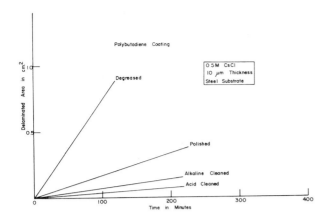

FIGURE 6 — The effect of surface preparation method on the rate of cathodic delamination of polybutadienne coatings from steel substrates immersed in 0.5 CsCl.

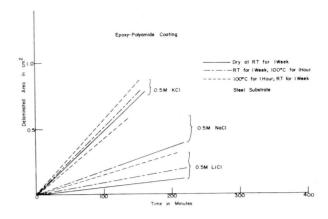

FIGURE 8 — The effect of curing procedure on rate of cathodic delamination of epoxy-polyamide coatings from a steel substrate immersed in 0.5 M alkali chloride solutions.

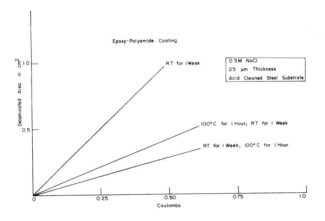

FIGURE 7 — The effect of curing procedure on the delamination parameter of epoxy-polyamide coatings from a steel substrate immersed in 0.5 M NaCl.

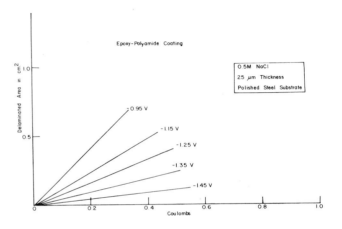

FIGURE 9 — The effect of the cathode potential on the delamination parameter for epoxy-polyamide coatings on steel substrates immersed in 0.5 M NaCl.

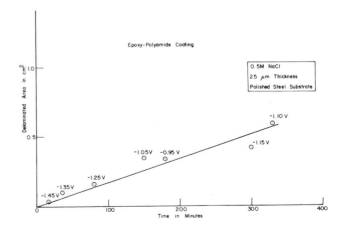

FIGURE 10 — The delamination of epoxy-polyamide coatings from steel substrates immersed in 0.5 M NaCl at different cathodic potentials.

negative. The rate of delamination apparently is independent of the applied potential as shown by the data collected in Figure 10.

An insufficient number of experiments was done to determine quantitatively the effect of coating thickness on delamination rate. The data in Figures 2 through 5 show that the rate of delamination was slower, when the coating thickness was 40 μm, than when it was 25 μm. Previous work with polybutadiene coatings showed that the delamination decreased as the coating thickness increased over a range of 10 to 20 μm.

Work in this laboratory has shown that dipping galvanized steel in 0.1M $CoCl_2$ solution decreases the ability of the surface to catalyze the cathodic reaction, $H_2O + 1/2O_2 + 2e^- = 2OH^-$. Several galvanized steel substrates were used for polybutadiene coatings (Figure 11) and a single substrate (Code E-12) was used with the epoxy-polyamide coating (Figure 12). In all cases, the pretreatment in $CoCl_2$ solution reduced appreciably the rate of delamination on cathodic treatment.

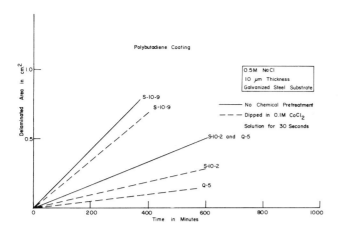

FIGURE 11 — The effect of pretreatment of galvanized steel in 0.1 M $CoCl_2$ on the rate of cathodic delamination of polybutadienne coatings immersed in 0.5 NaCl.

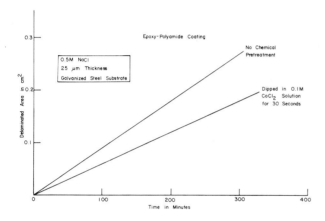

FIGURE 12 — The effect of pretreatment of galvanized steel in 0.1 M $CoCl_2$ on the rate of cathodic delamination of epoxy-polyamide coatings in 0.5 M NaCl.

Discussion

The data presented herein, and those reported previously, are supportive of the following hypothesis for the mechanism of cathodic delamination in neutral halide solutions of polybutadiene and epoxy-polyamide coatings from steel and galvanized steel substrates which contain a defect that exposes a small area of substrate metal.

The current that flows across the interface is the sum of two components: the current for the cathodic reaction, largely $2H^+ + 2e^- = H_2$, which occurs at the defect; and, the current for the reaction, $H_2O + 1/2O_2 + 2e^- = 2OH^-$, which occurs under the coating and adjacent to the delaminating region. The second reaction may also occur at the defect, because the concentration of dissolved oxygen ($2.3 \times 10^{-4}M$) is of the same order as the H^+ ion concentration, assuming that hydrolysis of Fe^{++} ions occurs in the vicinity of the defect. The appearance of visible gas bubbles at the defect is evidence that the $2H^+ + 2e^- = H_2$ reaction is an important consumer of electrons at the defect. The delamination parameter measures the relative activity as a cathode of the region exposed at the defect and the area under the coating, and the delamination rate is determined by the rate at which the reaction, $H_2O + 1/2O_2 + 2e^- = 2OH^-$, occurs under the coating. The rate of cathodic delamination in the presence of sufficient oxygen is largely determined by the rate at which cations reach the metal interface to serve as counter-ions for the OH^- generated. Under conditions where the substrate is inactive for the cathodic reaction under the coating, the rate of reaction may be limited by the catalytic properties of the oxide on the surface of the substrate metal. The comments that follow will provide support for this hypothesis.

There is overwhelming evidence that the predominant delamination reaction under the coating is $H_2O + 1/2O_2 + 2e^- = 2OH^-$. The pH is known to become high under the coating, both in our studies[3] and those of many other workers.[2] No evidence for hydrogen evolution under the coating has been observed in our work; therefore, we discard the reaction, $2H^+ + 2e^- = H_2$, as an important reaction under the coating in the delaminating region. At the defect, on the other hand, the hydrogen evolution reaction is the important reaction, as indicated by the evolution of gas bubbles during the cathodic treatment. The important role of oxygen in the delamination process is shown by the fact that the rate of delamination is negligibly low when the electrolyte is freed of dissolved oxygen.[1]

The most important variable in determining the delamination rate under cathodic treatment in oxygen-containing solutions is the nature of the cation. In all systems studied, the rate of delamination decreases in the order CsCl > KCl > NaCl > LiCl, although, under some conditions, the rate of delamination in KCl solutions tended to approach those in CsCl solutions. A rough parallelism exists between the rate of delamination and the diffusion coefficient of the cation in aqueous solutions of these four electrolytes. This parallelism is supportive of the concept that diffusion of the cation through the coating is rate limiting.

Information is available to calculate the rate of oxygen transmittal through polybutadiene coatings. The permeability of oxygen through polybutadiene is $19.2 \times 10^{-10} cm^3$-cm/second cm^2-cm Hg[4] and the solubility of oxygen in 0.125M CsCl solution at 25 C is 5.67 cm^3/l.[5] If it is assumed that the concentration of oxygen at the coating/electrolyte interface is 5.67 cm^3/l., and the concentration at the coating/substrate interface is zero, then the rate of oxygen passage through a 10 μm thick coating may be calculated to be 3.4×10^{-11} mole/cm^2-second. If the assumption is made that oxygen is consumed at the interface by the reaction, $H_2O + 1/2O_2 + 2e^- = 2OH^-$, then this amount of oxygen will consume 8.2×10^{13} electrons/cm^2-second equivalent to a current flow of 1.3×10^{-5} Amp/cm^2. This rate of oxygen passage through the film may be compared with the fastest rate of delamination observed in CsCl solutions, namely 2.2×10^{-4} cm^2/second, by making the assumption that there are approximately 10^{15} coating/substrate bonds per cm^2 of substrate surface, and that each electron flowing through the interface leads to rupture of one coating/substrate bond. Under such assumptions, the current flow is equivalent to 0.0035×10^{-5} amp/cm^2, a value approximately 1/400 of that obtainable if all oxygen molecules participated in the electrode reaction. Thus, these very crude calculations show that there is sufficient oxygen transmission through polybutadiene coatings, 10 μm in thickness, to support the most rapid delamination rate observed. The calculations also suggest that the rate of oxygen transmittal through the coating is not the rate controlling step in the delamination process.

These calculations also allow a separate determination as to whether the oxygen concentration is sufficient to support the $H_2O + 1/2O_2 + 2e^- = 2OH^-$ reaction at the rate at which it proceeds in the absence of a coating. Information is available from another study[6] to make such a calculation. Galvanized steel immersed in 3% NaCl (pH = 6.4) at room temperature under quiescent conditions exhibits a cathodic polarization curve that indicates that the reaction occurs at a rate equivalent to approximately 20×10^{-6} Amp/cm^2. This reaction rate should be compared with a delamination current of 0.0037×10^{-6} Amp/cm^2, calculated, as before, on the basis of a delamination rate of 2.3×10^{-5} cm^2/second observed with galvanized steel immersed in NaCl solutions. It is clear that the reaction rate itself is not limiting.

The data summarized in Figure 8 show that the method of curing the epoxy-polyamide coating has an effect on the rate of delamination. The slowest rate of delamination in LiCl, NaCl, and KCl solutions was obtained on coatings that were cured at room temperature, and the most rapid rate of delamination was obtained on coatings that were heated at 100 C for one hour before curing at room temperature. Coatings that were cured, at room temperature for a week, followed by heating at 100 C for one hour, had delamination properties not greatly different from coatings that were cured at room temperature. The tentative interpretation of these data is that the higher temperature treatment leads to a larger number of pathways of easy diffusion through the coating.

Two studies carried out in this laboratory indicate that water uptake by polybutadiene coatings and epoxy-polyamide coatings is rapid. Capacitance measurements[7] have yielded information on the rate of water uptake by polybutadiene coatings, and John Standish[8] has shown, by quartz microbalance studies, that epoxy-polyamide coatings on aluminum take up water rapidly. These data, unfortunately, do not allow calculation of the rate of water transmission but they do indicate that water uptake is rapid. Rapid water uptake suggests rapid movement of water through the coating.

The low rate of delamination of polybutadiene and epoxy-polyamide coatings from aluminum is in accord with service experience that indicates that epoxy-polyamide coatings are adherent to aluminum substrates under adverse environmental conditions. Our interpretation of the behavior of aluminum is as follows. Aluminum is a very reactive metal, and exposure to the atmosphere or to an aqueous medium, prior to application of the coating, results in the formation of a thin layer of aluminum oxide on the aluminum surface. This aluminum oxide is the surface that is in contact with the organic coating, and it is the surface on which any cathodic reaction must occur. The cathodic reaction requires three components: water, oxygen, and electrons. As shown previously, oxygen readily passes through the polybutadiene coating, and presumably it readily passes through the epoxy-polyamide coating, too. Information summarized above proves that water transmittal through the coating is rapid. The rate limiting constituent appears to be the electrons. The normal potential drop across the surface oxide during anodization of aluminum is of the order of 10^7 volts/cm. If the oxide film on aluminum is 3×10^{-7} cm in thickness, then it is apparent that a significant voltage is required to overcome the effective resistance of the aluminum oxide. Since electrons may flow through two parallel paths, one at the defect and the other below the coating adjacent to the defect, it is obvious that the electrons will follow the path of least resistance and flow, for the most part, through the defect where the aluminum surface does not have a continuous oxide film, since aluminum pitted at the defect during the experiment.

The experimental information and calculations summarized suggest that the rate of the cathodic delamination reaction under polybutadiene and epoxy-polyamide coatings on steel and galvanized steel substrates is not limited by the availability of oxygen, water, and electrons, the three reactants in the cathodic reaction. Since the OH$^-$ ions are being formed in a confined volume at the delaminating edge, space charge effects would rapidly prevent additional reaction. The charges of the OH$^-$ ions generated by the reaction can be balanced by the diffusion of cations through the film to the interface. The likely diffusing species is the alkali metal cation, rather than the H$^+$, because of the very large difference in concentration. The concentration of the cation in solution is 0.5M in a typical experiment, approximately 10^6 times greater than the H$^+$ concentration. The high pH under the coating[3] also is indicative of the fact that the alkali metal cations, and not the H$^+$ ions, are the major diffusing species. It is thus hypothesized that the rate limiting step in the cathodic delamination, of both polybutadiene and epoxy-polyamide coatings, from polished steel and galvanized steel substrates is the diffusion of alkali metal cations through the coating. It is recognized that metal cations can be supplied both through the coating and through the liquid that enters at the defect. The fact that the delamination rate decreases with increase in coating thickness, in the case of both types of coatings, is strongly suggestive that the cations are supplied through the coating and not from the liquid adjoining the delaminating region.

Data reported previously in a different form[2] allow the calculation of an activation energy for the delamination of polybutadiene coatings on galvanized steel substrates when cathodically treated in 0.5M NaCl. These data are given in Figure 13. The temperature

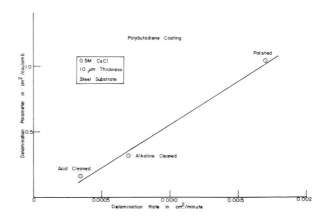

FIGURE 14 — The relation between delamination rate and delamination parameter for polybutadienne coatings on steel immersed in 0.5 M CsCl.

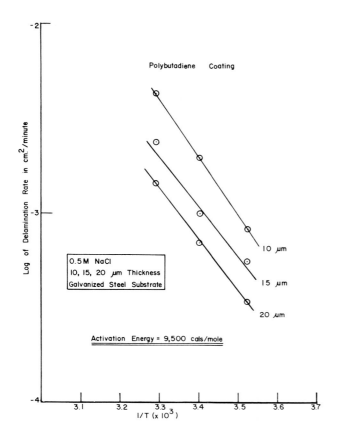

FIGURE 13 — The effect of temperature on the cathodic delamination of polybutadienne coatings from galvanized steel immersed in 0.5 M NaCl.

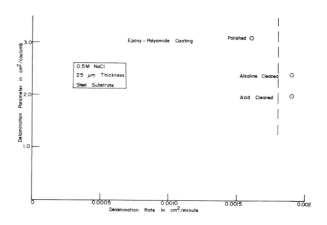

FIGURE 15 — Data showing the lack of a relationship between delamination rate and the delamination parameter for epoxy-polyamide coatings on steel substrates immersed in 0.5 M NaCl.

range, 11 to 31 C, is small, but nine data points are available for estimating the activation energy. These data yield an activation energy for the delamination process of the order of 9,500 cals./mole, a reasonable value for a diffusion process.

Data reported in Figures 11 and 12 show that predipping galvanized steel in 0.1M $CoCl_2$ for 30 seconds, prior to application of the polybutadiene or epoxy-polyamide coatings, results in a lower rate of cathodic delamination. These findings may be explained by a lower activity of the zinc oxide at the coating/substrate interface as a catalyst for the $H_2O + 1/2O_2 + 2e^- = 2OH^-$ reaction, since it is known that the introduction of small amounts of cobalt into the oxide film on zinc lowers the activity of the surface for this cathodic reaction.[9]

The delamination process has been studied in terms of two variables: time, and the number of coulombs passing through the interface. In the case of the polybutadiene coatings on steel and galvanized steel, the relative behavior under different experimental conditions was the same, whether the time or the number of coulombs passed was the variable. In some cases there was an approximately 1:1 relationship between these two variables, as shown, for example, in Figure 14. The behavior of the epoxy-polyamide coatings, however, was different from polybutadiene coatings in that a 1:1 relationship did not exist between these two variables under those conditions in which the surface pretreatment was varied. Figure 15, for the epoxy-polyamide coatings, should be compared with Figure 14. Our tentative interpretation of these two different behaviors is that the polybutadiene coating is very sensitive to the character of the oxided steel surface with which it is in contact, and that bond cleavage is at the polymer/substrate interface. The lack of sensitivity to the surface character, in the case of the epoxy-polyamide coatings, is considered to be indicative of the fact that bond cleavage is within the polymer itself. Evidence for bond fracture within the polymer during the delamination process has been obtained by Hammond et al.[10]

Acknowledgment

Appreciation is expressed to the Petroleum Research Fund for providing a fellowship to Wendy Wang during the early stages of this research. The research involving the epoxy-polyamide coatings received support from the Office of Naval Research, to whom the authors express their gratitude.

References

1. R.A. Dickie and A.G. Smith, *Chem. Tech. 1980,* No. 1, p. 31.
2. H. Leidheiser, Jr. and W. Wang, *J. Soc. Coatings Technol.*, in press.
3. H. Leidheiser, Jr. and M.W. Kendig, Corrosion **32**, p. 69 (1976).
4. Polymer Handbook, J. Brandrup and E.H. Immergut, Editors, Interscience (1966).
5. A Seidell and W.F. Linie, Solubilities of Inorganic and Metal Organic Compounds, 4th Edition, *Am. Chem. Soc.* **2**, pp. 1228-32 (1965).
6. H. Leidheiser, Jr. and I. Suzuki, Corrosion, **36**, p. 701 (1980).
7. R.E. Touhsaent and H. Leidheiser, Jr., Corrosion **28**, p. 435 (1972).
8. John Standish, Ph.D. Thesis, Lehigh University, October 1980.
9. H. Leidheiser, Jr. and I. Suzuki, submitted to *J. Electrochem. Soc.*, **128**, p. 242 (1981).
10. J.S. Hammond, J.W. Holubka, and R.A. Dickie, J. Coatings Technol. **51**, p. 45 (1979).

Cathodic Disbondment of Well Characterised Steel/Coating Interfaces

*J.E. Castle, J.F. Watts**

Introduction

Disbondment of polymer coatings from metallic substrates can be accelerated by the presence of water[1] or other hostile environments.[2] The problem may become more serious when the metal is polarised cathodically.[3] Such a situation can arise in practice around a defect in a can coating,[4] or a holiday in the coating of a buried pipeline.

Before application of any coatings system, it is necessary to treat the substrate to remove residual carbonaceous deposits. These may take the form of protective oils or greases, or may be due to handling contamination. These residues have been shown to have a deleterious effect on adhesion.[5,6] There are many types of pretreatment available to the coatings technologist, but they can broadly be divided into two types: those which involve mechanical removal of the surface, such as wire brushing or grit blasting; and solution treatments such as pickling or alkali cleaning. In this paper we report the characterization, by XPS, of three pretreatments.

Although some coatings may be applied to a cold substrate and cure cold (such as house paint), the use of "hot cure" coatings is becoming increasingly popular. These may be applied either to a preheated substrate to effect cure (as in powder coating), or applied to a cold substrate and subjected to a post-application cure (for example, can coatings and automobile finishes). The relationship between the surface chemistry before and after such stoving schedules has been investigated for two of the preparation techniques.

In a perfect situation, a coatings or adhesive system will fail in a cohesive manner, either of adherend or adherate.[7] In practice, adhesive failure occurs frequently. Bikerman,[8] explains this phenomenon by advocating weak boundary layers of either (or both) phases.

The type of bonding present at the interface cannot be examined easily using conventional techniques, because it is necessarily only one or two atom layers thick, but shielded from surface analysis techniques by an adhering layer of coating or substrate (usually coating). Usual profiling techniques, such as argon ion bombardment, cannot be used on polymeric materials due to sample degradation. By approaching the metal oxide/polymer interface through the oxide this problem can be reduced greatly.[9] A new method of achieving this has been devised using oxide stripping techniques.

Although cathodic polarisation is known to accelerate coatings disbondment,[10] the mechanism by which this occurs is not known. Uncertainty exists as to whether it is an acceleration of the failure mode operating under unpolarised conditions (that is, cohesive failure of the coating[11]), or reduction of the oxide leading to adhesion loss. XPS provides an ideal technique for such investigations, and has been used to good effect by Gettings and his colleagues to investigate failure modes and primer efficiency of adhesive joints, using mild steel and stainless steel substrates.[12–15] Hammond, et al, have recently used XPS to identify the failure mode of an automotive primer.[11]

XPS spectra were obtained in this present work using a vacuum generators ESCA 3 Mk II, operating at a vacuum of $\cong 10^{-9}\tau$, using AlKα radiation (excitation energy 1486.6 eV), and a 50 eV analyser energy. The spectra were acquired using a DEC PDP8e based data system; high resolution narrow scans (20 eV) were obtained for the C_{1s}, O_{1s}, and $Fe_{2p3/2}$ (and Cl_{2p} and $Na1s$ for cathodically disbonded samples) regions, in addition to a wide scan (1000 eV) to provide a survey of other possible impurities.

Surface Characterization

Efficiency of Surface Preparation Techniques

The steel used in this study was a sheet of standard mild steel cut into 10 × 6 mm coupons. Three different preparation techniques were investigated: 1) agitation in a commercial alkali cleaner[16] at 70 C for 5 minutes; 2) abrasion, under water, on 600 grade silicon carbide paper; 3) abrasion with grade F engineer's emery cloth (120 grit).

After preparation, the specimens were exposed to laboratory air for varying lengths of time to investigate any short term time dependency of the passivating process. Times of up to 100 hours were used.

Wide scans of alkali cleaned and emery abraded specimens are shown in Figure 1; narrow scans of the

---
*University of Surrey, Guildford, Surrey, England

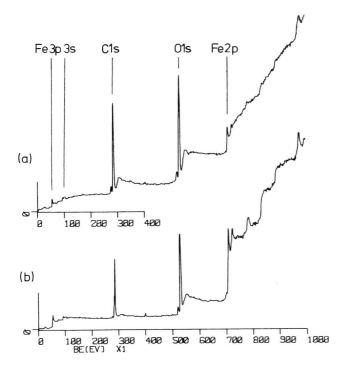

FIGURE 1 — Wide scan XPS spectra of mild steel atmospherically exposed for 72 hours after (a) alkali cleaning, and (b) emery abrasion.

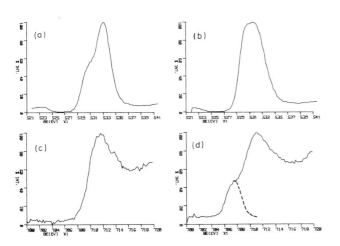

FIGURE 2 — (a) O_{1s} region of 1 (a) (b) O_{1s} region of 1 (b) (c) Fe2p3/2 region of 1 (a) (d) Fe2p3/2 region of 1 (b).

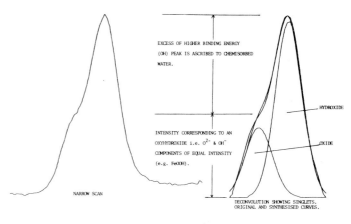

FIGURE 3 — Analysis of O_{1s} peak.

oxygen and iron regions are shown in Figure 2. The O_{1s} peak has two components; using the data system, the curve can be deconvoluted and quantified. The lower binding energy singlet can be ascribed to the oxide at approximately 530.2 eV ($C_{1s} = 285.0$ eV), while the higher binding energy peak (532.4 eV) is due to a hydroxide species.

On the emery prepared surface these two peaks are of approximately equal intensities. This is in accord with the findings of Roberts and Wood,[17] who showed that the surface of iron exposed to water vapour under UHV, clean conditions is passivated by the formation of an iron oxyhydroxide phase. Surfaces prepared by other methods consistently show an excess of the higher binding energy component attributed by Castle[18] to bound water associated with hydrocarbon contamination. Molecular or physisorbed water gives a characteristic peak at $\cong$ 533 eV,[19] and this was not observed as a distinct peak in our results, but contributing generally to the intensity of the higher binding energy shoulder. In practice, the O_{1s} peak was analysed, as depicted schematically in Figure 3, and the relative amounts of H_2O and $(OOH)^{3-}$ determined.

Both the emeried and SiC abraded specimens show a metallic iron component of the $Fe_{2p3/2}$ peak (Figure 2), which for it indicates that the passivating layer is thinner than the escape depth of the iron 2p electron, that is, $\cong 25\text{Å}$.[20]

A feature of some of the C_{1s} peaks is broadening on the higher binding energy side. Figure 4 shows an emeried specimen; the broadening is due to a component at $\cong 288.5$ eV which indicates the presence of C = O bonding. Such broadening is strongest on the emeried specimens, weaker on the SiC abraded examples, and never observed when alkali cleaning is employed. The broadening may be associated with bonding of the organic to the iron oxide layer.

Quantitative evaluation of these peaks on as prepared samples shows that alkali cleaning leaves the highest amount of residual carbon, and a high level of water; whereas, emery abrasion gives a much cleaner surface. The SiC abrasion treatment falls midway between the two extremes. Thus, we can say emery abrasion is the most effective form of contamination removal, and alkali cleaning (using conditions described here) is least effective.

The dependance of the various components of the peaks, on time of exposure to the atmosphere, is shown by the compositional plots in Figure 5. In contrast to earlier findings,[18] the level of water did not increase with time. However, the total amount of adsorbed water does correlate with the amount of residual carbon, Figure 6. A surface carbon of 40% en-

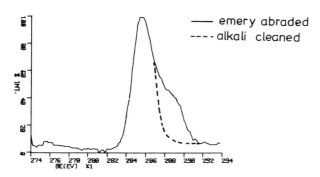

FIGURE 4 — C_{1s} region of emeried specimen exposed for 72 hours. Note high binding energy shoulder.

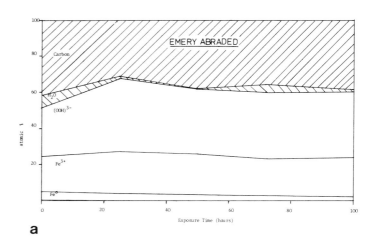

a

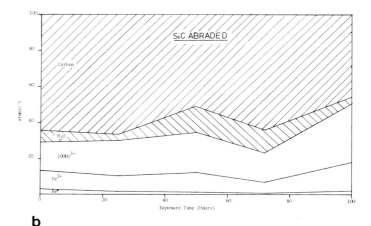

b

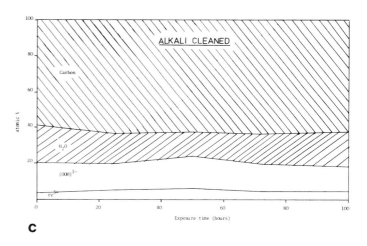

c

FIGURE 5 — Composition vs exposure time graphs for the three treatments investigated.

sures a satisfactorily low level of surface water and emery cleaning was, therefore, treated as a standard method in this work. Alkali cleaning, which leaves a higher level of adsorbed water, was adopted as a comparison.

Effect of Hot Curing Techniques

Although the foregoing conclusions regarding cleaning procedures are relevant for cold cure coatings, substrate heating effects must be considered when investigating hot cure systems. To investigate heating effects, emery abrasion and alkali cleaning techniques were chosen, followed by heating at 250 C for 1 hour (typical cure temperature for epoxy powder coating).

The effect of such heat treatment, together with surface composition, is shown in Figure 7 and Table 1 respectively. Most dramatic is the increase in oxygen and iron, the O_{1s} peak is now predominantly due to oxide (530.2 eV), while the iron peak is completely ferric. Although the carbon level is still greater on the alkali cleaned specimen, it has now reached an acceptable level of $\cong$ 40%. The oxygen peak is similar to that of the heat treated, emery abraded sample, showing a large excess of the lower binding energy peak.

Notwithstanding the general reduction of carbon levels, the broadening of the peak to higher binding energy levels actually is enhanced by heating, and becomes visible on the alkaline cleaned surfaces. Thus, this component may have particular significance in interface formation. Its distribution was, therefore, investigated by angular resolution of the electron spectra.[21] This technique is only applicable to smooth surfaces, and therefore cannot be used for emery or alkali cleaning techniques. To overcome this problem, a mild steel coupon was polished to a 1 μm diamond finish and heat treated. Analysis showed the surface composition to be close to that of the emeried specimen (Table 1). Figure 8 shows the variation in the C – H (285.0 eV) and C = O (288.5 eV) intensity ratio with take off angle, Θ. Low values of Θ enhance the near surface species, and results indicate the C – H group to be close to the surface. The C = O, therefore, lies below a carbonaceous contaminant layer, and probably adjacent to the oxide surface.

Polybutadiene Coating

The coating chosen for this investigation was a polybutadiene can coating.[22] This coating was applied to the mild steel substrate, using a wire wound bar coater giving stoved thickness of 17 ± 2 μm. Cure schedule was 200 C for 20 minutes. The disbondment of this material has been investigated by Leidheiser and Kendig using electrical techniques.[23,24]

Two surface treatments were employed prior to coating: emery abrasion and alkali cleaning. After

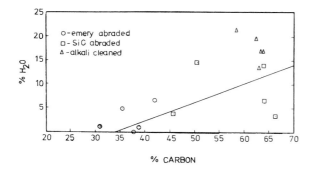

FIGURE 6 — Relation between surface water and hydrocarbon contamination.

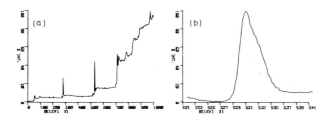

FIGURE 7 — (a) Wide scan of alkali cleaned mild steel heated in air at 250 C for 1 hour. (b) O_{1s} region of (a).

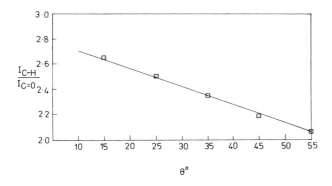

FIGURE 8 — Variation of $I_{C-H}/I_{C=O}$ ratio with photo-electron take off angle, Θ.

TABLE 1 — Surface composition of mild steel after various treatments and 1 hour at 250 C

	C	O	Fe
Alkali clean	42.0	37.9	20.1
Emery	32.4	49.7	17.9
1 μm diamond	30.6	49.2	20.3

stoving, the coating takes on a characteristic golden appearance; this color is due, in part, to thickening of the substrate oxide layer, which assumes a pale straw coloration on uncoated steel. This phenomenon will be discussed more in the next section.

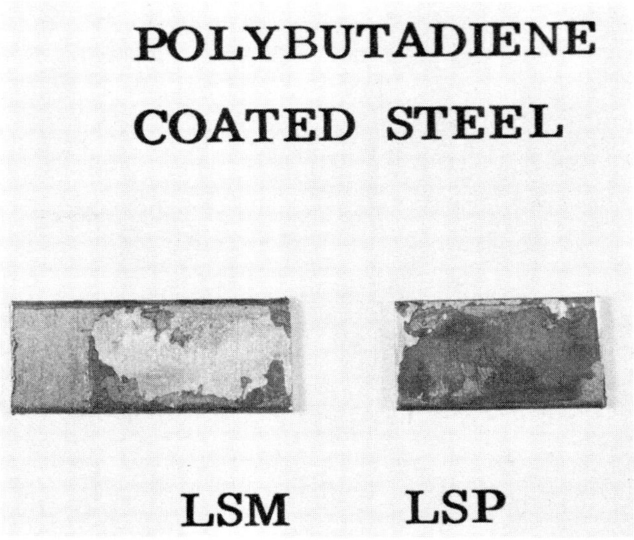

POLYBUTADIENE COATED STEEL

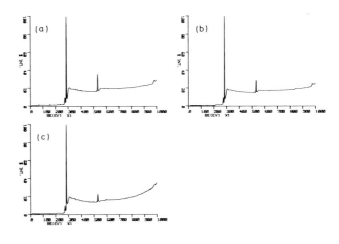

FIGURE 9 — "Metal" (LSM) and "polymer" (LSP) surfaces of a failed lap shear test specimen.

FIGURE 10 — (a) Wide scan of "polymer" surface (b) Wide scan of "metal" surface (c) Wide scan of polybutadiene standard.

The locus of failure of the coating under mechanical loading conditions was determined by carrying out lap shear tests on specimens produced from coated panels; bonded with cynoacrylate ester adhesive, the area of the lap was approximately 10 × 8 mm. Tests were carried out, using a Houndsfield tensometer; failure loads of approximately 2KN were obtained.

"Metal" and "polymer" surfaces of a failed, emeried specimen are shown in Figure 9. XPS analyses of these surfaces, together with the spectrum of a polybutadiene standard, are shown in Figure 10; they are all identical. This similarity of the three spectra shows the coating failure to be cohesive with the locus of failure in the polymer coating. Similar results were obtained on an alkali cleaned, coated panel.

a

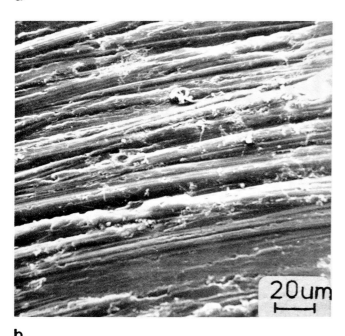

b

FIGURE 11 — Scanning electron micrographs of the oxide surface after stripping for (a) alkali cleaned specimen, and (b) emery abraded example.

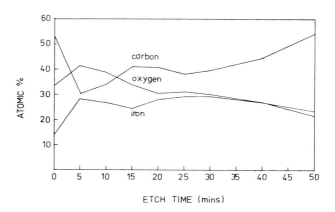

FIGURE 12 — Etch profile of a stripped oxide/polymer film.

Examination of Bond Characteristics by an Oxide Stripping Technique

As has been pointed out, it is not practical to approach the oxide/polymer interface through the polymer phase. Thus, an iodine/methanol oxide stripping technique has been used to remove the polymer/oxide layers as a duplex film. Oxide stripping was first described by Vernon, et al,[25] who performed gravimetric analyses of the films, and more recently by Dye, et al[26] who carried out electron diffraction studies. The application of the stripping technique to XPS analysis, where substantially larger, continuous films are required, has been developed in our laboratory, and full details of the technique, together with potential applications, will be published elswhere.[27]

Oxide stripping of the coated specimen is accomplished readily, and work of adhesion calculations show that the oxide/polymer bond is not disrupted by methanol.[28] It is possible that the film would fracture or spall from the coating on removal from the stripping rig. Scanning electron microscopy of the underside of the stripped material, Figure 11, shows the oxide in position, without disruption on either alkali cleaned or emeried coated panels. The oxide, identified by XPS, shows a replica of the original metal surace down to small intrusions and extrusions, present on the alkali cleaned example.

Several etch profiles through the oxide have been obtained, using 3KV argon ions and an ion gun calibrated to have an etch rate of $\cong 15$ Å minutes^{-1} on a stainless steel oxide. Times of up to 90 minutes have been used on stripped oxides, and the iron peak remains ferric until the interface is reached; thus, argon ion reduction of the iron species did not occur, contrary to the many observations made of oxide films on iron.

Because of the roughness of the interface, a sharp discontinuity is not observed in the etch profile, but a general increase in carbon level was taken as an indication that the substrate had been reached in certain regions (Figure 12). The oxygen level remained constant throughout the film, and only started to fall as the interface was approached; that is, as the carbon level rose. There also was found slight broadening of the oxygen peak. This broadening can be correlated with a broadening of the iron peak, due to the presence of a small amount of iron in the ferrous state (Figure 13). As stated before, it is unlikely that this is due to ion beam reduction, since the etch time is 50 minutes, compared with 90 minutes on thicker oxides, without reduction.

The broadening present in the C_{1s} peaks on uncoated panels is not evident. The depth of etch profiling indicates the thickness of the oxide on emeried

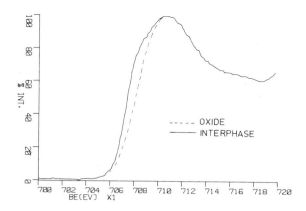

FIGURE 13 — Fe2p3/2 region at interface showing broadening due to Fe^{2+} component.

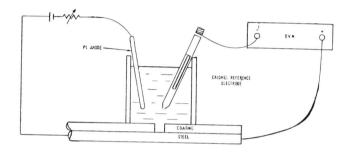

FIGURE 14 — Schematic diagram of cathodic disbondment test.

specimens is of the order of 300 Å, which is substantially thicker than after initial surface preparation (<25Å), indicating the dramatic oxide growth which takes place on stoving.

The Characteristics of the Surface and Interface

This work has shown that, although differing methods of surface treatment leave the metal in quite different surface states, these do not persist through a heat treatment. Adsorbed water and a large proportion of aliphatic carbon contamination is lost, but a more polar, organic group is enhanced. Angular work showed this to be located beneath superficial carbon at the oxide surface. However, attempts to locate it at the bonded interface were not successful. The interface may be marked by the presence of ferrous iron and by a residue of the high binding energy oxygen seen on the prepared surface. Our appreciation of these surface features was utilised then in a study of cathodic disbondment.

Cathodic Disbondment

Experimental

The experimental set up is shown schematically in Figure 14: a seal is made between the electrolyte ring

TABLE 2 — Identification and surface composition of specimens from disbondment experiments

SPEC	Pretreatment	Potential	Position
D4M1	Alkali clean	−1500mV	near defect
D4M2	Alkali clean	−1500mV	near limit of disbond
D4M3	Alkali clean	−1500mV	not disbonded
D5M1	Alkali clean	F.C.P.	near defect[1]
D5M2	Alkali clean	F.C.P.	near electrolyte ring[1]

[1] Not disbonded

SPEC	at% C	O	Fe	Cl	Na
D4M1	35.6	47.1	6.5	3.1	7.7
D4M2	36.7	43.8	6.5	3.8	9.2
D4M3	66.8	27.4	2.2	tr	3.6
D5M1	63.4	28.7	tr	1.3	6.7
D5M2	58.0	30.8	1.0	2.9	7.4

tr = trace.

and coated panel using silicone grease. A central initiating defect was made by drilling the coating. To obviate any effects brought about by defects in the coating (for example, dust particles), each panel was coated and stoved twice to give an effective thickness of $\cong$ 34 μm. The electrolyte employed was a 0.52M NaCl solution.

Tests were carried out with alkali cleaned, coated panels, polarised cathodically (−1500mV vs S.C.E.), or at the free corrosion potential for 250 hours. By this time, the disbondment front locally approached the electrolyte ring, in the case of the cathodic specimen. For the panel at rest potential, there was evidence of rusting at the defect when the disbondment front had advanced $\cong$ 1 mm.

XPS Results

A disbonded region could be visually identified and lifted, using a scalpel; the area between the disbondment front and electrolyte ring was adhering weakly, and could be removed with a scalpel. This weakness indicates some disruption of bonding outside the disbondment front.

XPS analyses were carried out for samples taken from both panels, and the peak areas were normalised to give atomic percent. Identification of specimens and atomic percentages are given in Table 2. Wide scans for two specimens are presented in Figure 15, showing the difference in carbon and iron levels for specimens inside and outside the disbondment front. Figure 16 shows oxygen on iron narrow scans.

Examination of surface compositions for the five specimens indicate a clear difference between the samples D4M1, and D4M2 (removed from within the disbonded region) and D4M3, D5M1, D5M2 (removed from outside the disbonded region). The latter group show a high level of carbon, and low oxygen and iron.

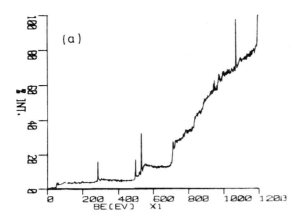

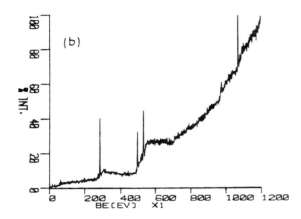

FIGURE 15 — Wide scan spectra for specimens (a) D4M2, and (b) D5M2.

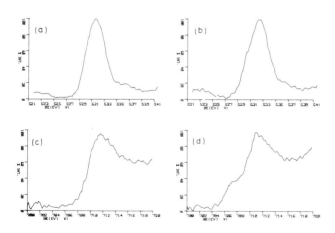

FIGURE 16 — (a) O_{1s} region of D4M1 (b) O_{1s} region of D4M3 (c) $Fe2p3/2$ region of D4M1 (d) $Fe2p3/2$ region of D4M2.

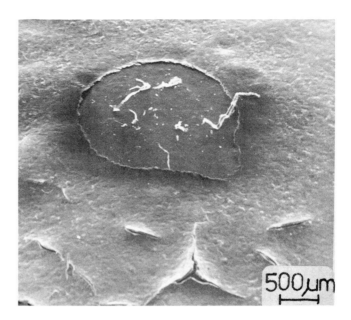

FIGURE 17 — Scanning electron micrograph of stripped oxide/polymer film showing disbondment halo.

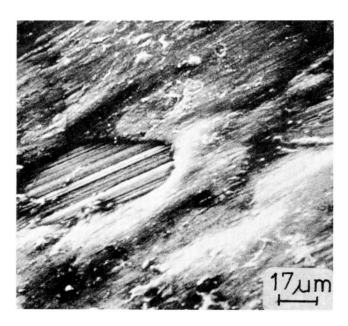

FIGURE 18 — Scanning electron micrograph showing elliptical region in oxide film within the disbondment halo.

The high levels of carbon indicate cohesive failure of the coating; the remaining polymer on the substrate attenuating the iron and oxygen signals. The O_{1s} peaks show some broadening (Figure 16), and can be deconvoluted to show that molecular water accounts for some 15% of the oxygen signal. Since the film was removed after dehydration in a vacuum desicator, the presence of water beneath the polymer indicates permeation of water through the coating.

Specimens D4M1 and D4M2 are similar, but inspection of the Fe2p3/2 peak for D4M2 (adjacent to the disbondment front), Figure 16, shows the presence of metallic iron, indicating some reduction of the oxide under the coating during cathodic treatment. Although reduction appears at the disbondment front, it is not apparent near the defect; that is, the area which was first disbonded, thus indicating some time dependency of the reduction process.

More important, possibly, than the question of oxide reduction, is the fact that the carbon levels in the disbonded region are remarkably low. Clearly, cathodic disbondment causes the total separation of the organic phase. On these samples, also, the water like oxygen component was less marked than on the samples prepared from outside the disbonded region. We note the high level of sodium, which is at a similar concentration inside and outside the disbonded area.

Film Stripping

To gain an insight into the initial stages of cathodic disbondment an alkali cleaned, coated panel was polarised cathodically for 3 hours; the disbondment front was estimated as ½ mm advanced. The area around the defect was stripped, using the iodine/methanol technique, and the duplex film examined from the oxide side by scanning electron microscopy. A halo around the defect is apparent at low magnification (Figure 17), corresponding to the visually identified disbonded region. At higher magnification, elliptical regions are seen, Figure 18, due to areas of localised oxide reduction distributed randomly in the disbondment zone.

The fact that the oxide is reduced only in localised patches indicates that oxide reduction is not a necessary precursor for cathodic disbondment to occur, but may occur after disbondment has taken place as a result of a high pH generated at the edge of the disbondment crevice. The XPS results indicate a thickening of the oxide once disbondment has been accomplished.

Cathodic treatment liberates hydrogen, but there is no evidence of hydrogen having been present at the polymer/oxide interface.

Discussion

When steel is prepared by abrasion or by chemical cleaning, a considerable hierarchy of layers remains. These layers are structured similar to those envisaged by workers in the decade preceeding surface analysis. We have demonstrated here that, up to the point of coating, there is a sequence: metal, oxide, hydroxide, polar organic, hydrocarbon, and water. Heating, in air, up to the temperatures used in curing the resin causes oxide growth, and a signficant reduction in adsorbed water and hydrocarbon. In practice, surfaces may be heated before or after coating, and in the latter case, the fate of the less strongly adsorbed material is uncertain. The question is important since some authors, (Bikerman, for example) have suggested that the locus of failure might be entirely within such a weakly bonded layer. We were unable to detect the presence of a high interfacial water concentration in etch profiles, even though we adopted the novel approach to the interface by way of the thin, adherent, oxide. Moreover, coatings on both abraded and chemically cleaned surfaces failed cohesively under shear stress, although they carried different thicknesses of water and organic contamination. We conclude therefore, that the contamination either dissolves in the polymer, or reacts with iron during heat treatment. In this context, it is worth noting that there is a considerable increase in the oxide thickness (20 Å → 300 Å) during stoving, and hence, the oxide/polymer interface would be extremely active. Attempts to control the oxidation by stoving in nitrogen failed, because the polybutadiene did not cure.

Two features were observed during the work which would be of significance in terms of the bonding of organic polymers to iron oxide. First, the formation of the polar carbon peak in the position corresponding to $C=O$. This is not a commonly observed feature of the carbon peaks arising from contamination; it has not, for example been observed in our extensive work on brasses. Also, the peak increased its intensity on heating, suggesting that it might arise as a result of reaction of hydrocarbon with the iron oxide to yield RCOOFe. It is disappointing that we were unable to observe this peak at the polymer/oxide interface; however, future work will have improved interface definition. The second feature of interest was observed during ion etching through the interface, and is the development of some Fe^{2+} character at the oxide surface. We are satisfied that this was not a beam induced artifact, and may well relate to the role of polybutadiene as a reducing agent for iron oxide.

When disbonding is induced by cathodic potential, we find a complete segregation between polymeric material and metal oxide surface. In the disbonded region, further from the initiating defect, the metal is bright, $Fe°$ appears in the spectrum, and there is a marked lack of corrosion product. These results lend emphasis to the segregation of organic molecules to the polymer side of the interface. Note that loss of organic molecules is not found on uncoated steel treated cathodically in sodium chloride. Thus, cathodic disbonding is an interphase separation. Within the disbonded region, the oxide is reduced only in localised patches, and visible disbondment extends well beyond these. Thus, oxide reduction is not a necessary feature of cathodic disbondment.

While disbondment was recognised easily, the analytical evidence was that both water and sodium ions had penetrated well beyond the disbondment front. However, the polymer had some residual adhesion, and its removal was accomplished only at the expense of having a considerable residue of organic carbon on the metal surface. The iron component of the surface analysis was so low that this failure would be described as cohesive. Yet, there was selective attack on the polymer in the proximity of the interface. This attack must be associated with the alkalinity evidenced by the excess sodium ion concentration. Note that in exposure of oxidized iron to sodium chloride solutions it is usual to record a surface excess of the chloride ions.[29]

We presume the altered zone ahead of the disbonded region to be active saponification, as proposed by Hammond, et al.[11] Alkaline conditions certainly would hydrolyse the FeOOCR salt which we proposed. Hydrolysis of the bond adjacent to the oxide surface would account for the interphase separation. Saponification, of a near interface polymer zone, prior to complete hydrolysis of the bond, would account for the low energy cohesive failure ahead of the disbondment front.

This work leaves the route by which water reaches the interface uncertain. The fact that it is accompanied by sodium ions does suggest, however, that it diffuses laterally from the defect zone. A future development of this program will be the tracing of pH contours by addition of specific cations such as magnesium and cadmium, of known solubility product, to the electrolyte.[30] Their deposition within the oxide/polymer interphase separation will be detected from the oxide side of the couple using electron microprobe analysis.

Acknowledgment

J.F.W. acknowledges financial support of the S.R.C. and British Gas Corporation through a CASE award.

References

1. C. Kerr, N.C. Macdonald, S. Orman, *Br. Polym. J.* **2**, pp. 67 to 70, (1970).
2. A. Kinloch, M.O.D. (PE), PERDE, Waltham Abbey, Tech Note 95, August, 1973.
3. W.A. Anderton, *J.O.C.C.A.* **53**, 181 to 191, (1971).
4. E.L. Koehler p. 117 to 133, Localised Corrosion, Ed R.W. Staehle, Pub. NACE.
5. R.W. Zurilla, V. Hospadaruk, Paper 780187, SAE Meeting, Detroit 1978.
6. E.L. Koehler, *Corrosion*, **33**, pp. 209 to 217 (1977).
7. K.L. Mittal, *J. Adhesion* **6**, pp. 337 to 8 (1974).
8. J.J. Bikerman, Science of Adhesive Joints, Academic Press, 1968.
9. J.S. Solomon, D. Hankin, N.T. McDevitt, p. 103 to 122 in Adhesion and Adsorption of Polymers, (1980), Ed. L.H. Lee, Plenum Pub. Corp. N.Y.
10. R.R. Wiggle, A.G. Smith, J.V. Petrocelli, *J. Paint Tech.* **40**, pp. 174 to 186 (1968).
11. J.S. Hammond, J.W. Holubka, R.A. Dickie, *A.C.S. Org. Coat. Plas. Chem.* **39**, pp. 506 to 11, (1978).
12. M. Gettings, F.S. Baker, and A.J. Kinloch, *J. Appl. Polym. Sci.*, **21**, pp. 2375 to 2392, (1977).
13. M. Gettings & A.J. Kinloch, *J. Mater. Sci*, **12**, p. 2511 (1977).
14. M. Gettings & A.J. Kinloch, *Surf. Interf. Anal.* **1**, pp. 165 to 171 (1979).
15. M. Gettings & A.J. Kinloch, *Surf. Interf. Anal.* **1**, pp. 189 to 195, (1979).
16. Solventol 3024 ex EFCO Ltd., Woking, England, Data Sheet 3024/1.
17. M.W. Roberts & P.R. Wood, *J. Electron Spec.* **11**, pp. 431 to 437 (1977).
18. J.E. Castle, pp. 435 to 454, Corrosion Control by Coating, Ed. H. Leidheiser, Science Press, 1979.
19. K. Asami, K. Haskimoto & S. Shimaidaira, *Corr. Sci.* **16**, pp. 35 to 45, (1976).
20. M.P. Seah & W.A. Dench, *Surf. Interf. Anal.* **1**, pp. 2 to 11 (1979).
21. C.S. Fadley, *J. Elect. Spec.* **5**, pp. 725 to 754 (1974).
22. Du Pont, RKY-662 Budium.
23. H. Leidheiser and M.W. Kendig, *Corrosion*, **32**, pp. 69 to 76 (1976).
24. M.W. Kendig and H. Leidheiser, *J.E.C.S.* **123**, pp. 982 to 989, (1976).
25. W.H.J. Vernon, F. Wormwell and T.J. Nurse, J. Chem. Soc., 1939, pp. 621 to 632.
26. J.G. Dye, O. Fursey, G.O. Lloyd & M. Robson, *J. Sci Inst. (J. Phys. E.)* 1968 Ser. 2, **1**, pp. 463 to 464.
27. J.E. Castle & J.F. Watts, to be published.
28. A.J. Kinloch, pp. 1 to 11, Adhesion 3, Ed. K.W. Allen. Applied Science. Pub. 1978.
29. J.E. Castle pp. 182 to 198, Applied Surface Analysis, Ed. J.L. Barr & L.E. Davis, ASTM STP 699, 1980.
30. J.E. Castle & R. Tanner-Tremaine, *Surf. Interf. Anal*, **1**, pp. 49 to 52, (1979).

Underfilm Corrosion Currents as the Cause of Failure of Protective Organic Coatings

E.L. Koehler

Introduction

Organic coatings have long been used to protect metals from corrosion, and good results have been achieved through their use. These accomplishments were the results of the work of practical investigators, relying almost entirely on long term service tests. Investigators of mechanisms involved have had a hard time trying to explain the practical accomplishments in the field. There have been a great number of opinions expressed, influenced by the varying backgrounds of the particular investigators. Much of what has been expressed has been speculation; some of it has been incorrect. It is a difficult problem to get "handles" on, and speculation has become a necessity. Nevertheless, it must be appreciated that there is a tendency to generalize on the results of limited investigations.

It must be appreciated, first of all, that the corrosion of organic coated metals is actually a corrosion problem. Any attendant phenomena occurring are important only as to how they affect the corrosion reactions. Being a corrosion phenomenon, corrosion of organic coated metals is susceptible to two important features of corrosion studies in general: (1) there is no one single mechanism that covers all situations; (2) results of short time tests, under forced conditions, are likely to reflect something other than practical service experience. A corrosion failure will occur by a mechanism which is dependent upon the conditions which we impose upon the system. We can not make a general statement to cover all possible situations. It is necessary to have some basic understanding as to the relevance of the particular experiments conducted. For example, the writer is aware of several papers, written lately, stating that the corrosion of organic coated metals is no different from the corrosion of bare metals, with the exception that the organic coating limits the area of metal exposure. It is felt that most of you will regard this as wrong. The investigators applied fast tests and obtained the results of fast tests.

Failure of a protective organic coating involves some mechanism of coating detachment. There are a number of ways in which an organic coating can be detached.[1] A great deal of attention has been given to what has been called "adhesion". Generally, we use this term when we mean adhesive strength. In many instances, there has been a loss of adhesion to the substrate under wet or high humidity conditions. Insofar as resistance to corrosion is concerned, it is in general only necessary that some degree of adhesion of the organic coating to the substrate be maintained. Excellent adhesion does not necessarily mean good performance, nor does relatively poor adhesion necessarily mean poor performance. It is not to be discounted that poor adhesion may result in local loss of coverage by the organic coating, as the result of mechanical abuse. Even in this case, however, the loss of local adhesion itself only promotes the initiation of the corrosion reaction. It is the ensuing corrosion reaction which results in the progressive failure of the coating.

Attention has been given almost exclusively to cathodically produced organic coating corrosion failures. Again, principal attention has been given to a mechanism involving possible saponification of organic coating constituents at the interface by cathodically produced hydroxide ion. Cathodic detachment is highly dependent on the substrate. It occurs very easily on untreated steel; less easily on tinplate; and in the writer's experience does not occur on aluminum. It has also not been observed on chromium oxide/chromium coated steel, but his experience may be somewhat more limited here.

While it is agreed that any of the possible mechanisms of adhesion loss may be involved in specific situations, it is preferred here to follow more generally the suggestions of Michaels and Bolger.[2] Adhesion loss occurs as a result of water displacement, and is intensified by pH values well removed from the isoelectric point of the oxide film, and as the result of the oxide having a much greater attractive force for the water than for the organic film. The particular metal oxide thus would be involved, as would the nature of the organic coating.

Forthright opinions have been expressed as to whether the rate determining step in the corrosion of an organic coated metal is related to the resistance of the organic coating to corrosion current flow, to the diffusion of water, or to the diffusion of oxygen. These opinions should be qualified better. An organic coating is principally effective in preventing corrosion of the substrate by acting as a barrier to corrosion processes. In this regard, as advocated by Mayne,[3] the resistance of the organic coating to corrosion current flow, as transported by ions, is most important. Ionic transport may explain why the organic coating works; it does not necessarily explain why it might fail.

We all like to make electrical measurements on organic coated metals, resistance, capacitance, polarization, or what have you, because such measurements are generally easy to make, and give results in a short time. We are all familiar with work such as that of Bacon, Smith and Rugg[4] which show that good organic coatings have resistances of greater than 10^8 ohms per square centimeter, while coatings which develop lower resistances will not be good. Unfortunately, such testing has limited value because coatings which have high resistances are also susceptible to corrosion failure. In a previous paper, it was pointed out that the corrosion of organic coated metal surfaces involves the flow of currents through a film of electrolyte beneath the organic coating.[5] In many instances, this flow of current through the interfacial water circumvents the high resistance of the organic coating. A different mechanism is involved, and the rate of oxygen diffusion through the film, as well as the electrical resistance of the film, may become less significant in determining the rate.

The earlier paper has reasoned, and it has recently been affirmed,[6] that a layer of water always exists between the organic coating and the metal oxide substrate. It should be readily acceptable that the loss of adhesion sometimes encountered, under exposure to water or to high humidity, is the result of the enhancement of this layer of water. Funke and Haagen have made weight measurements indicating the presence of such a layer of water under high humidity conditions.[7] It is the object of this paper to focus attention on this layer of water, be it anything from a monomolecular film to a liquid filled blister, from the standpoint of the flow of ionic current through this water layer along with the concurrent underfilm corrosion reactions.

Earlier Observations

With organic coatings of low resistance, the currents which can pass through the organic coating become quite significant. This may be illustrated by Figure 1, which has been used earlier.[8] This depicts the corrosion of organic coated steel, having a defect at which metal is exposed, in a corrosive solution. As is normal, the organic coated metal becomes cathodic to the exposed metal, and to the extent to which currents can pass through the organic coating, corrosion of the steel is stimulated. Conduction of current through the coating occurs by virtue of ionic transport. As indicated, transport of oxygen and water are also required. In general, it is the resistance of the organic coating which is rate determining in such a process; but in some cases, the corrosion rate may be limited by transport of oxygen or otherwise. Corrosion currents involved in this process pass out into the main body of electrolyte and are detectable in electrochemical experiments.

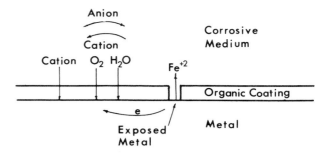

FIGURE 1 — Representation of processes through a protective organic coating, the coated metal acting as a cathode to stimulate corrosion of exposed metal.

About twenty years ago, the writer conducted experiments on organic coated aluminum with various types of coatings. Resistance and capacitance measurements were made, and polarization curves were measured. It was found, however, that none of these measurements related to whether the organic coating would break down, or how long it took to break down. Once the organic coating broke down by corrosion, it could be detected by any of these measurements, but it could also be seen visually. It seemed evident that the corrosion currents involved prior to the breakdown process did not pass out into the corroding solution, but were shielded from detection by the overlying organic coating.

Figure 2 indicates another type of observation made on organic coated tinplate, coated with a food

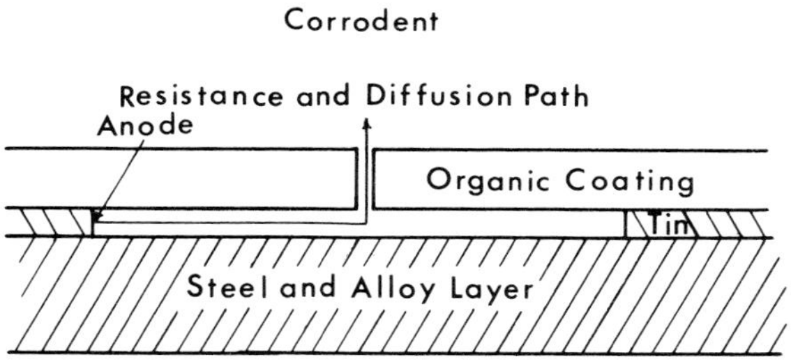

FIGURE 2 — Dissolution of tin between the organic coating and the alloy layer in tinplate coated with food can enamel.

can enamel. In a food product, tin is anodic to steel and affords it sacrificial protection. Accordingly, in the course of corrosion, the tin is dissolved out between the organic coating and the steel plus alloy layer. A small amount of such corrosion is normal, and is part of the means by which the steel of the can is protected. In an extreme case, such as represented by this laboratory test, the tin was dissolved out for quite some distance. In this case, the alloy layer covered steel established itself as the principal cathode for the dissolution of tin, the accompanying current not passing out into the main body of the corrodent. The corrosion here could not be detected by electrochemical means, nor could the corrosion be materially affected by applied currents. Anodically applied currents, for example, would simply cause pitting of the steel at the break in the organic coating, rather than speeding up the dissolution of tin. It was evident that, though detached, the organic coating still provided a significant degree of protection for the underlying metal.

The principal theme of this paper was advanced, but not emphasized, in the earlier paper. Figure 3,

FIGURE 3 — Atmospheric corrosion resulting from potassium chloride beneath the modified alkyd coating. One month at 38 C and 90% relative humidity.

taken from that work, illustrates three different types of corrosion damage taking place beneath the organic coating. In this test, a drop of 10^{-2} N potassium chloride solution, 3 mm in diameter, was placed on a steel sheet and allowed to evaporate dry. The sheet was then coated with the modified alkyd enamel. It was then stored in a room at 38 C and 90% relative humidity for one month. As mentioned in the paper, the central spot represents mostly local corrosion covered by rust. It is to be noted that it did not perceptibly spread by diffusion. The filiform corrosion was carried forward by an anodic process concurrent with the electrochemical transport of chloride ion beneath the coating. The dark ring represents cathodic detachment, promoted by the transport of potassium ions beneath the coating. This provided for at least part of the cathode for the corrosion of iron in the central spot. The contention here is that spread of corrosion beneath an organic coating involves ionic transport, and such ionic transport must be given consideration in any discussion of failure of organic coated metals.

Experimental

The coatings considered here were roll coated on steel and baked. Included are a modified alkyd coating at 8 µm thickness, baked for 8 minutes at 220 C and an epoxy-phenolic coating, at 7 µm, also baked for 8 minutes at 220 C. In some cases, these were covered with a vinyl top coat at 11 µm, baked for eight minutes at 170 C. Some mention should be made here of what to expect from the surface of such an enamel. Unfortunately, we can not expect a single coat to be perfectly uniform and defect free. Figure 4

FIGURE 4 — Extreme example of "microholes" in organic coated sheet. Width of the area shown is 0.55 mm; this example is not one of the specimens used in this paper.

shows a magnified view of the surface of a roll-coated enamel. Many defects, which we might call "microholes" are plainly visible. This represents a rather extreme case, and does not represent specimens used in this study. These microholes do not go to the base metal, but instead represent thin spots. When these are present, the majority of the cathodic current through the coating will take place through such microholes. A double coating, of course, will cover up most of these.

The present work consists of simple experiments involving the flow of currents in interfacial water. The type of cell used for most of this work is shown in Figure 5. It consists of half of a spherical "O"-ring

FIGURE 5 — Test cell used.

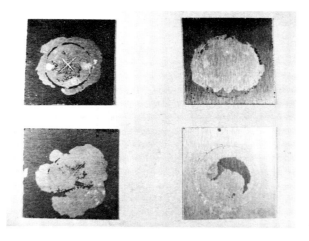

FIGURE 6 — Tested for 5 days with 0.1 M potassium chloride solution at 38 C. Nitrogen flushed. Cells in 100% relative humidity chamber. Top, modified alkyd; bottom, vinyl over modified alkyd; left, large X scratched anode; right, single point anode.

joint, sealed to the organic coated specimen by a buna-N "O"-ring. In most cases, this was a 15 mm "O"-ring joints; although in one case, a larger joint was used. A plastic back-up plate was used, and the assembly was held together with a joint clamp. In most cases, the test solution was 0.1 N potassium chloride solution. In some cases, the solution and the cell were preflushed with prepurified nitrogen. The solution was flushed for at least 24 hours, and by means of a three-way stopcock the cell was flushed and filled without admitting any air. The inlet and outlet tubes of the cell were then clamped off at the top.

Interfacial Currents in Cathodic Processes

In the test represented in Figure 6, the solution and the cells were flushed with the nitrogen, and the cells were held for five days at 38 C in a 100% relative humidity atmosphere. This arrangement placed the anode inside the cell, and the cathode outside the cell. The corrosion current between the two, then, must flow through the interfacial water. After the test, the specimens were scotch taped, and the detachment is visible. Immediately after taping, the surface of the metal was covered with a thin film of water, which evaporated quickly. The size of the "O"-ring is seen clearly in the specimen at upper left, since in this case the organic coating under the "O"-ring is not detached. It was found in earlier testing that there must be an exposed metal anode inside the ring for detachment to occur. Accordingly, in this test, anodes were provided in the form of scratched X's in the specimens to the left, and in the form of single points in the specimens to the right. The size of the anode made little difference. It is to be noted that detachment is generally more complete in affected areas outside the ring than in areas inside the ring. Applying phenolphthalein solution immediately after the test, to the areas outside of the ring, gave a strong pink reaction, indicating a basic solution. It was pointed out in the earlier paper that cathodic detachment is produced normally only by solutions of potassium or sodium salts, which are capable of producing strongly basic solutions. The area outside of the cell is thereby established as the cathode, as expected. The circular scratches outside the ring on the specimens to the right were made to see if such scratches through the enamel made any cathodic difference. Overall, they did not. The top two specimens were single coated with the modified alkyd enamel; the bottom two specimens were given the vinyl top coat. Surprisingly, the double coated specimens generally showed more cathodic detachment than the single coated. It is evident that effective oxygen goes through these coatings amazingly fast; and it is also evident, from the double coat vs single coat results, that the factors affecting underfilm conduction are more important here than other possible factors which would be related to the organic coating thickness. Evidence for this has also been indicated to be true for the epoxy-phenolic coating in the earlier paper, where it was shown that in 0.1 N sodium chloride solution, while the number of detachment spots was much greater for the single coat than for the double coat, the area detached for each spot was approximately the same.

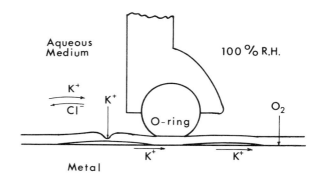

FIGURE 7 — Representation for the cathodic detachment of the coating on the outside of a cell at 100% relative humidity.

A diagrammatic representation for the situation is shown in Figure 7. Current is carried through the aqueous medium by potassium and chloride ions. It is also represented that potassium ions pass through the organic film at a point of ingress (in the usual case a microhole), and is electrolytically transported through the interfacial water to the point where potassium hydroxide solution is formed as the result of the reduction of atmospheric oxygen.

Figure 8 represents a test of vinyl over modified alkyd enamel, where the cells were not put in the 100% relative humidity chamber, but were tested at 38 C and ambient humidity. The anode scratched through the enamel was approximately 0.75 mm², and is seen at approximately average position, left. The

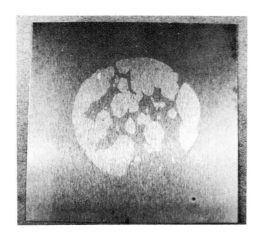

FIGURE 8 — Vinyl over modified alkyd coating. Nitrogen flushed. Test for 24 hours at 38 C and ambient relative humidity.

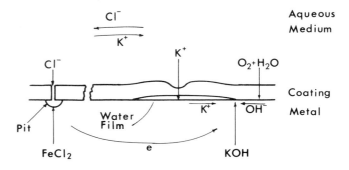

FIGURE 9 — Representation of detachment of the coating by potassium hydroxide solution formed as the result of the interfacial transport of hydroxyl ions from the area of oxygen reduction.

test area within the ring was 62 mm in diameter, as opposed to 21 mm in diameter for the regular cells. Test time here was 24 hours. Companion cells of the regular 21 mm diameter size were almost completely detached within the ring. This happened generally in short time for the modified alkyd base coat. In no case, in this test or in any other test, was there observable detachment outside of the ring when the cell was held at ambient humidity. Since these cells were also flushed with nitrogen, it would seem that the enamelled area at the periphery of the ring still serves as the cathode for the reduction of atmospheric oxygen. It is to be noted that these detached areas start from a point of origin and spread. As it is not conceivable that this point is actually a cathode for the reduction of oxygen, this point of origin must be only a point of cationic ingress. Likewise, since these points of detachment do not necessarily touch the edge of the ring, and since the mechanism of detachment is by the formation of potassium hydroxide, the only reasonable deduction would seem to be that current passes from the periphery of the ring, where oxygen is reduced, through the interfacial water by transport of hydroxyl ions. Such a mechanism for the formation of a detached area would lead to the situation depicted in Figure 9. The situation, as represented, indicates current as carried through the point of ingress by potassium ions. The cathode is represented as an area of undetached coating. Reduction of oxygen produces hydroxyl ions, which carry the current in the interfacial water to the point where potassium hydroxide is formed by the counter transport of potassium ions.

It is to be noted that there is some difference between these detached areas and those described by Leidheiser and Kendig.[9] Theirs were nucleated by a dark corrosion spot beneath the organic coating. Their description for the origin of such spots is probably correct. It would appear from the description that such spots start as small blisters which subsequently break. Under conditions of the present testing, where an exposed metal anodic area was deliberately provided, except for the area around the anode, detachment originated from a point of cationic ingress without a central corrosion spot.

Some of the tests were conducted at an ambient temperature of 22 C, rather than 38 C. This had the advantage of giving slower rates. Furthermore, it was not necessary in all cases to run tests on deaerated cells. Under such conditions, specimens of the modified alkyd coated plate, tested under 100% relative humidity conditions, are found to be completely detached within the ring within four days, along with considerable detachment outside the ring. The effect of the lower temperature did not seem to be as great as might be anticipated.

An argument might be presented that the spread of the detachment area is the result of the diffusion of potassium hydroxide under the coating, rather than by electrolytic transport processes. Experiments indicate that the rate of such diffusion is far lower than observed detachment rates. In Figure 10, specimens of the modified alkyd plate and the epoxy phenolic coated plate were tested for four days in 0.1 N sodium hydroxide solutions, after putting a scratch 5 mm long in the center of the specimens. There was no detachment at all observable on the epoxy-phenolic coated plate. Scotch taping indicated that enamel detach-

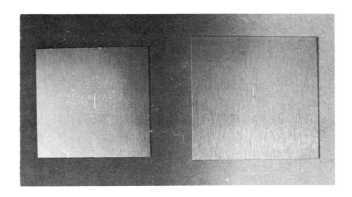

FIGURE 10 — Specimens with 0.5 cm scratches. Tested for 4 days at 22 C with 0.1 M sodium hydroxide solution. Left, modified alkyd; right, epoxy phenolic.

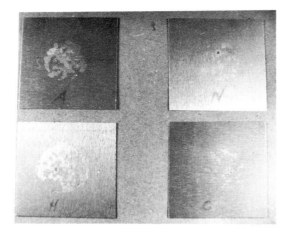

FIGURE 11 — Epoxy phenolic coating. Tested for 4 days at 38 C, 0.1 M potassium chloride solution. Top left, air saturated, ambient relative humidity; top right, nitrogen flushed, ambient relative humidity; bottom left, nitrogen flushed, cell in 100% relative humidity chamber; bottom right, nitrogen flushed, cell in sealed can flushed 4 hours with nitrogen.

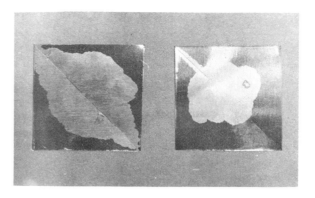

FIGURE 12 — Vinyl over modified alkyd coating. Tested for 5 days in 0.01 N potassium chloride solution at 22 C. Cell in 100% relative humidity chamber.

ment had spread only 0.1 mm on each side of the scratch for the modified alkyd plate. There were also small areas of detachment, apparently originating in microholes, which ran about 0.2 to 0.3 mm in diameter.

One consideration is that the spread of detachment in the nitrogen flushed cells may be related to some reduction process other than the reduction of atmospheric oxygen. To investigate this possibility, and to provide some additional confirmation for the description given in connection with Figure 9, the test depicted in Figure 11 was run. This test was run on the epoxy-phenolic coated plate to slow the rate. Tests were run at 38 C in duplicate in four conditions:

1. Upper left, air saturated solution, ambient relative humidity.

2. Upper right, nitrogen flushed solution and cell, ambient relative humidity.

This is seen to have materially reduced the number of detachment spots, but not their size.

3. Lower left, nitrogen flushed solution and cell, 100% relative humidity.

Relatively more detachment at the "O"-ring, with detachment spreading slightly beyond the ring.

4. Lower right, nitrogen flushed solution and cell, cell placed in an approximately two liter closed can, can flushed for 4 hours with nitrogen and sealed off.

While this did not eliminate detachment, the size of the detachment spots and the total detachment area were markedly reduced from the other cases.

Probably all of the oxygen in the can was not eliminated by flushing. At any rate, it does appear that detachment within the flushed cell (within the ring), is indeed related to the reduction of oxygen outside the ring, and that current must be carried between these areas, beneath the coating, by hydroxyl ions.

Two other types of experiments are presented here as of some interest in underfilm currents and cathodic detachment. The first of these is shown in Figure 12. These specimens were coated with the vinyl over modified alkyd enamels and held for five days at 22 C, with the cells at 100% relative humidity. The specimen at the left was scratched from corner to corner, and it is seen that the scratch directs the detachment pattern. This situation is somewhat complicated by the fact that pits develop in the scratch beneath the "O"-ring. In the specimen to the right, the scratch was brought up to the ring but did not go under it. On the bottom side of the specimen shown, the underfilm detachment did not even hit the scratch. The underfilm detachment did hit the top section of the scratch, although no part of the scratch was under the ring. It is to be expected that the surfaces of the scratch would be covered with potassium hydroxide solution, and it is indicated that, since a scratch has been shown to have no significant effect in speeding up the reduction of oxygen, the resistance of the electrolyte is much lower along this path than in the interfacial water between the organic coating and the metal.

In the test exemplified by Figure 13, the idea was to cut down on the ionic conduction in the interfacial water within the ring by reducing the thickness of the

FIGURE 13 — Vinyl over modified alkyd coating. Tested for 8 days at 22 C, 90% saturated sodium chloride solution. Cell in 100% relative humidity chamber.

interfacial water. This was done by using a solution with a high osmotic pressure, 90% saturated sodium chloride. The organic coating here was vinyl over the modified alkyd and the cell was held for eight days at 22 C and 100% relative humidity. Detachment rates within the ring are much lower than would be expected in this time with the 0.1 N potassium chloride solution. The principal point to be emphasized here, however, is how the underfilm detachment speeds up outside the "O"-ring, where the interfacial water, as dictated by the 100% relative humidity, is thicker. It can be seen to "mushroom" out once the detachment goes beyond the ring, as exemplified at the top, and especially at the lower right corner. This constitutes another example that the rate of this detachment is controlled by the interfacial conductance.

Anodic Detachment

So much for cathodic alkaline detachment. Interest here has been in exemplifying interfacial conductance, rather than in portraying the rapid type of coating deterioration which occurs under the conditions of these experiments. Practically speaking, if organic coatings performed no better than this, we would be in a bad way. Fortunately, any such rapid deterioration would be rapidly detected by our more practical minded protective organic coating friends who seem to be way ahead of us in providing useful protection. At any rate, practical cases of rapid cathodic detachment, encountered by this investigator in his field of work, have been rare.

Underfilm corrosion currents are dependent on both the nature of the organic coating and the nature of the metal substrate. As has been indicated, cathodic detachment has not been observed to occur on aluminum. For this metal, all cases of underfilm corrosion (personally observed) have been anodic in nature. On passage of currents between organic coated specimens of aluminum, corrosion and underfilm detachment have always been observed to occur on the anode.[8] What happens is illustrated in Figure 14. The corrosion observed has been a type of crevice corrosion, where the anode is in the fine crevice between the detached organic coating and the base metal. This type of crevice is sharper than the type of crevice which would develop if the corrosion penetrated more deeply into the metal; accordingly, this type of corrosion occurs preferentially at low rates, such as would be encountered in service. The thickness of the water layer between the organic coating and the metal shown in Figure 14 is an exaggeration for purposes of illustration. Actually, only a thin layer of surface metal is dissolved out from beneath the organic coating, and the layer of interfacial water is very thin. Protective organic coatings are meant to last a long time, and under practical conditions the cathodic activity provided to such an area, as in Figure 14, is low. The rate of such corrosion cannot be pushed beyond a certain rate. Investigators who are in a hurry to get their results will observe only local corrosion into the metal.

The same type of underfilm detachment on steel has been reported with beer as the corrodent. Again, under the application of a low current, underfilm detachment occurred only on the anode and not on the cathode.[10] Anodic undermining corrosion also involves flow of current through water beneath the coating, although likely only through a sensible water film.

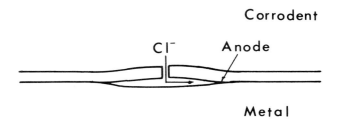

FIGURE 14 — Representation of anodic undermining corrosion, caused by the dissolution of a thin layer of metal beneath the coating. The thickness of the water layer is exagerrated for purposes of illustration.

Blistering

While it was not the intention of this work to go deeply into the subjects of blistering or enamel breakdown, it was given some consideration since interfacial water is involved. Liquid filled blisters which have been personally encountered in service have been anodic in nature; that is, corrosion of the metal beneath the blister was involved, along with the possible formation of corrosion product. Cathodic detachment produces thin films of liquid beneath the coating. In the present work, blistering was studied by using specimens coated with the epoxy phenolic enamel, unscratched. The test medium used was 0.01 N calcium chloride solution. A calcium salt was used here, since calcium is not capable of producing alkaline detachment. The specimens were made anodic at a galvanostatic current of 0.05 μAmp for the 3.5 cm^2 specimens. After seven days, a specimen had about six liquid filled blisters averaging about one mm in diameter. The liquid was rusty in the center and clear at the edges. Figure 15 indicates the appearance of these blisters. The center areas were corroded in a dished-out form. Peripheries were bright and unat-

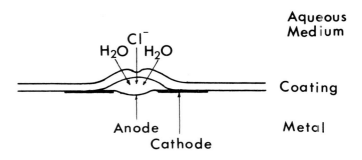

FIGURE 15 — Blister formation as the result of chloride ion transport through a defect.

tacked, giving the appearance that the primary detachment was cathodic. A faint ring around the blister had a slight discoloration. This disapeared after standing. The organic coating in this area was not removable with scotch tape. The double-lined area in the figure indicates the presumed cathodic area, although it extended relatively further out than indicated. At fifteen days, blisters were broken, exposing the metal to the outside corrosive conditions. At that time, the peripheries of the detached areas appeared etched.

The interpretation given here is as follows: Under action of the impressed current, chloride ions passed through the coating at some of the thin spots. Concurrent with this, iron was oxidized to form ferrous chloride. Water was drawn through the coating by osmotic action, to form the blister. The ferrous chloride solution is acid, and the center region established itself as anodic to the metal beneath the organic coating at the periphery of the blister. Since the calcium salt is not capable of producing alkaline detachment, and since the liquid in the blister must be presumed as acid, it would seem to follow that the primary detachment involved the autoreduction of ferric oxide in the oxide film according to the mechanism proposed by Gonzalez, Josephic, and Oriani.[11] According to this mechanism, oxidation of iron to the soluble ferrous state is accompanied at the cathodic regions by reduction of the ferric oxide in the film on the metal to the soluble ferrous state. As pointed out by these authors, in a noncomplexing medium pH of less than 4 is required for the ferrous iron produced to be sufficiently soluble in water that the organic coating becomes detached. An apparent cathodic ring beneath the coating around an anodic area is similar to the findings of G. M. Hoch who, by indicator techniques, showed the existence of cathodic rings around the anodic heads in filiform corrosion of aluminum.[12]

Similarly, blisters could be expected to originate if a salt contaminant of an acid anion existed on the surface of the metal beneath the organic coating. A small amount of chloride, for example, can be expected to do considerable damage in this regard. Once ferrous chloride is formed, the iron in solution will generally precipitate out some distance from the anode surface as insoluble corrosion product, releasing the chloride ion to form more ferrous chloride.

Discussion

There has been a great deal of attention given in the literature to the resistance of the organic coating to ionic transport, or to oxygen diffusion, or to water diffusion as rate controlling. Consideration has even been given to the possibility of replacing long term exposure tests by permeability measurements.[13] Surely, such things are important, but one schooled in the general discipline of corrosion science soon becomes wary of any short time test procedures which might be proposed as a generality. One might wonder, further, why the above things might be so important that he might ignore the importance of such things as the reactivity of the metal, the nature of its surface, the nature of possible surface contaminants, inhibitive pigments, coating discontinuities, or the nature of the corrosive environment, as the prime mover in determining what is going on. All these things are important.

Contrary to what may have been inferred, the nature of the corrosion processes beneath protective organic coatings is far from being settled. The present intent has been to give consideration to one factor which seems to have been largely overlooked. For convenience, we might arbitrarily separate sub-coating water into five separate categories:

1. That which always remains firmly adsorbed to a metal oxide surface.
2. Water adsorbed at the oxide-coating interface from the environment, below that level permitting mechanical detachment of the coating.
3. A layer of interfacial water of sufficient "thickness" to permit mechanical detachment of the coating.
4. A film of sensible water beneath the organic coating.
5. A liquid filled blister.

Categories 4 and 5 represent areas of corrosion produced failures. Category 2 might be regarded as "normal" for an organic coating.

Adhesive Strength

Involved here are areas of considerable uncertainty, and one could quickly get lost in speculation. Evidence would seem to indicate, however, that an organic coating should always be considered as bonded to a "layer" of water, rather than directly to the metal oxide or to the surface treatment film. The fact that water has an appreciable tensile strength has been popularized.[14-15] This can be appreciated readily by introducing a film of water between two smooth glass plates and trying to pull them apart. Loss of adhesion would seem to require a tearing action, rather than a straight pull. Such a stress could be expected to be resisted by the structuring of the interfacial water. It has been suggested that such structuring may "extend over considerable ranges, say tens to thousands of molecular diameters."[16] The same authors conclude later that the thickness of structured water at a solution/glass interface appears to range from 0.05 to 0.2 μm. More significance, perhaps, lies in what we call a water "layer". Almost certainly, this does not represent a uniform, planar layer. More likely, the organic coating has adhesion points or areas at which it comes in contact with the category 1 water, or even with the metal oxide film, with thicker layers of water in between. Loss of adhesive strength would then represent progressively more water in such regions.

Practically, the presence of a layer of water is attested to by the sensitivity of some organic coating/substrate combinations to loss of adhesive strength, as in category 3. It has been argued frequently that in loss of "adhesion," rupture always takes place within a material, not between the materials.[17,18] Evidence indicates that, in cases of poor wet adhesion, this rupture must be within the water layer. Loss of adhesive strength is, first of all, sensitive to the water environment. Typically, it is reversible, recovery taking place in anywhere from an hour to months. In a case of poor wet adhesion of an organic coating to tinplate, mentioned earlier, it took three months in a desiccator to recover good adhesion.[5] In that case it could also be mentioned that after removal of the coating by scotch taping there was no discernable residue of the organic coating on the surface. Also, potentiokinetic reduction of the tin oxide film after the coating removal indicated a mature oxide, unaffected by pulling off the organic coating.

Relationship to Oxygen Diffusion

It should be evident that flow of corrosion currents by ionic conduction beneath the organic coating is a requisite part of any corrosive deterioration of organic coated metals. In the case of atmospheric corrosion it is the only possibility. Evidence presented here indicates that such currents can be appreciable for even the firmly bonded coatings in category 2. One point to be noted is that, in spite of the rapid rate of deterioration in these tests, the rate of corrosion was controlled by the underfilm conductance properties. Oxygen diffusion through the coating could have supported even faster rates. It could be argued, of course, that these are thin coatings. Two points might be mentioned. The first is that prime consideration was the interface, not how much was on top of it. The second is that the vinyl over epoxy phenolic coating is actually a superior type of coating for its intended purpose, capable of limiting corrosion to low rates by virtue of its resistance to ion transport.

It is, of course, possible to restrict oxygen diffusion to a greater extent with a thicker coating. In view of the high rate of deterioration in the present case, it must be doubtful that this, in itself, could provide adequate protection. One difficulty in trying to relate measured oxygen diffusion rate through the coating to corrosion is that the cathode/anode ratio, as in Figure 1, can be large. With the situations depicted in this paper, as exemplified in Figures 7 and 9, we could currently have no idea as to the effective size of the cathode area. It is not intended here to discount the effects of oxygen diffusion. In any specific case, however, it must be put into its proper place relative to other factors.

Fortunately, this has been an extreme corrosion situation. It is reasonable to expect that the natures of both the oxide surface and the organic base coat, as well as the amount of water absorption at the interface, would affect ionic conduction processes. It is appreciated, for example, that a phosphate treatment has a major effect in retarding cathodic detachment.[19] While no evidence is currently available, it would not seem unreasonable to relate this to interfacial conduction. On a tinplate surface, we could expect easily to detach cathodically the modified alkyd coating in a sodium salt solution, but one would be hard-put to do this with the epoxy phenolic base coat. The limiting of underfilm conduction processes would also limit the area of cathode available for the reduction of oxygen.

Specific Ions

Specific ions are considered next. Potassium chloride and sodium chloride must be regarded as especially bad actors and not at all representative of what one might encounter in other environments. Anions and cations are required to carry the corrosion currents beneath the coatings, but their damaging effects are not limited to this. Cathodic detachment, except for that involving ferric oxide autoreduction, requires the development of a strongly basic solution beneath the coating. This pretty much limits the damaging cations to potassium and sodium. On the anionic side, the corrosion literature is full of examples of the effects of chloride ions in stimulating various types of corrosion. Beneath the organic coating, other anions can also serve to stimualte corrosion, even though not as effectively as chloride. The important consideration here is that a soluble iron salt is formed at the anode so that, even if there is subsequent formation of an insoluble corrosion product, it will be some distance away from the actual corroding surface.

Breakdown Processes

It would appear that interfacial corrosion currents must play a part in organic coating breakdown processes. These occur entirely beneath the organic coating. They also require the presence of damaging anions. How do they get there? The importance of cleaning metals before coating has long been recognized, as has the importance of such subcoating contaminents. The possibility of sulfur dioxide penetrating the coating as a gas, along with later reaction with water and oxygen to form sulfate ions, has been mentioned a number of times. While important, these do not appear to explain all observations.

Evidence has been presented frequently as to the high degree of impenetrability of high quality protective organic coatings by ions; and this, on the average, is certainly true. This is in line with the high resistance of such coatings, even including the relatively thin vinyl over epoxy phenolic mentioned here, to corrosion currents, which are carried by ionic transport. The difficulty here is that organic coating breakdown does not start from an average position. It starts from a specific point where the organic coating might be thin, or otherwise defective. As indicated in Figure 4, such spots can be very small, and it takes only a small amount of penetration at that point to get the process started. The discussion in connection with Figure 15 is one example of how blistering and breakdown might get started. The cited work of Leidheiser and Kendig also shows the breakdown processes as nucleated from discrete spots. It is unlikely that a few spots like this could be detected from macro measurements, but perhaps more local measurements being given attention lately may be informative.[20]

References

1. E.L. Koehler, Extended Abstracts, Boston Meeting, Electrchem. Soc. **73-2,** p. 164, (1973).
2. J. C. Bolger and A. S. Michaels, *Interface Conversion for Polymer Coatings,* Weiss & Cheever, ed. Elsevier, New York, (1968).
3. for example, J. E. O. Mayne, *J.O.C.C.A.,* **40,** 183, (1957).
4. C. Bacon, J. J. Smith and F. M. Rugg, *Ind. & Eng. Chem.,* **40,** p. 161, (1948).
5. E. L. Koehler, *Corrosion,* **33,** p. 209, (1977).
6. J. Castle, *Corrosion Control by Coatings,* Leidheiser, ed., p. 435, Science Press, Princeton, (1979).
7. W. Funke & H. Haagen, A.C.S. Coatings & Plastics Preprints, 173rd meeting **37,** n. 1, p. 275.
8. E. L. Koehler, *Proceedings, 4th Int. Cong Met. Corr.,* p. 736, (1972).
9. H. Leidheiser, Jr. and M. W. Kendig, *Corrosion,* **32,** p. 69, (1976).
10. E. L. Koehler, *Localized Corrosion,* Staehle et al, ed., p. 117, NACE, (1974).
11. O. D. Gonzalez, P. H. Josephic, and R. A. Oriani, *J. Electrochem Soc.,* **121,** p. 29, (1974).
12. G. M. Hoch, *Localized Corrosion,* Staehle et al, ed., p. 134, NACE, (1974).
13. H. Haagen and W. Funke, *J.O.C.C.A.,* **58,** p. 359, (1975).
14. P. F. Scholander, *Am. Scientist,* **60,** p. 583, (1972).
15. R. E. Apfel, *Sci. American,* **227,** p. 58, (Dec., 1972).
16. J. F. Schufle, C. Huang, and W. Drost-Hansen, *J. Coll. & Interface Sci.,* **54,** p. 184, (1976).
17. J. J. Bickerman, *J. Adhesion,* **3,** p. 333, (1972).
18. W. J. McGill, *J.O.C.C.A.,* **60,** p. 121, (1977).
19. R. R. Wiggle, A. G. Smith, and J. V. Petrocelli, *J. Paint Technol.,* **40,** p. 174, (1968).
20. J. Standish and H. Leidheiser, Jr., Corrosion **36,** p. 390 (1980).

Blistering of Paint Films

W. Funke*

The first indication of insufficient protection by organic coatings against corrosion is usually blister formation. Therefore, much effort has been given to elucidate the mechanism of this paint defect. Mainly four different mechanisms have been discussed in literature:

1. Blistering by volume expansion due to swelling.[1-3]
2. Blistering due to gas inclusion or gas formation.[4]
3. Blistering by osmotic processes, due to soluble impurities at the film/support interface.[3,5-8]
4. Electroosmotic blistering.[9-11,15]

It is doubtful whether blistering due to volume expansion of the film, as a consequence of water absorption or swelling, is really a common mechanism. Water absorption of organic coatings used for corrosion protection is usually in the range of 0.1 – 3 Wt. % (based on the vehicle fraction); and even part of this absorbed water does not really swell the vehicle, but accumulates at some internal interfaces, like the pigment/vehicle or the film/support interface.[12,13] Considering the viscoelastic properties of paint films, it is hard to understand why stresses derived from the absorption of such small amounts of swellant are not relieved by relaxation processes before local detachment and blister formation may occur. Moreover, it should be expected, that by volume expansion due to swelling wrinkling should result rather than blister formation, as it is observed when paint films are exposed to suitable organic solvents. There is also no reason why stresses due to volume expansion by swelling should not result in a delamination of the whole film, rather than in the formation of discrete local blisters.

Blisters may develop due to the presence or formation of volatile components in a paint film on film formation or in cathodic protection. Such components could be: air bubbles incorporated in the film during paint preparation; or application; or, also, perhaps some inclusions of solvents as a result of phase separation;[14] or volatile condensation products. Certainly, these blisters are not confined to some interface, but should be observed throughout the whole layer and thus be easily distinguished from blisters caused by other mechanisms. Blisters may also be formed by gas evolution due to electrolysis at the metal surface below a coating.

Electroosmosis has also been proposed as a possible mechanism of blistering.[9-11,15] In this case, water moves through a membrane or a capillary system under the influence of an electrical potential gradient, which may be caused by local corrosion cells. According to Helmholtz-Smoluchowski the electroosmotic permeability D_i [cm.Coul^{-1}] is given by

$$D_i = \varepsilon \zeta / 4 \pi \eta x_i$$

where ε = dielectricity constant, ζ = Zeta potential, η = viscosity and x_i = specific conductivity. According to Schmid,[16] this equation has to be modified in the case of membranes having pore radii in the range of 1 nm, which are also found in normal paint films. With increasing salt concentration, and increased conductivity, the water transport by this mechanism decreases.

Considering the arguments mentioned above, osmotic blister formation remains the most important mechanism. Recent result of van Meer-Lerk and Heertjes,[7,8] who compared permeation rates, measured by the rate of volume increase of blisters, with those directly obtained by permeation experiments, provide more evidence that the osmotic mechanism may explain most cases of blistering.

There is no doubt that most paint films are almost impermeable to inorganic anions, like Cl^- or $SO_4^=$, or at least greatly retard their permeation.[17,18] On the other hand, it could be shown that the osmotic pressure in blisters can be as high as 2500 to 3000 kPa (25 to 30 atm).[6] This pressure is much higher than the resistance of a film to deformation, which is estimated to be 6 to 40 kPa (0.06 to 0.4 atm).[7]

Whereas it is widely agreed that the osmotic mechanism is the most important for paint film blistering observed during practical performance, it is often speculative what special type of osmotic active substances are present at the film/support interface. The general information that water soluble impurities at the film/support are responsible for osmotic blistering, usually helps little to avoid this paint failure. It is well known, of course, that impurities, like salt residues from phosphating or from washing paint

*Institut fur Technische Chemie, Universitat Stuttgart, Forschungsinstitut fur Pigmente und Lacke e.V., Stuttgart

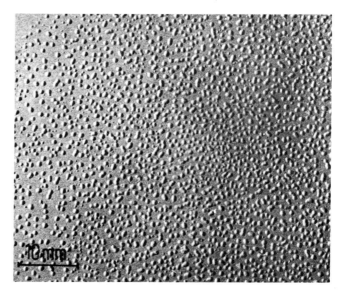

FIGURE 1 — Schematical representation of structures formed depending on the time of phase separation.

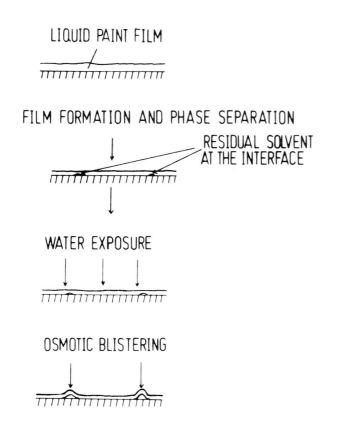

FIGURE 2 — Phase separation and osmotic blistering caused by solvent inclusions at the film/metal interface.

layers after wet sanding,[19,20] "sulfate nests"[21] due to atmospheric SO_2 or sebacious secretions may cause osmotic blistering.

However, at least in some cases it is difficult to understand, why contaminations should be localized and not also cover the area left between adjacent blisters; that is, why, on exposure to water, blisters are formed instead of a general delamination.

Blistering due to Phase Separation During Film Formation

As possible sources for osmotic processes, hydrophilic solvents, which are widely used in paint formulation, have to be considered also. Again, the

FIGURE 3 — Blister formation due to phase separation and solvent inclusions.

 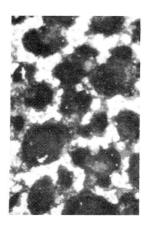

FIGURE 4 — Appearance of the steel surface (left) and the backside of the paint film (right) after exposure to water.

question arises why such solvents are localized, corresponding to the blistering pattern rather than uniformly distributed over the whole interface between the paint film and its support.

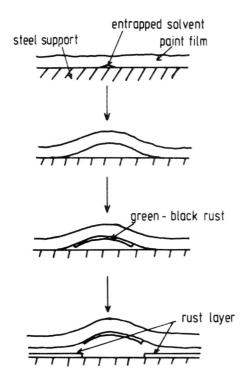

FIGURE 5 — Rust formation in blistering due to phase separation and solvent entrapment at the film/steel interface.

FIGURE 6 — Blister formation due to solvent entrapment on variation of film thickness.

In earlier studies, on phase separation during film formation, it could be shown[22,23] that on decreasing the fraction of a less volatile nonsolvent, phase separation shifts to later stages of film formation. Accordingly, the morphological structure of the film changes, from a layer of precipitated polymer with supernatant liquid, to films with open porous, closed porous structures and microcellular films. On approaching the formulation with the pure, more volatile solvent, the microcellular structure does not disappear uniformly over the whole section of the film, but gradually retreats from the surface down to the interface between the film and the support (Figure 1).

On exposure of paint films, containing such interfacial solvent inclusion, to high humidity or liquid water, osmotic blisters appear. The pattern of the blisters agrees perfectly with the cellular structure adjacent to the steel surface. The development of blisters due to phase separation (Figure 2) involves the diffusion of water into the solvent containing microcells adjacent to the steel surface. Obviously, the diffusion of the entrapped solvent through the film is much slower than its dilution by water due to the osmotic process. Consequently, the film lifts off locally, and blisters develop (Figure 3). If the film is detached shortly after blisters appear, no trace of corrosion products, like Fe^{++}, could be detected; but the blisters did contain substantial amounts of hydrophilic solvents. It is necessary, for this mechanism to operate, that the entrapped solvent is hydrophilic. High boiling solvents, like glycol ethers or esters, definitely tend to this kind of blister formation.

After some time, formation of green rust is observed at the dome of the blisters. Following the delamination over the cathodic areas, rust is deposited there at the steel surface, forming the characteristic pattern shown in Figure. 4. Rust formation is controlled largely by the permeability to oxygen.[24,25] Due to the relatively slow permeation of oxygen, it reacts with ferrous ions immediately, on arrival at the dome of the blister, and causes precipitation of rust over these areas. Therefore, oxygen is prevented from reacting at the steel surface below the blister, thus enforcing the anodic character of these areas. On the other hand, oxygen reacts at the cathodic areas with formation of hydroxyl ions, and delamination follows. The high pH value at these areas prevents the dissolution of corrosion products in the interfacial space. The depolarization of the cathodic areas by oxygen, therefore, causes insoluble corrosion products to be formed, which then are precipitated in contact with the cathodic areas of the steel surface (Figure 5.)

Blister formation due to phase separation during film formation depends significantly on film thickness. On stoving of paint films based on melamine-alkyd resins, blister formation on exposure in a humidity cabinet was more pronounced with increasing thickness (Figure 6). Likewise, with unpigmented films, underrusting becomes more distinct the thicker the films are (Figure 7).

Phase separation on film formation also explains these unexpected relations between film thickness and protective properties. In thinner films, the microcellular structure has already dissappeared, even adjacent to the film/support interface, while it is still present in the thicker film. Obviously, the less volatile hydrophilic solvent leaves the film more slowly the thicker the film is.

That blistering really is caused by entrapped solvent at the film/metal interface is also supported by the result, that no influence of film thickness on blistering and underrusting is observed if the coating

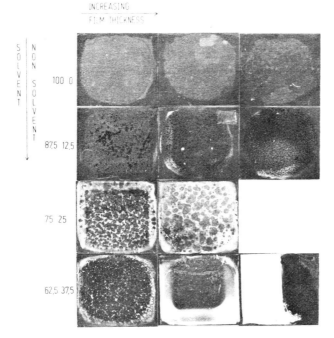

FIGURE 7 — Influence of the non-solvent fraction and the film thickness on blistering.

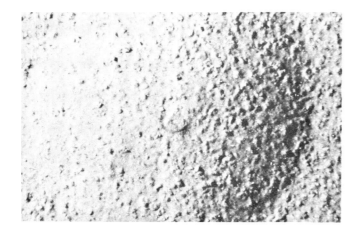

FIGURE 8 — Blister formation at thin pores and paint defects.

FIGURE 9 — Blister top enlarged.

FIGURE 10 — Steel surface beneath the blister.

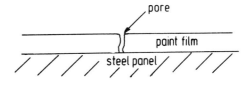

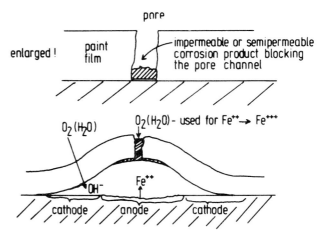

FIGURE 11 — Process of blister formation at small-pore paint defects.

is given a heat treatment at temperatures significantly above the glass transition point. Under this condition the entrapped solvent is released within reasonably short times even from the incoherent phase of microcells.

It may be argued that in normal coatings in general no distinct nonsolvents are used. However, on paint formulation the choice of solvents is mostly directed by the requirements of the dispersion and application process. Slowly evaporating, hydrophilic, latent or high boiling solvents may well become incompatible with the polymer at later stages of film formation especially because the chemical structure of the vehicle may significantly change as a result of the reactions involved in film formation.

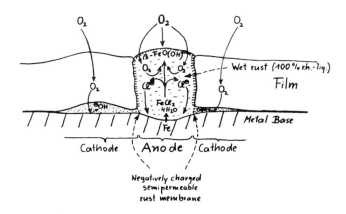

FIGURE 12 — Blister formation adjacent to scribes.

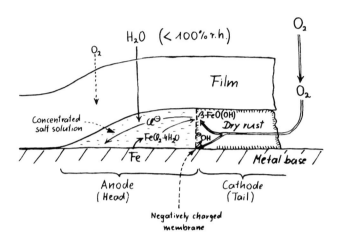

FIGURE 13 — Mechanism of filiform corrosion.

	Humidity of liquefaction % r.H./20°C
LiCl	15
$CaCl_2 \cdot 6H_2O$	32
NaCl	78
$Na_2SO_4 \cdot 10H_2O$	93
$FeSO_4 \cdot 7H_2O$	95
$FeCl_2$	56
$FeCl_3 \cdot 6H_2O$	47

FIGURE 14 — Humidity of liquefaction at 20 C of various inorganic salts (data from Bukowiecki[32]).

Blistering Adjacent to Scribes

Contrary to blistering at small sized paint defects, like pores or pinholes, if scribes are applied to a paint film covering steel, blister formation always is observed beside the scribe (Figure 12). In this case, the area of the unprotected steel surface is too large to be closed by a membrane of hydrated oxidic corrosion product. Rust is formed at the scribe, but a membrane covering this area is not stable. Due to the dynamic character of rust formation, the corrosion products formed are subjected to continuous rupture and healing processes. It may be assumed, therefore, that the difference between both blistering mechanisms is essentially due to the difference in the areas of unprotected steel surface.

Blistering Due to Paint Defects

The semipermeability of the polymer film to the osmotic active substance is an essential prerequisite for any osmotic process to operate. Films, having pores, pinholes, or similar small-sized paint defects, are not expected to show blistering, especially not at these paint defects. In testing a series of anticorrosive coatings by exposure to salt spray according to DIN 50021 and 53167, blisters were observed with some brownish spot at the center of their top (Figures 8 and 9). The steel surface beneath the blister is shown in Figure 10. It could be proven by chemical analysis that the brownish material was rust. Samples with small pores, artifically prepared by piercing the film with a thin steel needle, always exhibited blistering as a consequence of such paint defects. It remains to be explained why osmotic blistering may occur even at macroscopic pores. Obviously, the corrosion products formed at the bottom of the pore effectively close it, thus making this area semipermeable again (Figure 11). For this mechanism to operate, it is sufficient that a membrane of hydrated oxides of iron covers the hole. The permeability of such iron oxide membranes is being studied presently.

Filiform Corrosion

The mechanism of filiform corrosion has been discussed in a number of publications.[26-31] The essential conditions for this paint defect to take place may be summarized as follows

1. High humidity (65 to 95% r.h. at room temperature).
2. Sufficient water permeability of the paint film.
3. Stimulation by artifical or natural impurities like sulfurous dioxide or acetic acid.
4. Presence of film defects like mechanical damages, pores, insufficient coverage of points and edges, air bubbles, salt crystals or dust particles.

Despite proposals on the mechanism of filiform corrosion, the explanations given for the formation and growth of the filiforms are still somewhat unsatisfactory. There is no doubt that a filiform starts at some paint defect, and that the formation of the head is caused by osmotic processes similar to those on normal osmotic blistering. However, it is difficult to understand why an internal pressure may develop at all with the filiform tail open to the paint defect, and therefore also open to the air. As the tail section of the filiform contains dry rust, which is almost free from soluble inorganic salts, oxygen can diffuse comparatively free to the filiform head. The retention of the highly concentrated salt solution in the growing head is only possible if some membrane prevents the ions from migrating into the rust filled tail (Figure 13). This membrane may again be composed of hydrated corrosion products of iron, which are continuously renewed as the filiform head moves forward, leaving the older membrane to be transformed into dry rust.

Whereas the oxygen concentration in the tail containing porous dry rust corresponds essentially to that in the air (ca. 210 ml/l), its concentration in the concentrated salt solution of the filiform head, is certainly even lower than 31 ml/l, which is the oxygen concentration in pure water under comparable conditions. It is probable that this differential aeration cells directs the growth of the head, and also enables the continuous renewal of the hydrated oxide membrane when the filiform grows in length. Considerable evidence exists[32] that the relative humidity, above which filiform corrosion may occur, depends on the humidity of liquifaction of the salts (Figure 14) in the head of the filiform. Likewise, this relation may also be important for the humidity necessary for other types of blister formation.

References

1. N.A. Brunt, *J. Oil Col. Chem. Assoc.* **47**, p. 31 (1964); Verfkroniek **33** p. 93 (1960).
2. L. Bierner, *Farbe und Lack* **66**, p. 686 (1960).
3. D.M. James, *J. Oil Col. Chem. Assoc.* **43**, pp. 391, 658 (1960).
4. J.A. van Laar, *Paint Varnish Prod.* **51**, (8) pp. 31, 88; (9) p. 49; (11) pp. 41, 97 (1961).
5. W. Wettach, *Official Dig.* **33**, p. 1427 (1961); **32**, p. 1463 (1960).
6. J. L. Prosser, T.R. Bullett, *J. Oil Col. Chem. Assoc.* **45**, p. 836 (1962).
7. L.A. van der Meer-Lerk, P.M. Heertjes, *J. Oil Col. Chem. Assoc.* **58**, p. 79 (1975).
8. L.A. van der Meer-Lerk, P.M. Heertjes, ibid **62**, p. 256 (1979).
9. H. Grubitsch, K. Heckel, *Farbe und Lack* **66**, 22 (1969).
10. H. Grubitsch, K. Heckel, R. Saumer, *Farbe und Lack* **69**, p. 655 (1963).
11. H. Grubitsch et al, *Farbe und Lack* **70**, p. 167 (1964).
12. W. Funke, U. Zorll, W. Elser, *Farbe und Lack* **72**, p. 311 (1966).
13. W. Funke, *Werkstoffe und Korrosion* **20**, p. 12 (1969).
14. W. Funke, *J. Oil Col. Chem. Assoc.* **59**, p. 398 (1976).
15. W.W. Kittelberger, A.C. Elm, *Ind. Eng. Chem.* **39**, p. 876 (1947).
16. G. Schmid, *Chemie-Ing. Technik* **37**, p. 616 (1965).
17. M. Svoboda, D. Kuchynka, B. Knapek, *Farbe und Lack* **77**, p. 11 (1971).
18. M. Svoboda, J. Mleziva, *Progr. Org. Coatings* **2**, p. 207 (1974).
19. Anon, Water Treatment, *Chem. Processing* **6**, p. 861 (1960).
20. H. Nienhaus, *Ind.-Lackier-Betrieb* **27**, p. 281 (1959).
21. H. Schwartz, *Werkstoffe und Korrosion* **16**, pp. 93, 208 (1965).
22. M. Schmitthenner, W. Funke, Die Angew. *Makromol. Chem.* **67**, p. 117 (1978).
23. W. Funke, *J. Oil Col. Chem. Assoc.* **59**, p. 398 (1976).
24. W. Funke, H. Haagen, *Ind. Eng. Chem. Prod. Res. Dev.* **17**, p. 50 (1978).
25. W. Funke, E. Machunsky, G. Handloser, *Farbe und Lack* **84**, p. 493 (1978).
26. H. Kaesche, *Werkstoffe und Korrosion* **10**, p. 668 (1959).
27. W.H. Slabough, G. Kennedy, *Offic. Dig.* **34**, p. 1139 (1962).
28. J.F. Barton, *Paint Manufacture* Dec. 1964, p. 47; Nov. 1964, p. 53.
29. W.H. Slabough, E.J. Chan, *J. Paint Technol.* **38**, p. 417 (1966).
30. M.N.M. Boers, *Verfkroniek* **47**, p. 278 (1974).
31. W.H. Slabough, M. Grother, *Ind. Eng. Chem.* **46**, p. 1014 (1954).
32. A. Bukowiecki, Proceedings of the 8th Congress for Elektrodeposition and Surface Finishing 1972, p. 14, Forster Verlag AG, Zurich 1973.

Adhesion Loss of Organic Coatings
Causes and Consequences for Corrosion Protection

W. Schwenk*

Introduction

Adhesion loss of protective coatings may occur over the entire surface or locally as blisters. The mechanisms are based on mass transfer within or under the coating. The mechanisms can differ significantly and are related to the properties of the protective system. A proper understanding of these mechanisms is desired for the prediction of corrosion damage and for the choice of proper preventive measures. To investigate the different kinds of adhesion loss, we have carried out experiments with coatings on steel which are commonly used for protection in soil and sea water, especially of steel pipes.

Figure 1 describes the three possible corrosion interactions at a coated material in an aqueous environment containing Na^+ and Cl^- ions. In order to be realistic it is assumed that the coating contains holidays, such as mechanical damages or pores. Furthermore, the coating has a permeability for the molecules, H_2O and O_2, and for ions as well. The corrosion of the unprotected metal at the holidays occurs as a result of the formation of Evan's cells with a preferred generation of NaOH at the rim of the holiday. The alkalinity can reach $pH \approx 14$ and gives rise to alkaline peeling which is commonly known as cathodic disbonding.

In the case of coating with a marked DC conductivity (which means a sufficiently high permeability for ions) cell formation with the coated area acting as a cathode leads to pitting corrosion by enhanced anodic metal dissolution at the holiday. Furthermore, the cathodic reaction products at the steel/coating interface can give rise to an enhanced adhesion loss and to blistering. In general, permeating substances, such as O_2 and H_2O, can lead to a minimum underfilm corrosion yet a total loss of adhesion after long periods of time in service.

From this picture it follows that the DC conductivity of the coating material free of any holidays and pores, that is, the ionic permeability, is of a great significance. In most cases this conductivity is not exhibited in the as-received condition, but only after water absorption which depends in turn on immersion time, temperature, coating thickness and even polarization conditions. Figure 2 shows the effect of immersion

*Mannesmann Forschungsinstitut GmbH, Germany

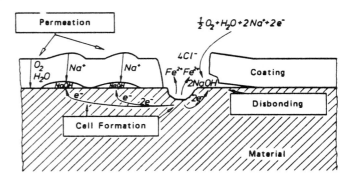

FIGURE 1 — Corrosion interactions at coated material.

time on the DC resistance of a phenolic resin in pure water at 200 F.[1] This type of coating is used for lining hot water tanks and often fails by pitting at pores and blistering due to cell formation. The cathodically enhanced decrease of DC resistance can be understood by electroosmosis which is responsible for an enhanced water take up.

For comparison purposes, it is of interest to note that the DC resistance of coatings with high quality

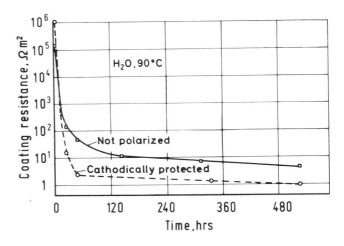

FIGURE 2 — Electric resistance (DC) of an over critically pigmented phenolic resin coating on steel plate in pure water.

coal tar epoxy is in the range of $10^6 \, \Omega \, m^2$, whereas 2 to 4 mm thick polyolefin coatings result in resistances of the order $10^{11} \, \Omega \, m^2$ that do not have any tendency to decrease with time.[2]

Table 1 summarizes the characteristic information relative to five different mechanisms of adhesion loss due to mass transfer. For a description of general mass transfer we have to consider both chemical and electrochemical forces:

$$v = D \left(\frac{1}{c} \, \text{grad} \, c + \frac{zF}{RT} \, \text{grad} \, \varphi \right) \quad (1)$$

where v = velocity of particles; D = diffusion constant; c = concentration; z = valency; F = Faraday's constant; R = gas constant; T = absolute temperature; and φ = electrical potential. Equation 1 is exactly valid for dilute solutions and involves the following Nernst-Einstein relationship between the mobility u and the diffusion constant:

$$u = \frac{DF}{RT} |z| \quad (2)$$

Electrical factors seem to have no influence in the case of corrosion in pure water. For practical conditions, however, there are two important aspects to be considered:

1) The potential of the coated steel surface is influenced by the corrosion of the bare steel at holidays or by foreign objects which are in electrical contact with the coated steel part under consideration.

2) Natural environments, water and soil, contain sodium and chloride ions. These ions may migrate through the coating in a similar way as molecules like H_2O and O_2 according to Equation 1. Potential gradients are highly effective. Furthermore, electrolysis takes place at the steel/coating interface:

$$Fe - 2e^- + 2Cl^- + H_2O = (Fe(OH)^+ + Cl^-) + (H^+ + Cl^-) \quad (3)$$

$$O_2 + 4e^- + 4Na^+ + 2H_2O = 4(Na^+ + OH^-) \quad (4a)$$

$$2H_2O + 2e^- + 2Na^+ = 2(Na^+ + OH^-) + H_2 \quad (4b)$$

Both sides of these equations are electrically balanced. The left hand side of each equation contains the polarization current which passes the metal surface and the equivalent rate of ionic migration through the coating. The right hand side of each equation contains the reaction products which give rise to characteristic values of pH (Table 1).

Osmosis is a well known subject. Surface preparation and passivating substances within the primer are preventive measures. Enhanced water permeation in a temperature gradient is of a high interest for vessels at elevated temperatures and for heat exchanger tubes as well. Blistering and total loss of adhesion are the consequences in general, with the exception of special resins which are over critically pigmented, and thereby have a microporous structure.

Cathodic electroosmosis is often believed to be the driving force of cathodic blistering. According to the principles of colloid chemistry, neutral water is transferred to the metal/coating interface cathodically. But field observations show that the cathodic blisters usually contain no neutral water but an alkaline solution. This alkalinity can be explained only in terms of migration of alkali cations and cathodic generation of OH^- by an electrochemical reaction according to Equations 4a, b. By a similar mechanism, anodic blisters are formed in chloride containing media according to Equation 3. In both cases the migration of special ions within the coating and their electrochemical reactions at the metal/coating interface are of fundamental interest. The reaction products, having a high solubility in water, give rise to osmotic water transport. This transport is the driving force for the growth of the blisters. Removal of the mentioned critical ions, for example, alkali compounds, and increase of DC resistance are preventive measures.

TABLE 1 — Mechanisms, characteristics and prevention of blistering.

Process	Causative factors	Contents and characteristics	Possible preventive measures
Osmosis	Hydrophilic metal contamination and corrosion products	Oxides, hydroxides (neutral water)	Inhibitive pigments in prime coat, metal passivation
Enhanced H_2O diffusion	Temperature gradient metal (cold) environment (warm)	(Neutral water)	Over critical pigmentation
Cathodic electroosmosis	Coating charged negative with respect to environment	(Neutral water)	High coating thickness (ions support deterioration by electrolysis)
Cathodic polarization	Migration of cations, generation of OH^-	(NaOH) High pH	Removal of alkali ions, high coating resistance
Anodic polarization	Migration of anions, anodic corrosion	Hydrolysed corrosion products, low pH	Removal of corrosive anions, high coating resistance

In the following some examples of these mechanisms are described.

Water Permeation in a Temperature Gradient

For internal protection of water tubes, a special polyamide coating material was recommended, even for elevated temperatures up to 200 F. Short time field tests were said to give excellent results. But due to the temperature gradient (warm water/cold tube wall) there is an existing danger of blistering. Figure 3 shows the effect of a temperature gradient "tube material/ water" on the incubation time for blistering.[3] The incubation time increases with decreasing gradient and with decreasing water temperature (140, 176 F). Isothermal tests showed blisters after 2300 hours in water of 176 F. Thus the curves join the abscissa, and for warm water tubes internal coatings are less suitable because of the danger of blistering, complete loss of adhesion and peeling off. The danger of peeling off can be markedly decreased in the case of thick coatings — because here we encounter the situation of a tube in a tube! The danger of peeling off does not exist in the case of external coatings!

Electrochemically Induced Blistering

In practice, adhesion loss and blistering are mostly caused by cathodic polarization because of the wide application of cathodic protection and also because of cell formation with bare steel corroding in the same environment. It was already mentioned that cell activity between the coated steel as a cathode and the bare steel at a holiday as an anode causes cathodic blistering below the undamaged coating and pitting at the holiday. Figure 4 shows a 2 mm deep pit in a pile coated with 300 μm coal tar epoxy after 5 years ex-

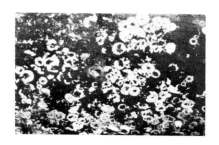

FIGURE 4 — Pitting (2mm deep) on a steel pile coated with coal tar epoxy resin (320 μm) after 5 years exposure in the North Sea (Heligoland)[4].

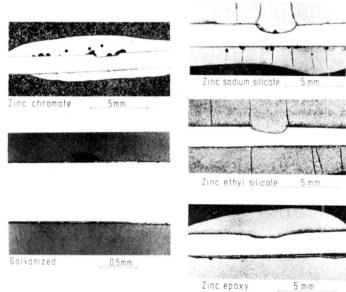

FIGURE 5 — Cross sections of steel samples with circular holidays in coal tar epoxy resin coating after 2 years immersion in 0.1 M NaCL.

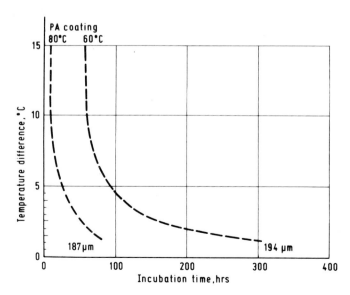

FIGURE 3 — Incubation time for formation of blisters in polyamide coatings on steel for various thermal gradients across a pipe wall coated on both sides (the cold side is permanently free of blisters)[3].

posure in the North Sea.[4] Other parts of the coated pile showed closed blisters containing NaOH. Piles with tar or asphalt coatings had been uniformly attacked without pitting. DC resistance of the coating has a significant influence on this type of cell activity. The current does not follow Ohm's law, and depends markedly on the nature of the prime coat. The effect of prime coat composition, as well as of coating resistance, can be demonstrated by the depth of the pits in the cross sections shown in Figure 5.[5] All specimens had been coated with 80 μm coal tar epoxy, and were exposed for 2 years to 0.1 M NaCl at ambient temperature. With the exception of galvanized specimens, all coated specimens showed a potential more noble than the free corrosion potential of bare steel in the same environment. This is the reason for the anodic polarization of the bare steel at

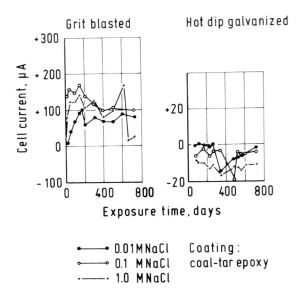

FIGURE 6 — Cell current between a coated steel sheet as cathode (300 cm²) and a bare steel sheet as anode (1 cm²)[6].

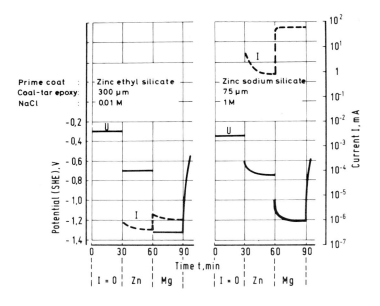

FIGURE 7 — Potential and current values measured on pore free coated steel specimens for determining coating resistance in the potential range between Mg and Zn. Calculation of the resistance according to Equation 5[6].

holidays, or of external electrodes in electrical connection. Figure 6 shows the cell currents measured between a coated specimen of 300 cm² and an iron electrode of 1.2 cm² during a 2 year exposure.[6] A cell current of 100 µA corresponds to a corrosion rate of 1 mm/year. Prime coats of Zn ethyl and Zn epoxy increase the resistance. Thus, the cell current is decreased. On the other hand, zinc metal, galvanized or metal sprayed, shifts the potential to more negative values. Therefore, the coated steel becomes the cathode.

The coating resistance increases, in general, with increasing thickness of the coal tar epoxy top coat. This effect is more than proportionate. Furthermore, it was found that the sodium silicate primer is unsuitable, and retains its negative influence even at a high thickness of the top coat. We assume that the DC resistance is predominantly an electrochemical polarization resistance, and to a lesser extent an ohmic resistance of the coating material itself. It was found for coal tar epoxy coating that thicknesses higher than 500 µm result in DC resistance values, which are generally high enough to prevent corrosion cell effects.

The DC resistance values of interest can be determined easily by electrically connecting the specimen with galvanic anodes of Zn and Mg, and measuring the differences of both the mixed potentials and the short circuit currents. The quotient of these differences corresponds to an integral polarization resistance value in the potential range from –0.7 to –1.3 volts (S.H.E.):

$$r = \frac{U_{Mg} - U_{Zn}}{I_{Mg} - I_{Zn}} A \qquad (5)$$

where U_{Mg}, U_{Zn} = mixed potentials of the specimen when connected either to the Mg or the Zn anode; I_{Mg}, I_{Zn} = short circuit currents; A = area of the specimen.

Figure 7 shows potential and current vs time curves. Polarization periods of 30 minutes were used to establish stationary conditions. The left hand specimen has a resistance of 10^9 Ω, whereas the right hand specimen has only 9 Ω.[6]

The assessment of test results showed a distinct relationship between coating resistance and susceptibility to blistering. Low resistance values are a clear indication of high susceptibility. Table 2 summarizes test results showing the influence of various factors on blistering. Zinc sodium silicate and zinc epoxy

TABLE 2 — Blistering after 2 years of exposure to NaCl solutions.

Prime coat	Zinc alkali silicate	Zinc epoxy resin	All specimens
Thickness of coal tar epoxy coating	Percentage of specimens with blisters		
75 µm	100	78	90
150 µm	94	33	68
300 µm	83	0	24
450 µm	67	0	12
NaCl concentration	Percentage of specimens with blisters		
0.01 M	100	17	44
0.1 M	100	33	51
1 M	61	22	46
All specimens	Geometric mean of the coating resistance in Ω m²		
With blisters	$5 \cdot 10^3$	$2 \cdot 10^3$	$4 \cdot 10^2$
No blisters	$6 \cdot 10^6$	$3 \cdot 10^6$	$3 \cdot 10^6$

represent the worst and the best prime coat, respectively. In the column "All specimens," the mean values obtained with seven different primers are given. The effect of top coat thickness is quite evident, and in line with the resistance values, except in the case of the sodium silicate primer. The effect of NaCl concentration of the environment on blistering displays a maximum at the intermediate concentration of 0.1 M. This maximum can be explained on the basis of an increase of Na^+ activity, and a decrease of H_2O activity with increasing NaCl concentration in line with the model that both Na^+ and H_2O are necessary for blister formation. The diminution of the relatively high susceptibility of specimens with sodium silicate primer at 1 M NaCl can be understood in terms of a constantly high Na^+ activity (maintained by the primer composition), and a decreased H_2O activity at high NaCl concentration of the environment. The effect of the coating resistance on the susceptibility to blistering is remarkable. The limiting value is about $10^4 \Omega m^2$.

The electrochemical processes that influence blistering have been examined systematically with the aid of thin coatings of pigment free alkyd resin on clean steel sheets.[3] Environments employed were near neutral aqueous solutions with the following ions: Na^+, Zn^{2+}, Cl^-, and acetate (CH_3COO^-). A pair of coated specimens was connected to an external voltage source of 1.5 volts, while a third specimen was immersed without any polarization for comparison purposes. In this case, the edges had been protected with extreme care to prevent steel corrosion at pores. In the course of immersion time, dark spots developed beneath the clear varnish, regardless of polarization and composition of the environment. Figure 8 shows these spots, which could be identified as iron oxide. Moisture, Cl^- and NaOH were not present. The spots

FIGURE 8 — Dark spots under a pigment free alkyd resin coating (200 μm). Osmotic water vapor diffusion and formation of Fe_3O_4.

FIGURE 9 — Anodic and cathodic blisters in a pigment free alkyd resin coating (40 μm) after ten days immersion in 0.2 M NaCl with an applied voltage of 1.5 between anode and cathode. Anode: Migration of Cl^- and anodic formation of $FeCl_2$. Cathode: Migration of Na^+ and cathodic formation of NaOH.

are observed below thin coatings of 40 μm more rarely than below thick coatings of 200 μm. It is concluded that residuals of the solvent promote their formation. Recognizing that the coating is free of corrosion inhibitors or pigments, it is assumed that the iron oxide spots are a result of water permeation and normal corrosion reaction. Besides these dark spots, all polarized specimens showed cathodic or anodic blisters in all the solutions which contained Na^+ or Cl^- ions. Figure 9 shows blisters formed in the NaCl solution. The cathodic blisters (also observed in sodium acetate solution, but not in zinc chloride or acetate!) display alkalinity with pH up to $\approx$ twelve. Initially, the steel surface is free of any rust. But in chloride containing environments, the steel becomes slightly rusted with increasing exposure time. This effect is accompanied by a decrease of pH and by chloride contamination of the rust. Thus, cathodic protection is not effective in closed blisters. The anodic blisters (also observed in zinc chloride solution, but not in sodium or zinc acetate solution!) display rust and hydrolysed ferrous chloride with a pH of about four. The size of the anodic blisters is markedly smaller than that of the cathodic blisters, explicable in terms of a superposition of osmosis and electroosmosis. By the latter, cathodes are enriched with water, whereas anodes are depleted of water. Hence, cathodic blisters are expected to be generally larger than anodic blisters.

Table 3 summarizes these experimental results. It follows from this investigation that alkali cations are responsible for cathodic blistering, and that "corrosive" anions such as chloride ions are responsible

TABLE 3 — Blistering in pigment-free alkyd resin coating (40 μm) after 4 months of exposure.

		Test environment			
		0.2 M NaCl	0.2 M ZnCl$_2$	0.2 M Zn(CH$_3$COO)$_2$	0.2 M NaCH$_3$COO
	Free corrosion	No blisters	No blisters	No blisters	No blisters
	Anode	Blisters (rust)	Blisters (rust)	No blisters	No blisters
	pH	≈ 4	≈ 4	—	—
External Polarization	Cl$^-$	Present	Present	—	—
	Cathode	Blisters	No blisters	No blisters	Weak blisters (bright)
	pH	Bright ≈ 12 / Rusty ≈ 10	—	—	8 – 10
	Cl$^-$	free / traces	—	—	—
	mA·m^{-2}	0.12	0.08	0.05	0.06

for anodic blistering. These results mean that besides polarization, there is a critical environmental effect. These results help to understand the effect of surface impurities and inorganic substances of the coating material, for example, pigments.

Anodic and cathodic blisters can be identified by analyzing their contents, especially by checking the pH. Both types of blisters are observed very often in practice. It is evident that anodic blistering is enhanced by anodic polarization, and cathodic blistering by cathodic polarization. Furthermore, one can assume that there are critical potential ranges for both anodic and cathodic blistering. These ranges can overlap. In this case, without any polarization, both types of blisters can coexist. This is in line with experimental results. With increasing resistance, the potential ranges diverge; hence, under these conditions, external polarization is necessary for blistering. The effect of polarization on blistering can be examined easily by use of chrono-potentiostatic tests. With respect to the realistic condition that steel corrosion at holidays results in cathodic polarization, one can conclude that slight cathodic polarization could be a useful test procedure for comparing the properties of different coating materials, and for developing qualification tests.[7]

Electroosmotic blistering without the additional influence of migrating ions is apparently uncommon. In the case of negative potentials and coatings with inert pigments, electroosmotic blistering can be expected. In one series of experiments, steel sheets were coated with 500 μm thick epoxy resin.[8] The DC resistance in 0.2 M NaCl was determined to be ≈ 10^6 Ω m^2 at about 3 volts. One specimen was provided with a very small pinhole, which did not markedly decrease the resistance. Cathodic polarization was carried out at 3 μA/m^2 in 0.2 M NaCl for 5 years. After that time, the entire exposed coating was found disbonded from the steel. Figure 10 shows the steel surface after removing the

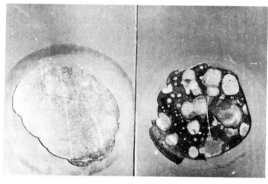

With intentional pin hole
Polarization:
–2 V/stainless steel
Bright surface

Without pores
Polarization:
–3 V/stainless steel
Rust stained

FIGURE 10 — Appearance of steel surface beneath an epoxy resin coating (500 μm) after 5 years immersion in 0.2 M NaCl solution under cathodic polarization with a current density of 3 x 10^{-3} mA m^{-2}.

disbonded coating. The specimen with pore free coating was slightly rusted and dry. Corrosion had occurred as a result of water transfer favored by electroosmosis. But the degree of corrosion was less than 1 μm metal loss. The specimen with the artificial pinhole in the coating was covered with an alkaline moisture, and had remained bright without any corrosion.

Cathodic Disbonding

The generation of high alkalinity by a cathodic partial reaction in the presence of alkali cations takes place not only in cathodic blisters, but also at the rim of holidays, with the consequence of a peeling off of

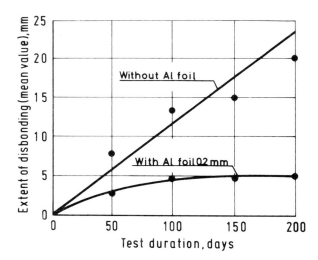

FIGURE 11 — Disbonding of a 3.5 mm thick asphalt coating on steel after immersion in 0.5 M NaCl solution saturated with oxygen. Disbonding started from corroding edges (influence of oxygen permeation)[8]

TABLE 4 — Disbonding in PE-coating (2.5 mm) in the case of free corrosion at bare steel edges.

Environment	Extent of disbonding in mm after an exposure of		
	30 d	100 d	180 d
0.1 M NaCl	50	50	50
0.5 M NaNO$_3$	30	50	50
0.1 M Li$_2$SO$_4$	50	50	50
0.1 M Na$_2$SO$_4$	50	50	50
0.1 M K$_2$SO$_4$	50	50	50
0.1 M NH$_4$Cl	4	15	30
0.1 M MgCl$_2$	0	0	0
0.1 M CaCl$_2$	0	0	1
0.1 M BaCl$_2$	4	6	6
0.1 M ZnCl$_2$	0	0	0
0.1 M ZnSO$_4$ + 0.1 M NaCl	15	50	50
0.1 M AlCl$_3$	0	0	0
0.1 M AlCl$_3$ + 0.1 M NaCl	45	45	50
0.1 M FeCl$_3$	0	0	0
0.1 M FeCl$_3$ + 0.1 M NaCl	15	40	40
0.1 M K$_2$CrO$_4$	0	0	0

TABLE 5 — Cathodic disbonding of a 2.5 mm thick polyethylene coating on steel.

Environment	Extent of disbonding in mm*
0.1 M NaNO$_3$	25
0.1 M NH$_4$NO$_3$	2 - 5
0.1 M CaCl$_2$	0
0.1 M ZnSO$_4$	0
0.1 M CdSO$_4$	0
0.1 M CdCl$_2$	0
0.1 M AgNO$_3$	0
0.1 M AgNO$_3$ + 0.5 M NaNO$_3$	0
0.1 M K$_2$CrO$_4$	25

*According to ASTM G8, radius of the holiday = 5 mm; test duration = 30 days; cathodic current density = 2mA·cm^{-2}

the coating from the metal (Figure 1). The process occurs under free corrosion conditions, and is enhanced by cathodic polarization. It is retarded by neutralization in the presence of acids, and by anodic polarization as well. Reaction of the hydroxide ions with polar bonding groups of the coating leads to disbonding, and to a consumption of hydroxide within the gap between metal and coating. This leads to a decrease in disbonding rate with time. On the other hand, hydroxide ions can be produced in the electrolyte within the gap by cathodic reaction of permeated oxygen. In this case, sodium ions have to migrate from the holiday for electrical neutralization, and the corroding steel at the holiday has to serve as an anode for cell activity. This mechanism can be demonstrated easily by stopping the oxygen permeation with the aid of an aluminum foil, which is fixed upon the coating.[9] Figure 11 displays the disbonding rate of asphalt on a steel plate caused by free corrosion of the bare edges in 0.5 M NaCl solution saturated with oxygen by bubbling. The disbonding of the specimen covered with aluminum foil is markedly reduced. Without the foil, oxygen permeation gives rise to a constantly high pH in the gap. Thus, the disbonding rate is not decreased.

The effect of disbonding has been the subject of many discussions and investigations in connection with pipeline coatings. Table 4 shows the effect of various chemicals on the disbonding of thick PE coating on pipeline steel under free corrosion conditions.[9] It is evident from these results that disbonding occurs only in the presence of cations of the alkali metals. These cations are well known to give rise to the formation of strong chemical bases. But there is really no correlation between disbonding and the aggressiveness of the environment with respect to steel corrosion, (FeCl$_3$). The resistance to disbonding in potassium chromate solution is due to the passive state of the steel in this environment. Thus, both anodic and cathodic reactions do not take place. If a cathodic reaction is forced to take place by external cathodic polarization, disbonding occurs quickly, (Table 5). The other results again demonstrate the effect of critical cations for disbonding. Ca^{2+}, Zn^{2+}, Cd^{2+} and Ag^+ are harmless. The inactivity of sodium ions in silver nitrate solution is surprising, but it can be explained by the electrochemical properties of cathodically discharged silver deposits, which have lower cathodic overvoltage than steel. Thus, the activity of hydroxide ions at the rim of the holiday is restricted.

On the basis of this mechanism of alkaline disbonding (Table 6), it is obvious that increased cathodic polarization, as well as increased concentration of alkali ions, increase the extent of disbonding. Under free corrosion conditions, the corrosion rate and the extent of disbonding may be controlled by ox-

TABLE 6 — Environmental parameters for disbonding and corrosion of coated steel at holiday.

PARAMETERS	COATING	BARE STEEL/HOLIDAY
Cathodic polarization	Increased disbonding	Cathodic Protection
Anodic polarization	No disbonding	Anodic corrosion
Alkali cations present	Disbonding	Aeration cell formation
Alkali cations not present	No disbonding	No cell formation
High oxygen concentration	Increased disbonding	Increased corrosion
Low oxygen concentration	Decreased disbonding	Decreased corrosion
Anodic passivation (CrO_3)	No disbonding	Passivated
Low pH (acidic)	Decreased disbonding	Increased corrosion
Neutral	Disbonding	Corrosion
Increased pH (~10)	Decreased disbonding	Inhibited
High pH (>12)	Disbonding	Passivated

ygen concentration. Increasing pH may inhibit corrosion and disbonding, whereas at high pH steel is passivated; however, disbonding is enhanced due to the high activity of hydroxide ions. In general, there is no correlation between disbonding and corrosiveness.[8,9] This leads to questions on the practical significance of disbonding and adhesion loss to corrosion protection.

Coatings for corrosion protection are to serve as barriers between metal and environment to keep the metal surface free of corrosive substances. Adhesion loss or disbonding does not invariably lead to an impairment in corrosion protection. Corrosion protection is lost neither by formation of closed blisters, nor by disbonding, as long as the solution within the blisters or beneath the disbonded coatings is not corrosive. Corrosion does not occur normally in alkaline environments, at low rates of oxygen permeation, or, generally speaking, at low rates of mass transfer through or beneath the coating. The protection is lost, however, if the coating is detached or peeled off from the metal surface, or if the blisters are open. In this respect, blisters seem to be generally harmful because of their susceptibility to mechanical damage. Furthermore, disbonding of thin coatings seems to be harmful because of mechanical peeling. Unlike internal coatings (as well as coatings on flat surfaces), thick external coatings for pipes and vessels are not adversely affected by disbonding, provided the coating remains tight on the metal surface and is not deteriorated. This is in agreement with long time experience in service.[10] Considering the fact that anodic polarization and acid environments enhance corrosion but prevent disbonding, whereas cathodic polarization and alkaline environments prevent corrosion but enhance disbonding, the disbonding test can hardly be used as a qualifying test for evaluating thick external coatings for pipe protection with a desired high impact resistance. Importance, however, is to be given to properties of the coating material such as aging resistance and strength, as well as to the tightness of the coating to the metal surface in service. Bond strength is of high practical interest in the as received condition for transportation and installation; but not for corrosion protection in long time service, provided the chemical and mechanical properties are satisfactory, and the coating and pipe surface are contiguous. Thick coatings differ basically from thin coatings whose protective properties depend on the behavior of the metal/coating interface, inhibiting pigments, and bond strength as well.

References

1. W. Schwenk, *polytechn. tijdschrift procestechniek* **26**, p. 345 (1971) German translation: *MW-Forschungsbericht Nr.* 548.
2. W. von Baeckmann, N. Schmitz-Pranghe, 2nd Int. Congr. Intern. Extern. Protection of Pipes, Canterbury 1977, Paper J1.
3. W. Meyer and W. Schwenk, *Farbe + Lack* **85**, p. 179 (1979).
4. E. Brauns and W. Schwenk, *Stahl & Eisen* **86**, p. 1014 (1966).
5. K. Meyer and W. Schwenk, *Schiff & Hafen* **26**, p. 1062 (1974).
6. H. Hildebrand and W. Schwenk, *Werkstoffe u. Korrosion* **30**, p. 542 (1979).
7. W. Goering, E. Koesters and R. Muenster, this volume.
8. W. Schwenk, *3R-international* **15**, p. 389 (1976).
9. W. Schwenk, *gwf gas/erdgas* **118**, p. 7 (1977).
10. P. Pickelmann and H. Hildebrand, *gwf gas/erdgas* **122**, p. 45 (1981).

Some Aspects of Cathodic Electrodeposition of Epoxy Latexes as Corrosion Resistant Coatings

C.C. Ho,* A. Humayun, M.S. El-Aasser and J.W. Vanderhoff**

Introduction

The process of electrodeposition of organic coating materials onto metal objects has been known since the early thirties. However, it was not until the beginning of the sixties that the process was widely adopted in the automotive and appliance industries to give prime and one coat finishes to a variety of products. A wealth of information on the mechanism and development of this process has been reported since.[1] However, most of the electrodeposition processes known hitherto are anodic. In contrast, development of the cathodic process has been relatively slow, primarily due to the lack of resins and polymers suitable for the cathodic process. Apart from a handful of polymer systems designed to deposit at the cathode, as described in some recent patents, little basic information of the cathodic systems for coating metallic substrates has been reported in the literature. Thus, the present fundamental studies on the cathodic electrodeposition of epoxy latexes on steel substrates represent a natural extension of the extensive work done on conventional anodic systems. This paper describes factors affecting the deposition behavior of cationic latexes, as carried out in our laboratory.

Cathodic Electrodeposition

The electrocoating process is based on the migration of macro-ions under the influence of an electric field, and the subsequent discharge and deposition at the appropriate electrode, which is the metal object to be coated. Thus, cathodic electrodeposition involves immersing an electroconductive object in an aqueous dispersion of a polymeric material, which contains cationic groups such as quaternary ammonium ions. An electric current is passed through the dispersion between the anode and the metallic object as the cathode to cause deposition of the polymer on the cathode.

Anodic electrodeposition is essentially the opposite of cathodic electrodeposition, in that negatively charged polymers are deposited on the anode, but there are distinct performance differences between the two types of coatings. Thus, the anodic process is usually accompanied by electrochemical dissolution of the anodic workpiece, and the subsequent incorporation of this dissolved metal into the finished coating gives rise to reduced corrosion resistance and discoloration of the anodic coating. A variety of methods have been developed since and used in circumventing these shortcomings of the anodic process. By contrast, cathodically deposited films are inherently corrosion protective, and devoid of staining and discoloration, as a result of the alkaline nature of the cationic polymer and the passivity of the cathode. The inherent performance qualities of cationic films enable these coatings to surpass those properties obtained with anionic films, specifically: corrosion protection; detergent resistance; gloss retention; and resistance to staining.

The electrodeposition process is an effective and proven technology amenable to complete automation and easy control. The equipment required to operate a cathode electrocoating system is not substantially different from that required to operate an anionic electrocoating system. With the incorporation of ultrafiltration technique in a closed loop rinsing system, the rinsed off "drag out" materials can be recycled into the system, thus assuring nearly 100% ultilization of polymer materials. Minimal pollution of the effluent water is maintained and the need for flocculating equipment is precluded. Coupled with the utilization of aqueous systems containing minimal volatile additives, the electrodeposition process provides a virtually nonpolluting coating process with the least impact on the environment.

Most of the fundamental studies of electrodeposition have been carried out with polymer resins that are solubilized with acids or bases, rather than with dispersed discrete particles, such as latexes and dispersions. Solubilized polymer resins, when electrodeposited, coalesce easily to form a cohesive film which is resistant to the passage of current. However, the process of film formation by electrodeposition of dispersions is more complex, and may be the result of an entirely different mechanism. Recent work by Wessling et al[2] has demonstrated the use of tertiary sulfonium and quaternary ammonium salts to impart

---
*Department of Chemistry, University of Malaya, Kuala Lumpur, Malaysia
**Emulsion Polymer Institute, Lehigh University, Bethlehem, Pennsylvania

positive charges to latexes for cathodic electrodeposition. Cationic aqueous dispersions of epoxy resins have been prepared previously in this laboratory using a mixed emulsifier system such as hexadecyltrimethylammonium bromide and hexadecane;[3] the particles of these dispersions are small, and thus they are ideal for the studies of cathodic electrodeposition.

Materials and Experimental

The epoxy resin used for this work was Epon 1001 (a solid at room temperature with an epoxide equivalent of 450 to 550; Shell Chemical Company) and the curing agent was Emerez 1511 (a highly viscous condensation product of a polyamine and a dibasic acid, with an amine value of 230 to 246; Emery Industries Inc.). Latexes were prepared from these resins by direct emulsification, according to the recipe and procedure given previously.[3] Briefly, the method prescribed dissolving the resin in a water immiscible solvent, then mixing hexadecane with the polymer solution, adding the solution to an aqueous solution of the emulsifier system (hexadecyltrimethylammonium bromide) to form a crude emulsion, and homogenizing by using the Manton-Gaulin submicron disperser. The emulsions produced were stable when subjected to steam stripping, under vacuum, to remove the solvents.

The particle size distributions of the Epon 1001 and Emerez 1511 latexes were examined by transmission electron microscopy using the cold stage technique. The substrate used was cold rolled steel plates (Bethlehem Steel Corporation). These were cleaned with soap solution, degreased according to ASTM method D609-73 using trichloroethylene, rinsed with deionized water followed by acetone, and then kept in a desiccator until use.

The electrodeposition was carried out at room temperature in a rectangular Plexiglas cell of dimension 2.7 × 3.8 (h) × 9 cm. Two equal-size carbon rods were used as anodes and these were positioned at the ends of the cell. An untreated cold rolled steel plate of 2.2 cm width was used as the cathode, which could be raised or lowered into the cell midway between the anodes by a motor drive (at a speed of 3 ft per minute) to ensure smooth entry. The experimental set up is illustrated schematically in Figure 1. A D.C. power supply capable of producing up to a maximum of 750 volts and 10 amperes was used. The circuit also included a strip chart recorder, to plot current as a function of time. The area under the current time curve was integrated using a Carl Zeiss MOP 3 analyzer, to obtain the coulombs of charge passed. All depositions were conducted at constant applied voltage with live entry. Typically, the electrodeposition bath contained a 10% solids dispersion of pH 7.5 to 8.0 for epoxy latex, and 9.3 to 9.7 for the mixture of latexes. Each plate was electrocoated for 60 seconds. After deposition, the plate was dip washed with deionized water, baked at

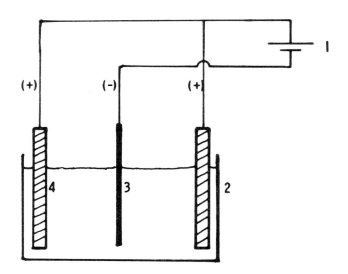

FIGURE 1 — Schematic illustration of the electrodeposition cell: (1) 750 volts, 10 amperes DC power supply; (2) rectangular plexiglass cell; (3) cold rolled steel cathode; (4) carbon anodes.

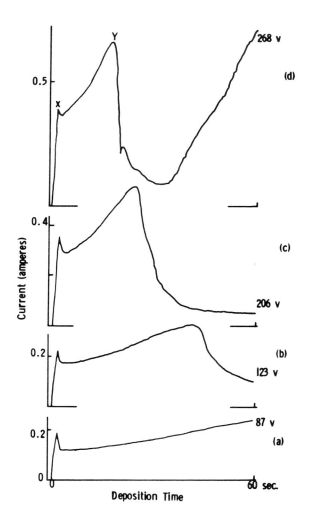

FIGURE 2 — Current time relationship for electrodeposition of Epon 1001 latex at various applied voltages.

90 C for 4 hours, and weighed to determine the amount of polymer deposited.

To study the morphology of the surface films after deposition, the plates were not baked, but were allowed to air dry, followed by vacuum drying at room temperature. The coated portion of the plate was sectioned, vacuum coated with a thin layer of palladium-gold alloy, and examined in the scanning electron microscope.

Results and Discussion

Particle size analysis of the Epon 1001 latex indicated that 95% of the particles were smaller than 110 nm and the number average particle size of the dispersion was 65 nm, whereas that of the Emerez 1511 latex was 20 to 30 nm.

Preliminary investigations of the electrokinetic properties of the epoxy latex showed that:

1. Electrophoretic mobility of the latex particles was constant down to a solids content of 10^{-3}% solids at pH 5.5 to 6.0, and then began to decrease on further dilution.

2. At fixed solids content, the mobility varied from $+3.5\,\mu m$ cm volt^{-1} second^{-1} at pH 3.5 to about $+2.4\,\mu m$ volt^{-1} second^{-1} at pH 10.7.

These results indicate that the emulsifier system used in the preparation of the epoxy dispersion can withstand dilution and variation in pH without any detrimental effect on the migration of the particles under the influence of an electric field.

Current-Time Voltage Relationship

Single-component system: epoxy latex. Typical plots of the variation of current with time at various applied voltages are shown in Figure 2. The deposition time was 60 seconds for each sample. The initial rapid rise in current within about one second (up to point X) indicates the completion of the immersion of the plate. The behavior of the current time curve beyond X depended on the magnitude of the applied voltage. At low voltage (Figure 2a), a conducting film was formed, as shown by the increasing current with deposition. Similar observations[2] have been reported for a quaternary ammonium stabilized water soluble polymer system deposited anodically at higher voltage of 200 volts. At intermediate voltage (Figure 2b), the current time curve was characterized by an initial rise in current, followed by a slow current cut off. The film retained considerable conductivity at the finish. At high voltage (Figure 2c), the current fell off rapidly after reaching a maximum value (point Y) to a low residual amperage, showing that the film was becoming insulating. At still higher voltage (Figure 2d), an uncontrolled increase in current was observed after a rapid current cutoff. This behavior was taken to indicate that film rupture had occurred. The rupture occurred at shorter times as voltage was increased. For example, the abrupt increase in current occurred 42 seconds after deposition started at 230 volts,

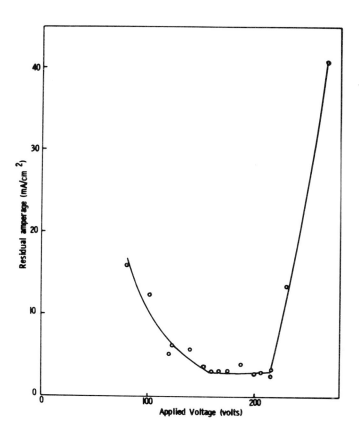

FIGURE 3 — The effect of applied voltage on residual amperage for Epon 1001 latex.

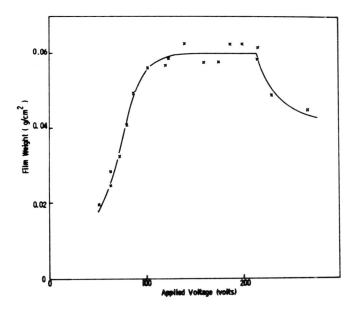

FIGURE 4 — Variation of film weight with applied voltage for Epon 1001 latex.

whereas it occurred after 33 seconds at 268 volts. These results show clearly that an optimum voltage range is required for rapid current cut off to give nonconducting films, and that film rupture is dependent upon the current voltage relationship of the deposited film.

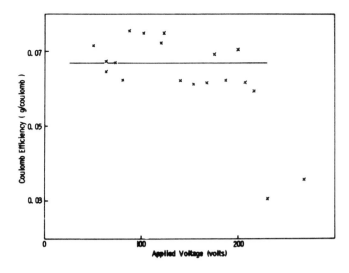

FIGURE 5 — Coulombic efficiency vs applied voltage for Epon 1001 latex.

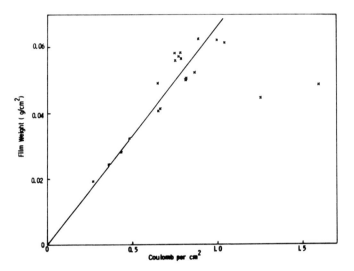

FIGURE 6 — Film weight vs coulombs for deposition at various applied voltage for Epon 1001 latex.

An illustration is given in Figure 3 in which the residual amperage (in terms of mAmp cm^{-2}) at the end of the deposition period of 60 seconds is given as a function of applied voltage. It can be seen that the residual amperage was at its lowest value of ca 3.0 mA cm^{-2} over the voltage range of 150 to 216 volts, beyond which a steep increase of residual amperage with a small increment of voltage was observed. Good nonconducting films were deposited within this voltage range and film rupture became evident above 216 volts. During film rupture, excessive bubbling was noted, and the bath temperature went up to 58 C at 268 volts. The appearance of excessive gassing was similar to boiling, and eventually the boiling action (of water) within the film did 'blow' part of the film off the substrate, as revealed by the morphology studies of the resulting deposited film and the lower film weight per cm^2.

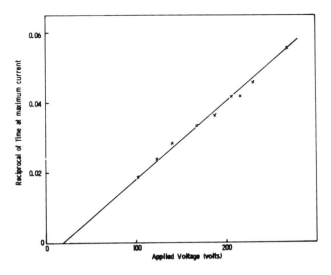

FIGURE 7 — Reciprocal of time at maximum current vs applied voltage for Epon 1001 latex.

The dependence of the deposition process on the applied voltage is also illustrated from a plot of film weight expressed as g cm^{-2} against the applied voltage in Figure 4. The film weight increased rapidly with increasing voltage up to ca 140 volts, when it reached a constant value of 60 mg cm^{-2}. It remained at this value with increasing voltage up to 216 volts. With another increase in the applied voltage, the film weight dropped abruptly. This result is explained by the loss of part of the deposited film to the bath, as a result of excessive bubble formation within the coating resembling boiling as outlined above. A decrease in film weight, at the onset of film rupture, has been reported previously by Mercouris et al[4] for anodic electrodeposition of polyacrylic resin on iron.

The coulombic efficiency, defined as the weight deposited per coulomb passed, was calculated for each applied voltage. The results are shown in Figure 5. Some scattering of data was noted, but the overall results suggest that the coulombic efficiency is essentially independent of applied voltage. An average value of 67.0 mg per coulomb for the coulombic efficiency was obtained with the omission of the two points for deposition at 230 and 268 volts. The coulombic efficiency could also be obtained by plotting film weight per cm^2 against the current passed (Figure 6). A straight line passing through the origin was obtained, showing that the weight deposited is directly proportional to the coulombs of charge passed for each deposition. The slope of the line gives the coulomb efficiency, which in this case is 66.5 mg per coulomb. Linearity of these data would be expected for a deposition mechanism by electrophoretic migration. Once again, the points obtained for deposition at 230 and 268 volts did not lie on the straight line. Lower film weights per cm^2 were obtained at these two applied voltages for the reasons already given. Similar observations and deductions have been made by Rheineck et al[5] for an anionic system.

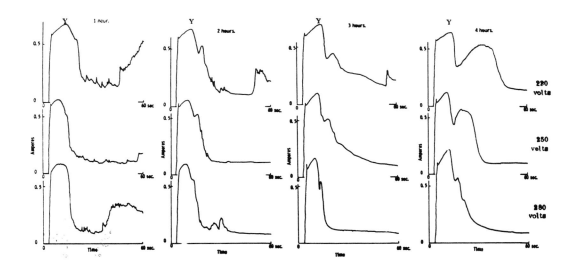

FIGURE 8 — Current time curves for stoichiometric mixtures of Epon 1001: Emerez 1511 (i.e., 2:1) aged for 1 hour, 2 hours, 3 hours, and 4 hours at room temperature before deposition at 220, 250, and 280 volts.

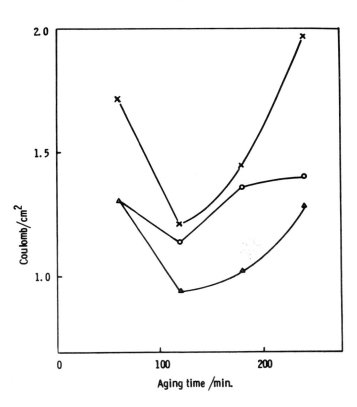

FIGURE 9 — Variation of coulomb of charge passed as a function of aging time before electrodeposition for stoichiometric mixtures of Epon 1001:Emerez 1511. Deposition at (x) 220 volts; (o) 250 volts and (Δ) 280 volts.

Further examination of the current time curves reveals that the time at which the current reaches a maximum value (Y), before it falls off, was strongly dependent upon the applied voltage. At high voltage, the value of Y was large and was reached earlier, and the current fell off almost vertically; whereas, at low voltage this maximum never developed and the film was conducting. If the reciprocal of the time at maximum current is plotted against the voltage, a linear relationship is obtained as shown in Figure 7. According to this plot, for a maximum to develop in the current time curve 60 seconds after deposition begins, an applied voltage of about 90 volts is required. This was found experimentally: below 90 volts, only conducting films were formed.

Two component system: epoxy resin-curing agent latex mixtures. The current time curves of the two component reactive system were much different from those of the epoxy latex. The shape of the curve depended, not only on the applied voltage, but also on the composition and the time of aging of the mixture before electrodeposition. It is therefore imperative to determine an optimum aging time prior to studying the other effects on the electrodeposition. Figure 8 shows the current time curves at various aging times for stoichiometric 2:1 mixtures of Epon 1001-Emerez 1511. Depositions at three different voltages for each aging time were done. For samples aged for one hour before deposition, there was a fairly rapid current cut off after the maximum current Y was reached; however the current then increased on further deposition, and the curve became 'wavy.' The same behavior was observed for all three deposition voltages used, 220, 250 and 280 volts. Increasing the aging time, to two hours, decreased the 'waviness' of the curves beyond Y and also decreased the residual current. At three hours aging time, the current time curves were smooth, and the residual current was reduced further. However, a small 'peak' appeared after Y. This peak seemed to be more marked at low applied voltage than at high voltage. Generally, the best current time curves were obtained for depositions at 280 volts, i.e., rapid current cut off with small residual amperage. Four hours aging time gave current time curves completely different from those of the shorter aging times. A broad maximum, as intense as the one at Y, ap-

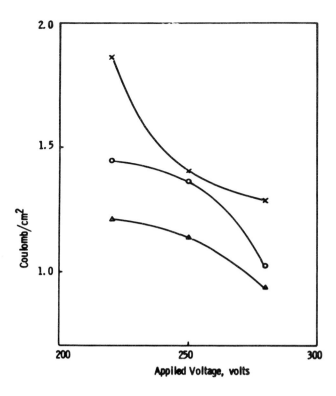

FIGURE 10 — Effect of applied voltage on coulomb/cm² for stoichiometric mixtures of Epon 1001:Emerez 1511. Mixtures aged for (Δ) 2 hours; (o) 3 hours; and (x) four hours before deposition.

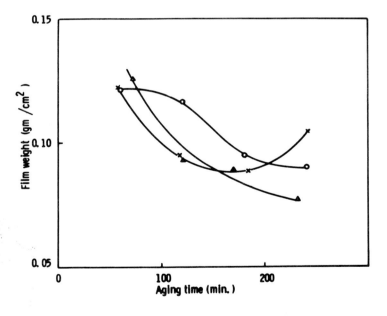

FIGURE 11 — Variation of film weight with aging time for stoichiometric mixtures (2:1) of Epon 1001:Emerez. Deposition at 220 volts (-x-), 250 volts (-o-) and 280 volts (-Δ-).

peared on further deposition. However, the residual current was comparatively low. It is further noted that, as the applied voltage increased, this second maximum was reduced in intensity and magnitude until, at 280 volts, only a small peak remained. This series of experiments was repeated for latex mixtures having Epon 1001-Emerez 1511 weight ratios of 3:1 and 4:1. Increasing the Epon 1001 weight fraction beyond stoichiometry with Emerez 1511 did not have any pronounced effect on the shape of the current time curves; essentially, the same coulombic curves were obtained as for the stoichiometric mixtures.

The effect of aging of the latex mixture was revealed more clearly when the variation of weight deposited per cm² and coulomb of charge passed per cm² were plotted against aging time. Figure 9 shows that the coulombs cm^{-2} decreased initially, with aging time, until a minimum was reached after about two hours aging time, and then increased again for longer aging times. An applied voltage of 280 volts gave the lowest values of coulomb cm^{-2} over the entire aging period. Higher values of coulomb cm^{-2} were obtained at lower voltages. This behaviour is illustrated clearly in Figure 10, which shows that samples aged for two hours gave the lowest values of coulombs cm^{-2} for all applied voltages. The film weight per cm² decreased with aging time for all applied voltages, except for 220 volts. At 220 volts, it decreased to a minimum, and then increased again (Figure 11). In fact, an inverse relationship seems to exist between the weight deposited per cm² and the aging time for deposition at 280 volts, for different ratios of Epon 1001 and Emerez 1511 as shown in Figure 12.

Based on these results, all subsequent electrodepositions were conducted using the stoichiometric mixtures of Epon 1001 and Emerez 1511 aged for 2¼ hours at room temperature before deposition. Figure 13 shows the current time curves of such mixtures deposited at various voltages. At low voltage (100 volts), the coating was conducting, as indicated by a monotonic increase in current with deposition time. At intermediate voltage (100 to 200 volts), a second broad maximum appeared after the

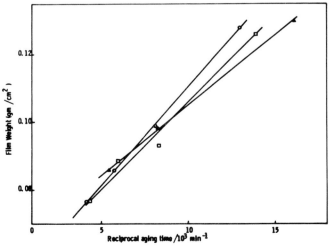

FIGURE 12 — Variation of film weight/cm² with reciprocal aging time. Weight ratio of Epon 1001:Emerez 1511 at (-□-) 2:1 (-o-) 3:1 and (-Δ-) 4:1. Mixtures deposited at 280 volts.

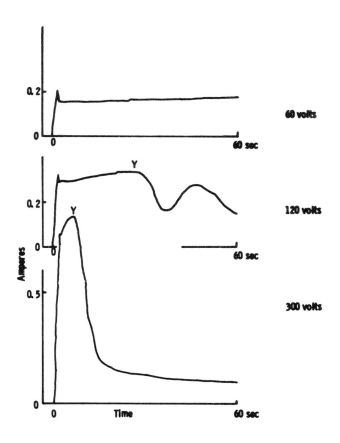

FIGURE 13 — Current time curves for stoichiometric mixtures of Epon 1001:Emerez 1511 aged for 2¼ hours at room temperature before deposition at various voltages.

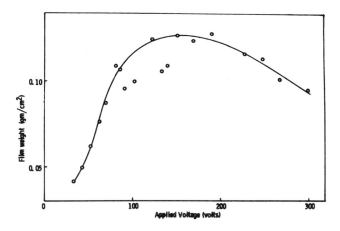

FIGURE 14 — The effect of applied voltage on the film weight for stoichiometric mixtures of Epon 1001:Emerez 1511 aged for 2¼ hours at room temperature before deposition.

first (Y), followed by a gradual decrease in current. At high voltage, a rapid current cut off was observed. Gassing at the cathode was noted. However, the phenomenon of film rupture was not as obvious as it was in the case of the single component system. From the above results, better insulating films were formed only at the highest voltage used.

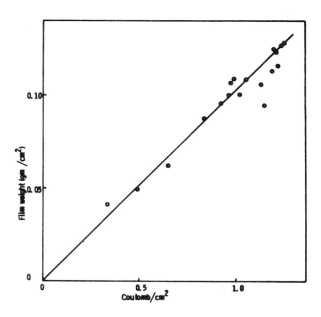

FIGURE 15 — The variation of film weight with coulomb/cm² for stoichiometric mixtures of Epon 1001: Emerez 1511 aged for 2¼ hours at room temperature before deposition.

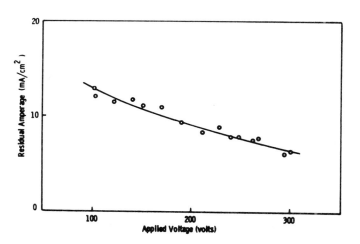

FIGURE 16 — Effect of applied voltage on residual amperage for stoichiometric mixtures of Epon 1001:Emerez 1511 aged for 2¼ hours before deposition.

The coating weight increased almost linearly with increasing applied voltage (Figure 14), until a maximum value of about 0.125 g cm^{-2} was reached; then it decreased gradually with an increase in applied voltage. The rate of weight increase with voltage was almost twice as fast as for the single component system. The maximum weight deposited was also about double that of the single component system. The weight per cm² was directly proportional to the coulombs of charge passed as indicated by a straight line passing through the origin shown in Figure 15. The slope of the line gives the coulombic efficiency of 103 mg per coulomb for this system.

The residual amperage for the mixture as a function of applied voltage is shown in Figure 16. In contrast

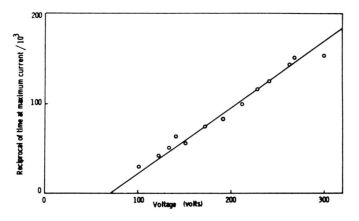

FIGURE 17 — Reciprocal of time at maximum current vs applied voltage for stoichiometric mixtures of Epon 1001:Emerez 1511 aged for 2¼ hours before deposition.

component system was also observed for the two component system as shown in Figure 17.

Nature and Morphology of Coating Films

Single-component system: epoxy resin latex. The unbaked films were white in color, opaque, bubbly, and rough in appearance. The water content of the wet film was about 20 to 30%. The deposited film was thick (circa 28 mil maximum) at high voltage. These thicker films showed smooth surface regions along the edges of the plate. Electron microscopy revealed a high degree of coalescence and fusion of the latex particles into a smooth film at these regions resembling the original bulk epoxy solids. The scanning electron micrographs of these are shown in Figure 18. In contrast, discrete aggregates in which individual latex particles are still discernible were observed in films deposited at low voltage (Figure 19). All depos-

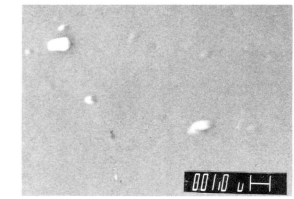

FIGURE 18 — Scanning electron micrographs of: a) electrodeposited coating of Epon 1001 latex showing smooth surface regions along edge of plate. Electrocoated at 200 volts. b) Solid Epon 1001 epoxy resin.

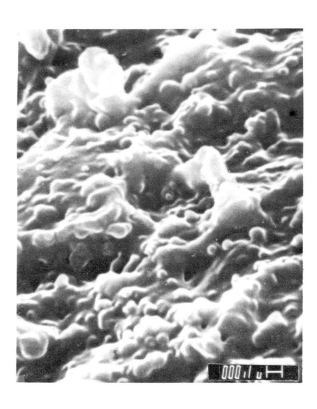

FIGURE 19 — Scanning electron micrograph of electrodeposited coating of Epon 1001 latex at 80 volts.

with the single component system, the residual amperage decreased continuously with increasing applied voltage up to 300 volts. It is, therefore, not possible to tell from this plot whether film rupture had occurred. On the other hand, a similar reciprocal relationship between the time to reach maximum current at Y and the applied voltage exhibited by the single

ited films cracked upon air drying at room temperature. Some cracked right through the deposit, revealing the metal substrate underneath. Latex films are known to crack at low humidity and low temperature below the film formation temperature of the latex polymer. The melting temperature of Epon 1001 is 65 to 75 C according to specifications of the manufacturer whereas the deposition and air drying were done at room temperature. Partial melting of the latex polymer did take place, however, at the high voltage

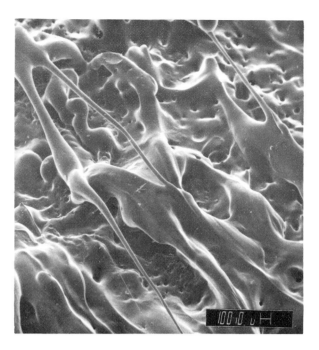

FIGURE 20 — Scanning electron micrograph of electrodeposited coating of Epon 1001 latex at 268 volts showing film rupture phenomenon.

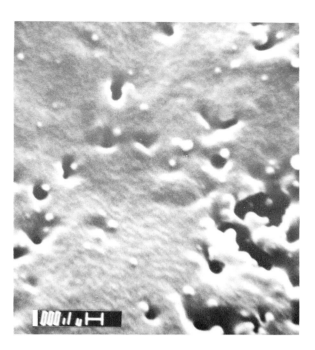

FIGURE 22 — Scanning electron micrograph of electrodeposit of Epon 1001 latex showing surface of deposit that was originally in contact with the substrate. Deposited at 140 volts.

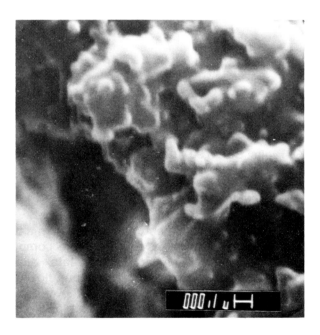

FIGURE 21 — Scanning electron micrograph of electrodeposited coating of Epon 1001 latex showing interior of coating. Applied voltage was 173 volts.

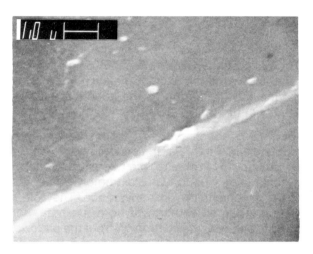

FIGURE 23 — Scanning electron micrograph of baked electrocoating of Epon 1001 latex. Deposited at 173 volts.

(e.g. 268 volts) where film rupture occurred. The top layer of the deposit was blown off leaving behind a fused layer adhered to the substrate with the badly torn surface clearly visible (Figure 20). The temperature of the water vapor within the deposit could easily have exceeded 100 C, causing partial melting of the epoxy and better adhesion to substrate.

The interior of the deposit (Figure 21), as observed through the cracked region, revealed that it still contained aggregates similar to those shown in Figure 19; however, the contours of the aggregates were less sharp, indicating that a certain degree of coalescence and fusion had occurred. The coating substrate (polymer-metal) interface revealed that a high degree of coalescence and fusion of the latex particles had already taken place, giving a smooth, flat interface; nevertheless, some embedded individual latex particles in the continuous film could still be seen (Figure 22). The adhesion of the film to the substrate was, however, poor.

Can Failures Still Occur When the Correct Coating (For a Given Environment) is Selected and Applied Properly?

Kenneth B. Tator*

The application of organic coatings is perhaps by far the most widely used method of corrosion control. When you consider the definition of corrosion, "Corrosion is a deterioration of a substance (usually a metal) or its properties because of a reaction with its environment," it can be seen that the role coatings play in corrosion protection is indeed vast! Coatings not only protect against corrosion, but often are considered an architecturally "finished" surface also designed to be aesthetically appealing. As such, coatings are highly visible, and are a part of our everyday life as they protect bridges, ships, steel mills, chemical factories, automobiles, stores, and even our houses and lawn furniture from the ravages of the environment. Much research and development work has gone into the development of the sophisticated coatings on the marketplace today. Vinyls, epoxies, chlorinated rubbers, zinc riches, urethanes, etc., are all generic materials that involve complex, and sometimes not understood, chemical reactions. However, for the most part, raw materials suppliers and coating manufacturers have reduced these complications to a working product that is taken for granted by many specifiers and users. There is much published information available from technical societies, raw material suppliers, and coating manufacturers that states, for example, that "vinyls are highly resistant to strong acids and alkalies, waters, alcohols, aliphatic hydrocarbons, fats and oils — but are swelled by aromatic hydrocarbons"; that "polyester coatings are tile-like, hard, nonporous, and durable, capable of withstanding dilute alkalies, organic solvents, hot water, and a high order of nuclear radiation — as well as mechanical abuse, impact, and abrasion"; that "amino resins (urea- and melamine- formaldehyde compounds) promote a fast bake curing, and have prolonged resistance to elevated temperatures, superior alkali resistance, maximum hardness and outdoor durability." Crosslinked epoxy resins are noted for their "outstanding adhesion, flexibility, and chemical resistance (especially to alkalies), leading to their use as industrial maintenance paints, architectural paints and enamels, spar varnishes for marine applications, and coatings for floors, furniture, concrete, and metal." Similar good words are written for virtually all generic classes of coatings, expounding on their wide resistance to chemicals, their toughness and durability, and special use for given situations. In fact, such information is generally so widespread and readily available that for most services, the appropriate selection of a coating for corrosion protection is not difficult. Coating salesmen, relying on their experience and familiarity with their product line, are excellent sources for advice in coating selection. The United States Government has an excellent series of coating formulations for use on a variety of surfaces in numerous environments, as well as some excellent standard test methods. Other governments, notably the Canadian, British, German, and Swedish, also have excellent source material for coating specifiers. Lastly, there is always the age old question, "Shall we call in a specialist — or louse it up ourselves?" Consultants are available to assist when problems arise.

Because of the high technology of many coating materials available today, there has been increasing awareness of the need for proper coating application. Principally as a result of nuclear quality assurance and its spinoff into other industries, quality coating application and knowledgeable competent inspection agencies are increasingly used to apply coatings, and to assure that the application is proper. In fact, there is a trend developing for some coating applicators to specialize in the application of coatings for use as tank linings, immersion, and other specialized services, such as nuclear plants, shipboard cargo, and ballast tanks, etc. Correspondingly, other applicators specialize solely in the application of baked phenolics or epoxies, zinc rich or urethane coatings. These applicators exhibit a great deal of technical expertise and familiarity with the coating material being applied, the material being protected, or the environment of intended use.

Thus, while one usually pays more (often much more), it is increasingly possible to select the correct coating for an environment, and have it "perfectly" applied.

But what if such a coating should fail? Is it then the correct coating, or was it really properly applied? The following four case histories illustrate this paradox.

---

*KTA-Tator Associates, Inc., Corapolis, Pennsylvania

The Case of The Blistered Coating

When coating in confined spaces, the flash point of solvents in a coating is always a major consideration. Ideally, a coating with little or no solvent is to be preferred. If, at the same time, the coating must resist fresh and salt water immersion, as well as diesel and fuel oil, selection is even more limited. Finally, it must be possible to apply using conventional equipment (brush, spray, and roller) by a relatively unskilled work force that is supervised intermittently. These requirements faced the coating specifier when writing the specification for coating the interior of compensating fuel oil storage tanks at a large naval construction facility. As these tanks also comprised the outer hull of an ecological ship design, there were numerous small compartments (particularly at the double bottoms) where ventilation would be poor, and workspace cramped.

Thus, the choice of a ketimine cured epoxy coating system was ideal. The coating is better than 95% solids, and is spray applied easily. Furthermore, it cures to a hard, glossy, resistant epoxy coating with outstanding resistance to water and oil.

The shipyard did their own application, and inspection and quality control was tight. All surfaces were abrasive blast cleaned to white metal; after cleaning and inspection, the coating was applied. Contamination between coats was kept at a minimum, and thicknesses were those recommended in the coating manufacturer's literature. A total of three coats was applied to all internal surfaces. Ambient temperatures ranged from the mid sixties to the high eighties, and the tanks were well ventilated during and after coating. Prior to filling the tanks, the coating had cured in some areas as long as four or five months, in others, as little as two weeks or so. Regardless, everything seemed just right.

However, after the shake down cruise, some blisters in the coating of one tank were observed. Upon draining other tanks, it was seen that blisters were widespread among all tanks examined. They were, for the most part, relatively small, and usually occurred between coats. However, in a number of areas, the blisters were larger, and went down to bare metal. When opened, they were filled with liquid; however, there was no corrosion of the underlying steel. In fact, the steel surface appeared just as bright as the day it was blasted.

The owner of the ship was understandably unhappy about the blistering. There was particular concern because the blisters had occurred so soon (the expected life of the coating was estimated at 10 to 15 years). Because of the potentially large cost of recoating the tanks, the writer was asked to investigate and determine the cause of failure. Upon examination, it was found that there were no obvious application deficiencies (the surface preparation was excellent, with the required cleanliness and anchor pattern). Failure was occuring in tanks that were painted both in winter and summer, so that temperature was ruled out as the cause. Both thick and thin areas showed signs of blistering (conversely, in some areas, there was no failure of thick and thin areas) so that coating thickness was ruled out as the cause, also. The only potential pattern observed was that occasionally some of the more sheltered areas, or pockets, within a tank had more blisters than the more exposed surfaces. Also, last minute touch up, while comprising only a small percentage of the total area, generally showed the most blistering. These observations, plus knowledge of the ketimine curing mechanism, unraveled the mystery.

Chemistry of a Ketimine Epoxy

In this case, a standard epichlorhydrin-bisphenol A epoxy resin was used, of the general formulation:

$$\underset{H}{\overset{H}{C}}-\underset{H}{\overset{H}{C}}-\underset{H}{\overset{H}{C}}-O-\underset{H-C-H}{\overset{H-C-H}{\bigcirc}}-O-\underset{HOHH}{\overset{HHH}{C-C-C}}-O-\underset{H-C-H}{\overset{H-C-H}{\bigcirc}}-O-\underset{H}{\overset{HHH}{C-C-C}}$$

n = 0 to 15

The crosslinking reaction mechanisms of this epoxy are well known and thoroughly described in the literature.

Ketimine, on the other hand, is less well known. Ketimine is essentially a ketone blocked polyamine, which under dry conditions reacts slowly, if at all, with the epoxy. However, in humid conditions the ketimine decomposes to form an amine and a ketone:

$$\underset{R_2}{\overset{R_1}{C}}=N-R-N=\underset{R_2}{\overset{R_1}{C}} + 2H_2O \rightarrow \underset{H}{\overset{H}{N}}-R-\underset{H}{\overset{H}{N}} + 2R_1-\overset{O}{\overset{\|}{C}}-R_2$$

Ketimine + Moisture → Polyamine + Ketone
R, R_1 and R_2 are Alkyl Groups

The polyamine produced by the decomposition of the ketimine then reacts with the epoxy, in the normal manner, to form a crosslinked, cured film.

However, as atmospheric moisture is required, there is a tendency for the coating to surface cure, and entrap and retard subsequent moisture access to the underlying coating. This same surface curing retards solvent evaporation of the ketone by-product. This evaporation is further retarded by the polar nature of the resin system. Even though only a small amount of the ketone was retained by the coating, it was enough to cause the blistering failure.

Essentially, the failure mechanism was one of hydrogen bonding attraction to the water molecule, resulting in a permeation of the water into the coating to sites of retained solvent. The affinity was that of the carbonyl group to the water hydrogen:

Hydrogen bonding attraction to the electronegative carbonyl oxygen

Ketone   Water

That this was the failure mechanism was also confirmed by the observance of bright clean metal beneath blisters that originated at the steel interface. Such an observation can occur only if one of the elements necessary for ferrous corrosion is not present. In this particular case, free oxygen (O_2), present as dissolved oxygen in the ballast water, did not penetrate through the coating because of its relatively larger molecular size. Accordingly, within the blister on the blast cleaned steel, there was only water with trace amounts of the ketone solvent. Since oxygen was not present, rust could not form, and the steel remained bright and shiny.

Whose fault was it? That question is still unresolved. However, it was decided not to replace the coating, on the legal premise that no corrosion failure had occurred; therefore, the coating was still protecting and doing its job. To date, there have been no new blisters, and those blisters already formed have not appeared to grow.

It should be noted that this problem is not unique with ketimine cured epoxies. Any highly polar resin system, such as vinyl, chlorinated rubber, epoxy, or other synthetic resins that have a tendency to retain solvents, may blister in a similar manner when placed in water immersion service. This tendency is considerably higher as the purity of the water increases, and is a real problem in nuclear power plants where the use of deionized water is commonplace.

The "Peeling" Zinc Rich Primer

Within the last five to ten years, there has been an overwhelming abundance of literature and technical papers extolling the virtues of zinc rich primers. The protective advantages of the galvanic nature of the zinc-steel coupling have led many specifiers to use zinc rich primers whenever possible. Likewise, topcoating zinc rich primers, when properly done, seals and protects the zinc, minimizes the rate of galvanic sacrifice, and synergistrically increases the protective life of the system. In this case, the primer and topcoats were applied in strict accordance with the manufacturer's literature. This particular problem was investigated by the author on two separate occasions: first, in Florida, with a peeling of the zinc rich primer beneath a water base acrylic topcoat; and later, in West Virginia, with the zinc rich primer peeling when topcoated with a polyamide cured epoxy.

As might be imagined, the owner and the general contractor were quite concerned because if extensive repair was required, the resulting time delays would set back an already tight construction schedule. Perhaps more importantly, however, the field painting contractor balked when he saw the failure, and:

1. Refused to touch up or repair the existing damage unless paid an "extra."
2. Refused to guarantee the performance or longevity of any coatings applied over the failing coating.

Upon investigation, it was soon determined that the zinc rich primer (a solvent base, ethyl silicate inorganic zinc) had been applied in a fabricating shop to a properly cleaned steel surface, but had been topcoated as soon as possible. The topcoat application, although in conformance with the minimum topcoating time intervals, temperature, and humidity intervals permitted by the manufacturer's literature, had been done as soon as possible after the application of the zinc rich primer to hasten the flow of the steel through the fabricating shop. In both cases, the manufacturer's literature and advertising indicated the entire coating system (primer and topcoat) could be applied in one 8 hour work day. This was done for the steel supplied to both job sites, and within a few days after coating, the steel was shipped from the fabricating shop to the erection site. On arrival, the steel looked to be in good condition, except for what was considered, at the time, slightly excessive abrasion damage during transit. However, upon erection, it soon became apparent that the coating had poor abrasion and handling resistance. Once a score or cut had been made through the coating, it would readily peel and lift from the original damage point in strips, sometimes as large as one foot or more in length. Shortly after delamination occurred, rusting commenced, and by the time the field painter was called to the job, the amount of field touch up was far in excess of that originally estimated. Accordingly, the painter refused to begin work.

Investigation revealed that, indeed, the field touch up work required was far in excess of that normally encountered and expected with an inorganic zinc rich coating system. Furthermore, even though over six weeks had elapsed since the onset of the problem, large areas of the coating system could still be delaminated from the underlying steel. The delamination occurred immediately adjacent to the steel/zinc rich primer interface, with only a slight zinc dust residue remaining within the anchor pattern of the blast cleaned metal. For an instant, immediately after cutting into and delaminating the coating, one could discern a strong alcohol smell. The physical appearance of the failure, the alcohol smell, additional observations, and laboratory work led ultimately to the conclusion that the zinc rich primer was not cured sufficiently prior to topcoating; the failure occurred as a direct result of this.

In order to understand the mechanism of the failure, one must have a brief familiarity with solvent base self curing ethyl silicate coatings. These are the most widely used inorganic zinc rich primers in the United States, and usually do not fail by peeling. The zinc rich primer consists of finely divided metallic zinc dust, and perhaps minor amounts of lead, iron oxide, or other pigments disbursed in a partially hydrolyzed ethyl silicate binder. This binder, generally an oily yellowish liquid, must, after application of the paint, undergo further hydrolysis with moisture from the air to achieve its final properties. Thus, while the inorganic zinc rich primer becomes dry and hard, often within twenty minutes or so after application (and at that time is able to resist mild weathering and even rain showers), it still is not cured. While the time required for curing will vary depending upon the temperature, relative humidity, and other atmospheric conditions, recent work (including some by the author) indicates that the time required for complete curing is much longer than many people had thought necessary.

The curing involves a reaction with moisture from the atmosphere to complete the hydrolysis of the ethyl silicate, liberating ethanol; premature topcoating prevents access of the atmospheric moisture required to complete the cure. The ethyl silicate zinc rich primer curing reaction is:[1]

$$2 \begin{bmatrix} OR \\ RO-Si-OR \\ OR \end{bmatrix} + H_2O \rightarrow RO-\underset{OR}{\overset{OR}{Si}}-O-\underset{OR}{\overset{OR}{Si}}-OR + 2\,ROH \text{ (Ethyl Alcohol)}$$

tetra ethyl ortho silicate ($R = C_2H_5$)   Partially Hydrolyzed Teos

$$n \begin{bmatrix} OR & OR \\ RO-Si-O-Si-OR \\ OR & OR \end{bmatrix} + n\,H_2O \text{ (Atmospheric Moisture)} \rightarrow \begin{bmatrix} \cdots-Si-O-Si-O-Si-\cdots \\ O \quad O \quad O \\ \cdots-O-Si-O-Si-O-Si-O-\cdots \\ O \quad O \quad O \\ \cdots-O-Si-O-Si-O-Si-\cdots \\ O \quad O \quad O \end{bmatrix}_n + n\,ROH \text{ (Ethyl Alcohol)}$$

Crosslinked Silicate Binder

This explanation of the cause of the failure was accepted ultimately by all concerned parties; consequently, the coating manufacturer made satisfactory financial arrangements with both the shop fabricator, who supplied the steel, and his field painting subcontractor. Shortly afterward, promotional activities and literatuare regarding the rapid topcoatability of certain of their zinc rich primers disappeared from the suppliers' coating product manuals and advertising literature.

The Case of The Tank Lining That Didn't

This case involves the massive and extreme failure of a glass filled polyester coating applied to the interior of some large outside storage tanks used in a drug manufacturing process. The case was litigated, and, at the conclusion of the trial, the coating manufacturer was declared liable on technical grounds.

---

[1]Tator, Kenneth B., from an article in the *American Painting Contractor;* **55,** No. 8; August 1978, pp. 10-13.

However, in the author's opinion, although the coating manufacturer misunderstood and misrepresented his coating product, the real culprit was the person(s) who authorized use of this particular lining system inside the tank.

The tanks were "standard" carbon steel storage tanks similar in size and configuration to the "fixed roof" tanks commonly seen in petroleum storage yards. The tanks were used to hold aluminum and magnesium hydroxides, which form the basis of a well known stomach antacid medicine.

A coating company had just developed a new (at that time) product that, they said, was similar to sprayed on fiberglass. Their laboratory testing indicated that the coating had good resistance to alkalis and water. Although the coating had never been used before as a lining for these type of tanks, and in fact had only limited field use, the fiberglass simile appealed to the buyer. As a result, all the tanks were sand blasted to white metal and coated with approximately 60 mils of the new coating.

The coating was applied to a properly cleaned surface, cured out, and appeared excellent in all respects. Shortly afterward, the tanks were placed in service. Within a month, restricted flow through the tank piping system resulted in the partial draining and inspection of one of the tanks. The lining was found to be hanging in large sheets (maximum dimension of 5 or 6 feet or more) off the walls and tank ceiling. The discovery that the five other tanks in the tank farm were affected similarly led to near pandemonium, as the contents of the tank had to be discarded, all medicine manufactured from product that had been in the tanks had to be recalled, and the plant shut down until relining of the tanks could be done.

The coating resin was an isophthalic polyester resin.

$$\underset{H}{\overset{O}{OH-C}}\overset{O}{=}\underset{H}{\overset{}{C}}-\overset{O}{C}-\left[-\underset{H}{\overset{H}{O}}-\overset{O}{C}-\underset{H}{\overset{}{O}}-\overset{}{\bigcirc}-\overset{O}{C}-\underset{H}{\overset{H}{O}}-\overset{O}{C}-\underset{H}{\overset{H}{C}}=\overset{O}{C}-\overset{O}{C}-\right]_n-\underset{H}{\overset{}{O}}-\overset{O}{C}-\underset{H}{\overset{H}{C}}-OH$$

n = 3 to 6

This particular resin system had only been adapted recently for spray applied coating use, and all testing, thus far, both in the laboratory and preliminary field, revealed that it should be entirely satisfactory. In these tests, the resin had excellent acid and solvent resistance, and although slightly swelled by alkalis, did not fail or otherwise show any signs of distress. As the resin system had had extensive field usage as an FRP laminate, it was almost a foregone conclusion that the resins used for lining the drug manufacturer's tanks should be successful. However, reconstruction of the failure mechanism revealed that the polyester, despite the glass flake reinforcement, exhibited considerable shrinkage upon application. Thus, while adhesion at the time of application was satisfactory, the coating film itself was under considerable internal stress. Furthermore, because of the small size of laboratory test panels, the effect of the stress had not been duplicated in the labortory testing.

Continuous immersion in the alkaline solutions resulted in penetration of the alkali with resultant softening and swelling of the coating system. The chemistry of the softening reaction is the same as for saponification, except steric hindrance of adjacent pendant side chains prevents total failure, but softening and swelling still occurs. The softening reaction for saponification is:

$$R-\overset{O}{\underset{\|}{C}}-O-R' + MOH \rightarrow R-\overset{O}{\underset{\|}{C}}-OM + R'OH$$

Ester group + Metal Hydroxide → Soap + Alcohol

R and R' are Alkyl or Aryl Groups

Aiding this chemical attack (it was later discovered) were localized areas of the coating where curing had not taken place properly, probably because of an improper mixture of the polyester: MEK-peroxide catalyst mixture. When all of this was combined with mechanical flexing of the tanks' sides and bottom during the filling and draining operations, as well as annual and daily thermal expansion and contraction — the failure was inevitable!

In retrospect, the coating manufacturer realized the error of its ways shortly after the first signs of failure were discovered, and the coating was quietly removed from the market soon thereafter.

The Case of The "Cracking" Coating

In recent years, one of the real challenges to the coating industry has been the selection of an appropriate coating to line and protect scrubbers, scrubber ducts, bypass ducts, and carbon steel chimney liners from corrosion. An alternative to coatings is the use of rubber linings (scrubber only), acid resistant brick, or stainless steel and other high cost materials of construction. However, the protection of carbon steel by organic coatings in these services has been highly desirable. Numerous coating systems have been tried, with varying degrees of success. However, the extremely strong acids resulting from acid dew point condensations of flu gases, as well as high temperatures in the bypass condition, or moderate temperatures with high moisture content in the scrubbed condition, has presented an overly severe environment for most coatings. Conventional vinyls, epoxies, chlorinated rubbers, etc., do not stand a chance, and isophthalic polyesters only fare somewhat better. Fluoropolymers, while resistant, are too expensive (applied cost of 35 to $50 per square foot).

Thus, when a resin becomes available that had a dry heat resistance of 300 F, resisted a 70% sulfuric acid concentration at temperatures approaching 200 F, and could be spray applied, it seemed the ideal coating resin. Furthermore, this resin, a vinyl ester, had been used extensively in the plastics industry, and its characteristics were reasonably well known. The chemical structure of the vinyl ester is (contrast with an isophthalic polyester shown previously).

[Chemical structure diagram of vinyl ester]

$n = 1$ or 2

Chemical attack on both of these types of resins occurs through the hydrolysis of ester groups, or the splitting of unsaturated carbon to carbon double bonds that remain unreacted during polymerization. In the vinyl ester structure, note that the molecule is terminated by ester groups adjacent to carbon unsaturation. Such an arrangement makes the carbon double bonds extremely reactive, and during polymerization, the double bond is split to form a more stable saturated link. The polyester, on the other hand, is not terminated by ester groups, and during polymerization, it is more likely that unsaturated carbon double bonds will remain in the main chain. Furthermore, there are many more ester groups comprising the chain backbone of the polyester, making it more susceptible to acid cleavage, or alkaline hydrolysis. Accordingly, as polyesters had been used with some success in scrubbers and chimney linings, it was felt that the vinyl ester should provide an ideal, relatively low cost protective coating. In fact, *in situ* tests indicated that this was the case, as coatings formulated from the resin clearly outperformed all competitor products when placed on test panels and exposed in chimneys, ducts, and scrubbers.

A three package coating was formulated by two or three innovative coating manufacturers, and offered for corrosion protection in severe environments. The formulation is similar to that for many polyesters, and involves: the separate packaging of the vinyl ester resin and additives to retard air inhibition; a pigment; and, in the third package, a peroxide blend for free radical polymeric initiation. After testing, the coating was applied to steel vessels in a number of environments. Two situations the author is personally familiar with: that of tanks in a reactivated carbon waste water treatment facility, and a chimney liner and scrubber ducting in a coal burning power plant. In both cases, the coating was applied under rigorous independent inspection, with surveillance and advice by the coating manufacturer. The coating system was deemed the best available for the environment, and the coating application was done in strict accordance with the manufacturer's application instructions under controlled ambient conditions. Again, everything seemed perfect!

However, within two years of service time, massive

failure had occurred in both the waste water treatment tanks and the scrubber ducting. It should be noted that while service conditions were somewhat different (acid water immersion within the tanks, and wet acid water vapor within the ducting), failure was similar in both cases, manifest as a flaking and peeling down to the underlying blast cleaned steel. There was, however, no evidence of chemical attack or deterioration of the coating material itself.

The reason for the failure? Well, there is not agreement on this either. As to date, there is no definitive reason as to why the coating system should have failed. However, it is the author's opinion that the failure resulted essentially from accumulations of stress within, and applied to the coating that ultimately caused its demise. The fact that there is no evidence of chemical deterioration of the coating, and the underlying steel after the coating disbonds is clean, bright, and unaffected by the environment is a further argument against chemical attack or permeation. The perfectly clean disbonding, and the almost mirror like replication of the anchor pattern on the underside of the disbonded paint chips would indicate that initial adhesion, at the time of application, was poor. However, this is known to be untrue, as the coating had been personally tested on some of the steel by the author at the time of application, and the coating was known to be very tightly adherent. Later, coating at the same area failed extensively by disbonding. Thus, it would appear that some force or stress would cause a uniform disbonding of large sheets of the coating. This is difficult to postulate unless it is considered that moisture from the environment did indeed permeate the coating (perhaps affected to some degree by hydroxyl groups in the resin). Such moisture take up would swell, and slightly expand, the coating and reduce polar adhesion forces at the coating/steel interface.

If the scrubber or tank was taken out of service, then the environment would no longer be moisture saturated, and the coating could dry out. The coating would shrink, due to moisture loss, with an initial surface drying, and ultimately, a shrinkage from the coating surface down. Such a shrinkage would diminish adhesion, and induce shrinkage stresses on the coating. If such a condition were repeated a number of times, it is likely, in the author's opinion, that the observed disbonding failure would result. In fact, because of concern about the aggressiveness of the environment, and uncertainty of the coating's ability to protect, the equipment (in both cases) had been taken out of service and inspected. The scrubber duct, although in what was considered a less severe moisture environment, was cycled more often because other scrubbers would often take up the load, allowing any one scrubber to be out of service for inspection, maintenance, or repair.

It was felt that the act of taking the unit out of service for inspection caused the disbonding failure. These were sufficient to cause the failure: the additional shrinkage stresses, coupled with diminished adhesion at the coating/steel interface; in conjunction with all other stresses on the coating, such as those resulting from curing of the resin (vinyl esters normally shrink considerably upon curing); different coefficients of thermal expansion/contraction of the coating versus the steel; and any mechanical or vibrational stresses.

In essence then, a combination of inherent physical factors, and the sensitivity of coating resin to them, caused the failure. Such a failure mode is unique in coatings; for the most part, a properly applied coating adheres well, and failure or deterioration results from environmental attack or deterioration.

The fact that coating failure occurred in each of the above cases precludes the possibility of a "perfectly applied coating that was appropriate for the environment." However, at the time of selection, there was every reason to believe that the coating was the best available, and at the time of application, everything was proper. Thus while most coating failures can be explained by application deficiencies or improper coating selection or specification, these failures do not fit into that category. Rather, an unanticipated anomaly, discovered during the post mortem, was believed responsible for the failure.

Film Application Method as Related to Corrosion Control

*Herbert J. Schmidt, Jr.**

Formulation Latitude are key words of corrosion control in the manufacturing and application of protective coatings. But the best and most inert coating will not provide the desired life expectancy if it is applied to an improperly cleaned and prepared surface. Before addressing ourselves on how to achieve the most advantageous method of controlling corrosion by the selection and application of a coating film, let us first consider briefly surface preparation.

The first consideration of any protective coating application is to insure that the substrate has been properly prepared. Depending on the protective coating selected, the substrate should be warmed above the ambient dew point, and the surface to be coated should be cleaned of all deleterious material. The optimum etch or anchor pattern for the best results was believed previously to require a profile depth of approximately one-third of the coating thickness. Recently, it has been concluded that several factors, such as texture of the surface, degree of cleaning and type of abrasive, may be of greater importance than the maximum profile depth, as long as a certain critical thickness of the coating over the peaks of the anchor pattern is attained. Evaluation of surface cleaning methods is discussed in another paper in this book. A clear and definitive discussion of surface preparation was presented at the Plenary Lecture during NACE, CORROSION/80 by A.N. McKelvie of the Paint Research Association, Teddington, England. This paper is entitled — "Can Coating Successfully Protect Steel, What Are the Ingredients of Success". It was reprinted in the May, 1980 issue of Materials Performance.

If a decorative fusion bonded powder coating is desired and/or a thin film coating from 0.5 mil (0.013mm) to a few mils thick is desired, three to five detergent wash and rinse cycles followed by an iron phosphate treatment (to improve powder adhesion) is recommended.

Consideration should also be given to three important criteria for a successful, long life protective coating application. They all relate to the coating material: 1) selection; 2) application; 3) cure.

Selecting the optimum coating material, for a specific environment, to control corrosion is no easy task. Many times, chemical resistant coatings are not flexible. Loading or filling chemical resistant coatings may produce excellent laboratory results of the coated test panels, but they may not be suitable for field or job conditions. On the other side of the spectrum, the most flexible protective coating normally has limited chemical resistance. Therefore, Formulation Latitude is an important key to a successful protective coating application.

The method of coating film application will, in most instances, dictate the performance and success of the protective coating film. When powder or fusion bonded coating is applied, essentially the same components of the powder coating film are found after application as were originally formulated into the powder coating.

Because of the necessity of liquifying solids to permit a liquid coating to be applied, the solvent in the coating has to evaporate. Because of these and environmental considerations, coating suppliers are reformulating many of their coatings to high solids. A high solids coating usually is described as a coating that is a solvent system containing over 70% solids. High solids coatings consist of two component coatings, waterborn emulsions and waterborn solution type coatings, among others. It is estimated that by 1990, the average volume solids content of a liquid coating will be at least 50% higher, as compared to 1980.

Protective coating films must be properly cured to achieve the results that were planned by the chemist or coating formulator. Improper cure can result in entrapping solvents in a liquid coating film or producing a sponge like fusion bonded coating film. Each of these conditions will result in poor bond to the substrate and will cause early failure of the coating. When the cure of a coating film is over accelerated, the results are brittleness, charring, poor bond, and degradation of the coating film.

Many articles and papers have been written on surface preparation and the selection of protective coatings. There is significantly less information on coating application and how properly to cure a coating film.

The first consideration in curing a protective coating film is to insure that the coating film is in-

---
*MCP Facilities Corporation, Glen Head, New York

timately bonded to the substrate. The substrate should be of a temperature and cleanliness that permits the coating to flow into and fill all the interstices. The coating film should cure from the substrate outward through the coating film to its surface. Before the coating film surface develops a skin, the underlying coating should cure by turning into a solid mass, or be in that process. This can be achieved in different ways, but the most common practice is to have the substrate at the coating interface warmed or heated sufficiently to permit the coating film to cure from the substrate outward through the coating film. Powder coatings go through a liquid or gel phase from the powder form to a solid homogeneous mass, which becomes the resultant coating film. The surface of a cured coating film should present a highly resistant chemical and physical barrier to the outside environment.

Often, the choice of a protective coating material depends on the nature of the material to be coated. Assuming that the material or item to be coated is of a shape, weight, and size that would permit it to be coated either by a liquid coating system or a powder or fusion bonded coating system, the choice, in almost every instance, would have to be made on which coating system is most cost effective.

Intercoat adhesion of multiple coats of liquid coatings has been a problem since the advent of these coatings. Today, there are liquid coatings formulated to be applied in a multi-pass, one coat application which results in a high build film. A powder coating can be applied to a cured film of 10 mils. A market survey has shown that the total application of protective coatings is increasing at the rate of approximately 5% per year. Liquid coatings, which now command the major market share, are projected to increase at the rate of 3 to 4% per year, while powder or fusion bonded coatings are projected to increase at approximately 15% per year from the present small market share.

Liquid coatings have been used for many years. As with powder coatings, liquid coatings should be selected properly and applied to produce long life coatings finishes. To achieve optimum results using liquid coatings, if a liquid coating can be applied using electrostatic spray equipment, the results are often desirable. In many instances, airless spray equipment has many advantages over conventional liquid spray equipment. In line heaters can reduce or eliminate the need for additional solvent in the application of liquid coatings. Excellent results can be obtained with a liquid coating application by strict attention to achieving maximum coating deposition with uniform thickness and the minimal use of solvents.

Powder coatings offer energy savings and the reduction or elimination of pollutants. Pollution control laws and worker safety regulations are causing organic coating finishers and applicators to evaluate their present technology and application methods and consider alternatives. High solids liquid coatings, water base liquid coatings and fusion bonded powder coatings offer attractive alternatives.

Powder coatings are either thermosetting or thermoplastic. Thermosetting powders are cured and permanently fixed upon exposure to heat, and account for approximately 90% of the powders used to finish steel. These include epoxies, most polyesters, acrylics, and polyurethanes. Thermoplastic powders can be remelted and include vinyl, nylon, polypropylene, and some acrylics and polysters. Significant progress is being made to reduce the energy cost to cure powder coatings. Today's standard powders cure in 8 minutes at 350 F (177 C). Five years ago, thermosetting powders were cured 15 minutes at 400 to 450 F (204 to 232 C). There are slow cure powders available that will cure at 290 F (143 C), and fast cure powders which will cure sufficiently, in 10 seconds, for the coated materials to be handled.

Uniform film thickness is an advantage of fusion bonded coatings. When these coatings are applied using electrostatic powder spray equipment, the charged powder particles are uniformly attracted to all surfaces of the substrate. Today, present powder application system technology achieves well over 90% material utilization efficiency. The remaining 10% is overspray. This is powder that does not adhere to the substrate. Existing technology has produced equipment that can recover and recycle up to 98% of powder over spray. For tomorrow, equipment is being designed that will apply the exact amount of powder needed, virtually eliminating over spray.

Quality assurance by effective quality control of each step will produce the best coating and finish. Finally, and most importantly, to insure that the optimum coating film is obtained, attention must be paid to each step of the process: handling, surface preparation, coating application, and curing.

Environmentally controlled coating plant facilities provide the most desirable condition for surface preparation and application of protective coatings. For liquid coatings, the primer, intermediate, and finish coats should be shop applied. Only when aesthetics are a major concern, should the finish coat be applied in the field.

The results of paying strict attention to the many details that are involved will produce a high quality, long life protective coating application.

Corrosion Protective Properties of Paint Films

Isao Sekine*

Introduction

Coal tar epoxy paint is an excellent coating which has been used widely for the protection of metal from corrosion. The quality of the coating is attributed to uniting the best qualities of coal tar, the rust preventing ability such as excellent water resistance, brine resistance, chemical resistance etc., and the epoxy resin's ability to adhere to a variety of substrates. The coal tar epoxy paint is the best in combining all these properties. Such properties and deterioration of coal tar epoxy paint have been studied by many investigators.[1,2] However, it is not clear how the corrosion protective property of this paint varies depending on the various mixing proportions of coal tar and epoxy resin.

In this paper, the properties of the coal tar epoxy paint were studied as a function of the different mixing ratios of these ingredients. Various electrochemical methods were applied to studies of brine resistance. The results are discussed by comparison with other kinds of paints.

Experimental

Coal tar epoxy paints used are shown in Table 1.

TABLE 1 — Composition of coal tar epoxy paints (wt %)

Paint	Coal tar	Epoxy resin	Curing agent	Accelerator	Ratio of T:E
(1)	41.1	41.1	16.4	1.4	1:1
(2)	49.3	32.9	16.4	1.4	3:2
(3)	32.9	49.3	16.4	1.4	2:3
(4)	54.8	27.4	16.4	1.4	2:1

Other paints employed for comparison with coal tar epoxy paint are given in Table 2.

**Department of Industrial Chemistry, Faculty of Science and Technology, Science University of Tokyo, Noda, Chiba 278, Japan*

TABLE 2 — Paints for comparison

Paint	Name
(5)	Modified epoxy paint
(6)	Alkyd paint
(7)	Oily anticorrosive paint
(8)	Oily anticorrosive paint (Red lead)
(9)	Rust preventing agent + CIE[1] paint

[1] Coal tar epoxy, ratio of T:E = 1:1

Cold rolled steel (SPCCB) plates, 7 × 15 cm, were polished with emery paper up to No. 1200, and cleaned successively by immersion in perchloroethylene, and ethanol, then stored in a desiccator. These plates were coated with the paints and dried for a week. The thickness of the painted films was 180~200 μm. The back and edges of the specimens were masked thickly in paraffin wax so that a surface area of 50 cm² was exposed. The specimens were immersed in 3% NaCl, and the temperature was maintained at 30 ± 0.5 C.

The natural electrode potential was measured by a potentiostat (Hokuto Denko, PS-500B). A saturated calomel electrode (SCE) was used as a reference electrode.

The tan δ values were determined by calculating the parallel capacitance C_p and parallel resistance R_p measured by a phasesensitive detector (PSD, Fuso Seisakusho, model 332) as has been described previously.[3] The reference signal of the sinusoidal wave in PSD was within 9 mV (RMS) over the frequency range of 0.18~9 kHz. The pulse polarization and corrosion current were measured by a corrosion rate meter (Nippon paint).[4] The permeability of water was determined by gravimetry.[5] That is, the amount of water in a cup put into a desiccator decreases by permeating through the free film of paint.

Results and Discussion

The Figure 1 shows the natural electrode potential against time curves for the electrode coated by coal tar epoxy paints. These potentials were held constant at -0.25~-0.35 V for some time after immersion. However the potentials of paint (4), (3), and (2) began to

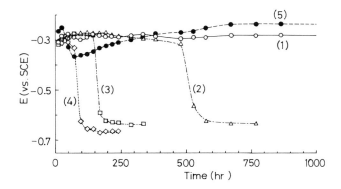

FIGURE 1 — Natural electrode potential vs time curves.

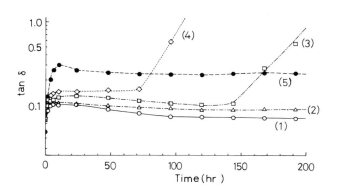

FIGURE 2 — tan δ vs time curves.

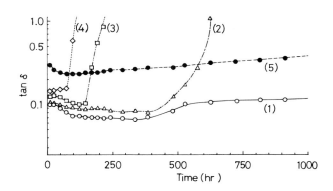

FIGURE 3 — tan δ time curves.

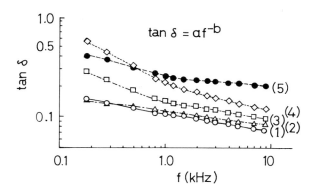

FIGURE 4 — Frequency dependence of tan δ at 48 hours after immersion.

shift rapidly to less noble potentials after the lapse of about 70, 145, and 470 hours, respectively, and approached the steady value of about −0.65 V. Whereas paint (1) did not show such a potential shift even after 1000 hours.

Figure 2 shows the tan δ against time curves when C_p and R_p are measured up to 200 hours at 3 kHz. In each of the paints (1)∼(4) the tan δ value increases rapidly after immersion, and subsequently shows a steady value after 10 hours. Such a rapid increase of tan δ is considered to be due to a swelling process caused by water absorption. Tan δ, which showed a steady value, may also be attributed to a swelling process resulting from elution of the soluble components of paint film as has been discussed by Ohyabu et al.[6] The tan δ of paints (1) and (2) remained constant for 200 hours, as shown in Figure 2.

It is further obvious from Figure 3 for measurement up to 1000 hours that tan δ in paints (4), (3), and (2) increases steeply at about 70, 140, and 450 hours, respectively, after immersion. On the other hand, in paint (1) such behavior is not observed, even after 1000 hours. Consequently, a steady value of tan δ of paint (1) is about 0.1, and this value corresponds to the standard of an excellent paint, as has been reported by Okamoto et al.[7] Interestingly, the rapid shift of the corrosion potential to less noble potentials takes place at approximately the same time as the steep increase of tan δ. It can, therefore, be presumed that at this point in time the paint film begins to deteriorate, and the elution of iron as ferrous ions increases rapidly.

The corrosion protective property of these tar epoxy paints is verified by comparison with nontar epoxy paint (5) as shown in Figure 1 through 3. Namely, even in measurement after 1000 hours, the behavior of the potential and tan δ of paint (5) is similar to that of paint (1), and rapid change is not observed. The tan δ value of paint (5) is generally somewhat greater than that of paint (1). Consequently, the protective ability of paint (5) on the basis of these measurements may not be better than paint (1).

Figure 4 shows the frequency dependence of tan δ at 48 hours after each specimen was immersed. Whichever paint is studied, tan δ shows a greater value at low frequency, but decreases with increasing frequency. Assuming that these points fall on a straight line, and an equation, $\tan \delta = af^{-b}$, applies to this case, paints (1), (2), and (5) which did not deteriorate in the long term, exhibit small values of the slope over a wide range of frequency. In these paints, the b values shown as a slope of the straight line are (1) 0.200, (2) 0.136, and (5) 0.233. On the other hand, the b values of paints (3) and (4), which deteriorated in a

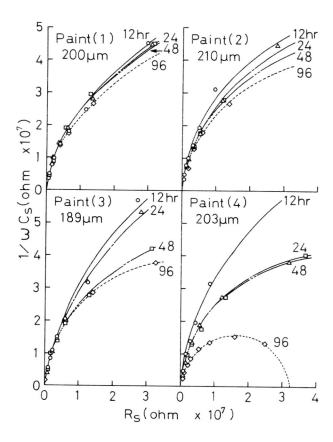

FIGURE 5 — Complex impedance plane plot.

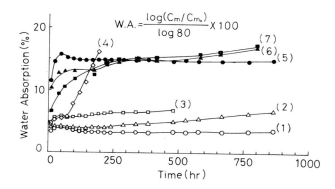

FIGURE 6 — Water absorption vs time curves.

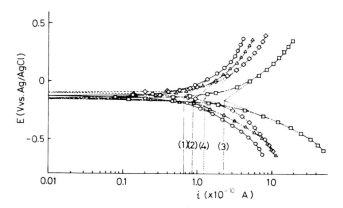

FIGURE 7 — Current vs potential curves at 120 hours after immersion.

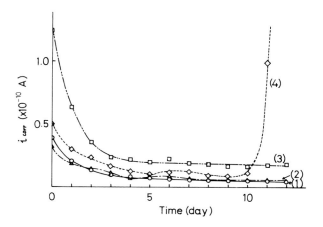

FIGURE 8 — Time dependence of the corrosion current.

relatively short time, show a greater decline ((3) 0.417 and (4) 0.540), especially at a lower frequency than 1 kHz.

From these facts, the protective property of paint film can be inferred from the frequency dependence of tan δ.

Figure 5 shows the results of the complex impedance plane plot. It is considered that most of these curves can not be shown as a complete semicircle, but as an arc in this frequency range. At the paint film free electrode/solution interface, the charge transfer resistance R_θ and double layer capacity C_{dl} are obtained from the diameter of semicircle and the ω_{max}, corresponding to the top of the semicircle, respective-

ly. The value of R_θ, that is, the size of circle, generally decreases with immersion time. Since the circle of paints (1) and (2) does not decrease with immersion time, the paint film, apparently, does not deteriorate. However, the circle of paints (3) and (4) decreases appreciably with time. For example, the value of R_θ and C_{dl} at 96 hours of paint (4) is 7.9×10^8 Ω·cm² and 27.8 pF/cm², respectively. The value of C_{dl} obtained at the interface which consists of metal, paint film, and solution, is abnormally small compared to the value obtained for the film free electrode. Therefore, it would not be reasonable that this value is regarded as that for C_{dl}. The decrease in resistance of the paint film, and the increase of water absorption occur in parallel, as will be mentioned later. If the change of R and C of paint film is appreciable, the values of R and C obtained as R_θ and C_{dl} reflect the resistance and capacity of the paint film. Therefore, a small value is clear evidence that the paint film has deteriorated.

The amounts of water absorption by paint films are obtained from the electric capacitance using impedance measurements. Figure 6 shows the water absorption against time curves. The water absorption of paint (1) shows the lowest value. In other paints this

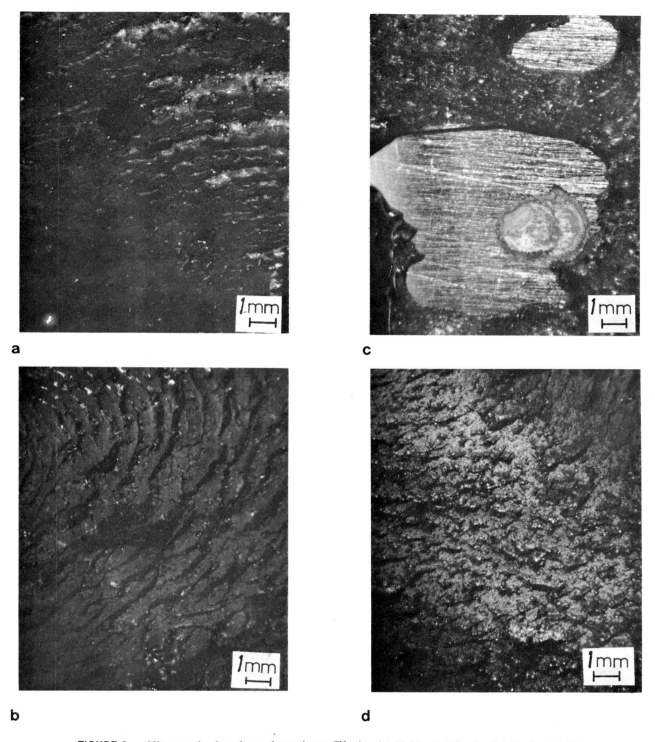

FIGURE 9 — Micrograph of surface of specimen, 7X. a) paint (1); b) paint (2); c) paint (3); d) paint (4).

value is high, in the order of (5)>(6)>(7)>(4)>(3)>(2), after about 100 hours. In particular, the value of paint (4) increases rapidly at the same time that the paint film begins to deteriorate. Since the amount of epoxy resin in paint (4) is less than that of coal tar, such a paint film easily forms many pinholes, the adhesion force on the metal surface becomes weak, and the paint film peels off from the metal as corrosion progresses. Consequently, the steep increase of water absorption in paint (4) seems to be due to water collected at the interface between the metal and paint film.

Figure 7 shows the current vs potential curves at 120 hours after immersion. These values are obtained when a specimen is polarized at 100 mV/minute to each side of the anode and cathode from the natural electrode potential. The value of corrosion current was the smallest for paint (1).

Figure 8 shows the time dependence of the corrosion current determined from the polarization curve.

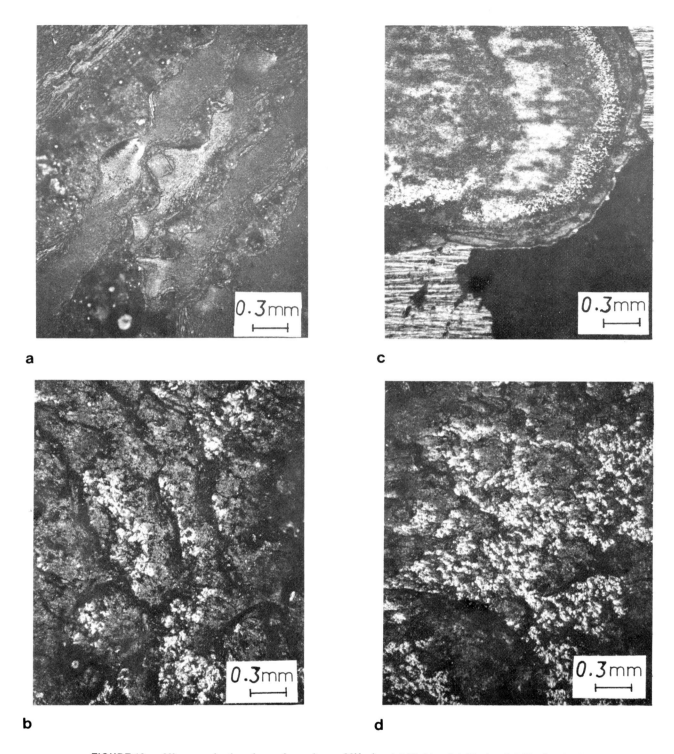

FIGURE 10 — Micrograph of surface of specimen, 33X. a) paint (1); b) paint (2); c) paint (3); d) paint (4).

From the time the specimens are first immersed, i_{corr} of paint (3) is generally greater than that of other paints. The i_{corr} value of paints (1) and (2) is relatively small, but in the case of paint (4), the i_{corr} value increases steeply after about 10 days. As can be seen from this figure, the corrosion current of paint (4) is comparatively smaller than that of paint (3) after 2 or 3 days immersion. However, according to many tests, the corrosion protective properties of paints (3) and (4) do not differ greatly.

Figures 9a, b, c, and d show a 7 times magnification of a micrograph of the surface of each specimen when the paint film was made to peel off from the metal substrate 2 months after immersion. Much cohesive destruction of paint can be seen on the upper surface of the film in paint (1), and at the center of the film in paint (2). But the metal substrate is not seen at all. Consequently, the corrosion protective property of paints (1) and (2) is similar. In paint (3), the interfacial exfoliation of the paint and substrate can be seen

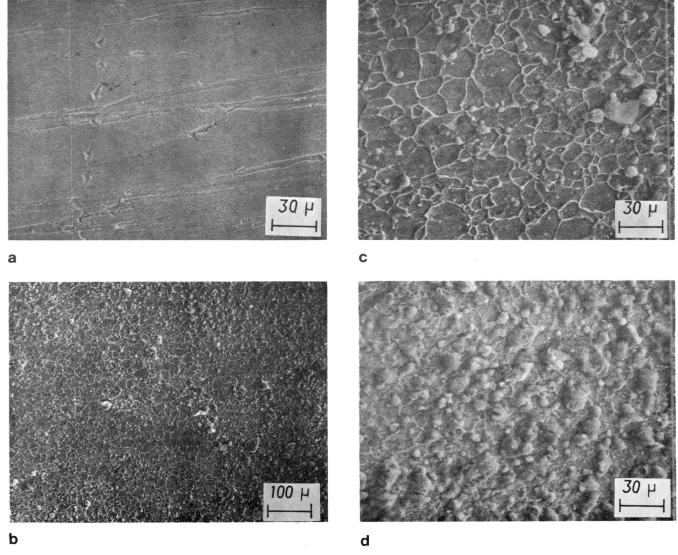

FIGURE 11 — SEM for adhesion failure of paint (3): a) non-corroded area; b) corroded area; c) central area of corrosion; d) peripheral area of corrosion.

simultaneously with the cohesive destruction of paint. The substrate of these exfoliated areas was partly corroded. In paint (4), the cohesive destruction of paint takes place at the upper adhesive interface. Figures 10a through d show a 33 times magnification of a micrograph of the surface of a specimen.

An enlarged photo of the corroded area of paint (3) is shown in Figure 11. The central area of corrosion corresponds to the lower left side of, and the peripheral area of corrosion to the upper right side. A reticulate pattern seems to be the grain boundaries of iron, and the small stone like substance is the corrosion product. On the other hand, the peripheral area of corrosion is covered, generally, by a corrosion product similar to general corrosion without pitting.

The amounts of water which permeated the free film of various paints are listed in Table 3. As can be seen from this table, in the coal tar epoxy paints, the water permeabilities are clearly less in paints containing larger quantities of coal tar. In particular, the value of paint (4) agreed closely with that obtained by Guruviah.[8] On the other hand, the value of permeability of paints which were free from coal tar is relatively high. Consequently, the time dependence of the natural electrode potential, tan δ, charge transfer resistance, water absorption, corrosion current, and water permeability are closely correlated to each other.

Table 4 shows the transference number of chlorine anions. These values were obtained from measurements of the membrane potential of the free paint film. Especially in paint (7), the oily anticorrosive paint, the film shows a strong anion selectivity, because the transference number at pH 2 and 12 is relatively high. All the coal tar epoxy paints exhibited similar values of the transference number. Therefore, these paint films would not have a strong anion selective permeability.

The corrosion mechanism of iron by chlorine anions is shown in Figure 12. If a paint film shows

TABLE 3 — Water permeability of paint films

Paint	Thickness (μm)	Permeability (mg/cm^2/day)	Gradient of relative humidity
(1)	100	1.3	100%/31%
(2)	100	1.4	100 /31
(3)	100	1.9	100 /31
(4)	100	1.1	100 /31
(5)	100	2.4	100 /31
(6)	100	2.7	100 /31
(7)	100	1.9	100 /31

TABLE 4 — Transference numbers of Cl$^-$ ions

Paint	pH 2.0	pH 5.5	pH 12.0
(1)	0.54	0.50	0.65
(2)	0.51	0.54	0.61
(3)	0.58	0.53	0.52
(4)	0.54	0.53	0.53
(5)	0.50	0.48	0.56
(6)	0.56	0.50	0.58
(7)	0.64	0.50	0.77

$$E_m = (1 - 2t_-) \frac{RT}{F} \ln \frac{C_1}{C_2}$$

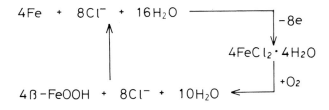

FIGURE 12 — Corrosion mechanism by chlorine anions.

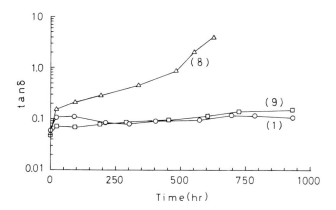

FIGURE 13 — tan δ vs time curves.

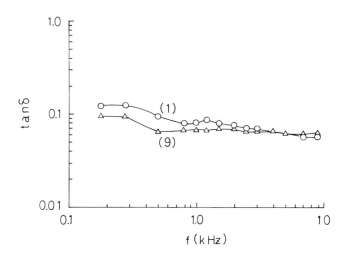

FIGURE 14 — tan δ vs frequency curves.

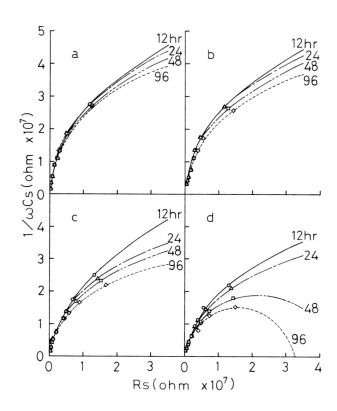

FIGURE 15 — Complex impedance plane plot: a) RPA + Paint (1), 75 + 197 μm; b) RPA + Paint (2), 64 + 207μm; c) RPA + Paint (3), 64 + 217 μm; d) RPA + Paint (4), 59 + 274 μm.

anion selectivity, the corrosion mechanism of iron is considered to be caused by chlorine anions with water and oxygen as described by Funke.[9]

Figure 13 gives the tan δ/time curves which paint (9) shows when applied to a fully rusted substrate, compared with paints (1) and (8) on the original polished substrate. The tan δ values of both paints (1) and (9) are nearly the same. Furthermore, these paints were superior to paint (8). In the case of paint (8), tan δ increases steeply after about 500 hours, and the paint film is observed to deteriorate.

Figure 14 shows the tan δ vs frequency curves. The value of tan δ of paint (9) is also close to that of paint (1).

Figure 15 shows the complex impedance plane plot for a rust preventing agent with various ratios of coal tar epoxy paint. Even if a rust preventing agent was used, it was found that the size of an arc or semicircle, that is, the resistance value of each paint film, is similar to that of Figure 5. From these results, removing the rust from the metal substrate is unnecessary if paint (9) is used. This paint can, therefore, be considered economically advantageous.

Conclusion

The corrosion protective properties of coal tar epoxy paint, depending on the different mixing ratios of coal tar and epoxy resin, were studied by various electrochemical methods.

1. From the values of the natural electrode potential, tan δ and charge transfer resistance on immersion time, the mixing ratios of coal tar and epoxy resin show excellent brine resistance in the order of (1)>(2)>(3)>(4).

2. In the frequency characteristics of tan δ, the paint in which the slope of curve and the tan δ value are smaller, that is, (1)<(2)<(3)<(4) resisted deterioration for a much longer period.

3. Water absorption and corrosion current against time was smaller, in the order of (1)<(2)<(3)<(4), but water permeability against time was (4)<(1)<(2)<(3).

4. The optimum mixing proportion of coal tar and epoxy resin was found to be 1:1 as in paint (1).

5. In comparison with the coal tar epoxy paint and other paints, the use of a rust preventing agent was effective.

Acknowledgement

The author wishes to thank Mr. Shunzo Miyazaki, Hokkai Seikan Company, for getting the data of pulse polarization and making the SEM pictures.

References

1. H. Haagen and W. Funke, *J. Oil Col. Chem. Assoc.*, **58**, p. 359 (1975).
2. R. Fernandez-Prini and H. Corti, *J. Coating Technol.*, **49**, p. 62 (1977).
3. I. Sekine and H. Ohkawa, *Bull. Chem. Soc. Japan*, **52**, p. 2853 (1979).
4. T. Yamamoto, H. Amako and Y. Ohyabu, *Shikizai Kyokaishi*, **48**, p. 352 (1975).
5. D.Y. Perera and P.M. Heertjes, *J. Oil Col. Chem. Assoc.*, **54**, p. 313 (1971).
6. Y. Ohyabu, H. Kawai and S. Ikeda, *Shikizai Kyokaishi*, **37**, p. 90 (1964).
7. G. Okamoto, T. Morozumi and T. Yamashina, *Kohgyo Kagaku Zasshi*, **61**, p. 291 (1958).
8. S. Guruviah, *J. Oil Col. Chem. Assoc.*, **53**, p. 669 (1970).
9. W. Funke, ibid., **62**, p. 63 (1979).

Performance of Marine and Industrial Coatings in the Arabian Gulf

Kenneth I. Rhodes and E. Michael Moore, Jr. *

Introduction

In this paper, the performance of selected coatings used in the Arabian Gulf area over the last thirty years is reviewed. Marine and industrial coatings are dealt with in the most detail; however, a short discussion of residential coatings also is included. The service lives of the various types of coatings are compared, and the effects of surface preparation and application techniques on coating performance are discussed.

The data presented in this paper come from two sources: (1) actual service data; and (2), panel testing programs. Records of the service performance of marine coatings used by ARAMCO go back to 1949. Service performance data on industrial coatings in refinery and tank farm areas date from the mid 1950s. Systematic panel testing was begun in the mid 1950s, discontinued in the mid 1960s, and restarted in 1974. Extensive data on marine, industrial, and residential coating performance was gathered during the panel testing programs and is reported below.

This paper also presents a number of selection criteria that experience has shown to be important when choosing coatings for use in the Arabian Gulf area. An important criterion is the shelf life of the coating. Remote areas and high summer temperatures make it desirable for coatings to have a shelf life of at least one year. The shelf life problem has been somewhat alleviated in recent years with the establishment of several local factories which can manufacture a wide range of products on an as needed basis.

The paper concludes that the most critical factor affecting the service performance of coatings in the Arabian Gulf is surface preparation. The widespread practice of using desert sand as a blasting abrasive results in reduced performance of high build coatings such as epoxies. The advantages of commercial abrasive blasting grits over desert sand are discussed. Conventional alkyd coating systems are recommended for use instead of more "exotic" paint systems in many cases where adequate surface preparation with abrasive blasting grit cannot be achieved.

Performance of Industrial Coatings

Table 1 shows panel test data on the industrial coatings used by ARAMCO in the Arabian Gulf area.

One of the interesting test results is that after 30 months continuous exposure, a three and a four coat alkyd system have performed as well as a three coat amine cured epoxy system in coastal, atmospheric exposure. A chlorinated rubber system has also shown excellent performance. This four coat alkyd system is in widespread use on exterior structural steel, piping and tanks; it consists of 3 mils DFT (dry film thickness) of zinc chromate plus 2 mils DFT of aluminum. Actual service records extending back about 20 years show that the zinc chromate/aluminum system has a typical service life of 8 to 10 years in external tank coating usage in a coastal refinery when good surface preparation and application techniques are used. On the other hand, poor surface preparation and application have resulted in service lives of 1 to 3 years for the same system.

This information is particularly significant in connection with maintenance painting in industrial and marine areas where sandblasting is not feasible or not permitted. There are many "exotic" coatings (epoxies, inorganic zincs, polyurethanes) that will give longer service than conventional alkyd systems if the surface preparation and application are superior. It is a complete waste of time to apply exotic coatings over mediocre surface preparation, or with poor application technique. High performance coatings require a minimum height of surface profile which, in turn, is based upon the thickness of paint to be applied and the size and type of abrasive used. If possible, the pattern depth should be approximately 1/4 to 1/3 the thickness of the coating system, except in high build systems of over 300 microns (12 mils) where normally a profile of 75 to 100 microns (3 to 4 mils) should be used.

When desert sand is used as the blasting abrasive, the maximum anchor pattern obtainable is about 25 to 40 microns (1 to 1½ mils). The total DFT of epoxy or inorganic zinc primer/epoxy topcoat systems should be on the order of 375 microns (15 mils). It is clear that even the optimum use of desert sand will yield an anchor pattern only about 1/10 of thickness of the coating. The result is that the theoretical "superior performance" of high build, high performance coatings cannot be achieved even under otherwise optimum conditions when desert sand is used as the abrasive. It is, therefore, frequently the case that conventional systems outperform the "exotic" systems in the Gulf area.

---

*Arabian American Oil Company, Dhahran, Saudi Arabia

TABLE 1 — Summary of Paint Exposure Test Results

Coating System	Thickness (Micron)	Exposure time (mos)	Condition (% Failure)				Comments
			Mild Atmosphere	Refinery	Marine Area	Submerged in Gulf	
ZnChromate/ Aluminum Alkyd System	125	36	0	0	0	—	Three systems tested. A fourth system OK after 28 months.
Zinc Chromate/ Alkyd Enamel System	125 to 150	36	0	0	0	—	Five systems tested. Four have phosphating pretreatment.
Chlorinated Rubber	125 to 150	36	—	0	0	100 (1)	(1) Failed after 4 months.
Catalysed Epoxy System	375	28	0	0	0	0 (1)	Six systems tested. Two have inorganic zinc silicate primers. (1) Only one system exposed in Marine rack.
Catalysed Epoxy System	250	36	0	0	0	5 (1)	Special epoxy for application to damp steel surface. (1) Primarily mechanical damage resulting from barnacle removal.
Catalysed Coal Tar Epoxy System	400	28	0	0	0	—	Three systems tested.
Epoxy Splash Zone Compound	2500	30	—	—	—	0	Three systems tested.
Surface Pretreatment/ Epoxy System	125	26	75	80	95	—	Proprietary phosphating pretreatment on wire brushed panels.
Vinyl Mastic	150 to 175	28	0	0	0	—	—

There are several areas where the "exotic" systems should be specified, however, in spite of the insufficient anchor pattern problem. These areas are the following:

1. Industrial areas routinely deluged to test the fire fighting system.
2. Areas subject to splash and spillage of acids or alkalis.
3. Areas exposed to excessive blowing sand (wind girders on storage tanks are an example).
4. Steel in cold service.
5. Areas subject to repeated mechanical abrasion.
6. Marine atmospheric areas where sand blasting is permitted.
7. Immersion service.

Of these areas, immersion is probably the most severe service. The authors have experienced quite good results (where surface preparation was adequate) with epoxies and coal tar epoxies as internal coatings in tanks and vessels in water and petroleum product service. The maximum service temperature of coal tar epoxies should be limited to 71 C (160 F). There are numerous epoxy systems suitable for service up to 93 C (200 F). Service lives in excess of 10 years are readily obtainable when the paint systems are correctly applied. Water tank internal coatings should be supplemented with cathodic protection.

In the authors' opinion, the major limiting factor on the service performance of coatings applied to the interiors of tanks and vessels is the use of desert sand rather than abrasive grit for surface preparation. Because of its roundness, wind blown desert sand cuts slowly, and results in a shallow, rounded anchor pattern. The sand has a high tendency to shatter into dust upon impact with the surface being blasted. In confined areas (even large confined areas such as storage tanks) the sand blaster usually cannot see the surface he is working. Consequently, the blasting process itself is slow, and the results are frequently spotty. Commercial abrasive blasting grits are technically superior to desert sand. They cut faster, visibility during blasting is better, and the anchor pattern is better (deeper and sharper). Other factors being the same, the use of commercial blasting grits will improve the service life of the coatings now in use, reducing future maintenance requirements. The initial cost of commercial abrasive blasting grits is higher than that of desert sand. Most commercial grits are reusable, however, bringing the actual cost into line with that of desert sand, which is not reusable.

Performance of Marine Coatings

The earliest available records of marine coatings used by ARAMCO in the Arabian Gulf show that hot applied coal tar enamels were specified for both splash zone and atmospheric zone. These were first applied between 1949 and 1950. Spot repairs were made with cold applied bituminous coatings. In the splash zone, and in areas exposed to direct sunlight, the applications were not long lived. However, in the atmospheric zone not in direct sunlight (for example, under the decking on piers) the original coating provided protection for 20 to 25 years. In recent years, the remaining coal tar enamels were removed due to random disbonding which could not be located readily by sight. Most of the coating had be be laboriously removed by mechanical means. The corrosion protection provided was excellent.

In the mid 1950s a red lead primer/chlorinated rubber topcoat system was tried, based primarily on reported good results elsewhere in the Gulf. This system was short lived. Panel testing in the last 3 years has verified the inappropriateness of chlorinated rubber coatings in the splash and immersion zones due to rapid, total destruction by barnacles in about 4 months. As shown in Table 1, however, the atmospheric exposure panels are performing well. It is not known whether the earlier poor performance of chlorinated rubbers in the atmospheric zone was due to defective material or deficiencies in application.

No effective splash zone coating was used until 1963 and 1964, when four different two pack epoxy putty type splash zone compounds were tried. On the basis of a two year test, "Cooks Splash Zone Compound" was selected, and this coating (or equivalent) continues to be the standard splash zone coating whenever monel sheathing is not used. On this subject, the authors prefer to specify monel sheathing in the splash zone whenever possible, because of its excellent performance in this service. Tubular constructions are particularly suitable for the application of monel sheathing because sheet metal forming operations are minimal. The putty type splash zone compounds should be used in preference to monel sheathing for spot maintenance repairs and for H-beams and angle irons where monel sheathing is not feasible.

In April, 1958, a three coat, 11 mils DFT catalysed epoxy coating was test applied in the atmospheric zone of a pier. To improve edge coverage on this H-beam structure, glass cloth was embedded around the edges. The DFT at the edges approached 40 mils. The test was monitored for 4½ years. The results were excellent, and this coating has been in general offshore use since that time. Service life of catalyzed epoxy coatings in Arabian Gulf atmospheric service depends largely on the quality of the surface preparation prior to painting. With a white metal sandblast and proper application, service lives of 5 to 10 years are obtainable. On the other hand, the authors have seen numerous cases of complete system failure within one year, when the epoxy paint was applied over wire brushed surfaces. The authors are not aware of a single case where epoxy paint in marine atmospheric service has been applied over a wire brushed surface and lasted longer than one year.

In the authors' experience, the submerged zone of offshore structures is readily protected by cathodic protection without the use of coatings. Tests in the 1950s in the Arabian Gulf showed that once an offshore structure had been polarized, the current requirements needed to maintain polarization were unaffected by whether the structure was coated. We define the submerged zone as the region below mean sea level plus one foot. This definition is empirical, and is based on observations taken over a period of about 15 years. Above MSL + 1 the cathodic protection system does not prevent corrosion; therefore, splash zone compounds or monel sheathing are applied.

A category of submerged facilities that does require coating in addition to cathodic protection is pipelines. Concrete weight coated, glass cloth reinforced asphaltic and coal tar enamel coatings have been used successfully since 1948 in the Gulf by ARAMCO. External corrosion leaks have been rare.

Performance of Residential Coatings

In recent years, several paint companies have established local factories on a joint venture basis. Most, if not all, of these factories produce a wide range of good quality residential paints. In the authors' experience, the water based paints perform better than the enamel based paints, especially on new construction where the cement blocks and the plaster may not be thoroughly dry. Although quite detailed purchase specifications are available for evaluating latex paints, the authors have found that a few simple tests suffice. Basically, a residential paint must perform the following functions: (1) adhere well to concrete, plaster, and wood; (2) be easy to apply; (3) resist weathering and intentional washing; (4) have sufficient hiding power so that color changes can be accomplished by overcoating with one (or at the most two) coats. Drawdown tests to determine hiding power, and exposure tests on concrete blocks to verify adhesion, ease of applicability, weatherability, and washing resistance suffice to eliminate unsuitable paints. Most of the "failures" that the authors have encountered have been due to insufficient hiding power on interior coatings. Exterior and exterior/interior grade coatings usually have superior hiding power compared to the strictly interior grades, and the authors prefer their use as interior coatings for that reason. In addition, the use of exterior or exterior/interior grade latex coatings for the interiors of buildings reduce the variety of paints that need to be kept in stock for maintenance use. Accurate service performance figures for latex coatings in residential use are not readily obtainable because of frequent

TABLE 2 — Comparisons of typical properties and service performance of several generic types of coatings. (data taken from reference 1).

Property	Alkyd	2 Pack Expoxies	Latex	Inorganic Zincs	Chl. Rubber	2 Pack Urethane	Vinyl
Adhesion	3	1	3	2	3	2	4
Hardness	3	2	3	1	3	2	3
Flexibility	3	3	1	4	2	3	1
Resistance to							
Abrasion	3	2	4	3	3	1	3
Acid	4	3	3	5	3	3	2
Alkali	4	1	2	4	2	2	2
Heat	4	3	4	2	4	3	5
Strong Solvents	4	1	3	3	4	2	4
Water	3	2	3	3	2	2	2
Use on							
Wood	3	4	2	6	5	4	6
Fresh (dry)							
Concrete	6	2	2	6	2	3	3
Metal	2	2	4	2	3	2	2
Interior use	3	3	2	3	3	3	3
Exterior use							
Rural Areas	3	3	2	3	3	3	3
Seashore	4		3	4	2	2	1
Industrial Areas	4	1	3	4	2	2	2
Submerged Zones	6	2	6	5	4	3	2

KEY: — 1 Outstanding, 2 Very Good, 3 Good or Average, 4 Fare, 5 Poor, 6 Not Recommended

color changes by the inhabitants. Service lives on the order of 3 to 5 years should be expected, however.

Selection of Coatings for the Arabian Gulf Area

Keeping in mind that there is no such thing as a "perfect" coating, in a practical sense we cannot expect a coating to give perfect protection. Thus, practical coatings are a compromise between the maximum protection that can be expected from a system and how much is available to pay for them. Paint selection depends on where the coating is to be used, the condition of the surface and substrate, and the reason for painting. Table 2, taken from the literature, gives a general overview of coating performance in various services.[1]

No one maintenance paint product is best for all environments or substrate conditions. The basic problem is to select the product which will provide adequate performance under a given set of conditions. To select this product with some confidence, it is best to test it under conditions equivalent to those it will face in service. Laboratory evaluations, no matter how sophisticated, should serve only as guidelines. Too often, a coating is chosen only after weighing evidence accumulated in accelerated laboratory tests. None of these accelerated tests consistently provide more than a reproducible means of discriminating among a group of competitive materials. This selection process is usually specific to the type of accelerated test, and frequently does not predict how a given coating will perform in service. Wherever possible, coatings should be tested in the actual environments. Long term panel tests are preferable to large scale tests, because they are much more economical, and they are much easier to monitor. The authors' experience with large scale tests shows that monitoring the application conditions, and following up the test for a period of several years, varies from the difficult to the impossible. The usefulness of these large scale tests is, therefore, largely negated by the minimal information they eventually yield. Large scale tests are not particularly beneficial to the paint vendors, either. Since new systems usually require some getting used to by the painters, technically suitable coatings can fail prematurely (and, consequently, not be accepted by the customer) simply due to painter inexperience with the system.

Selection of a coating from among the many types available can be made intelligently on the basis of the known performance of coatings for similar applications, the known chemical composition and physical properties of the paints, and the results of exposure tests on coatings under consideration. Of these factors, the best basis for critical judgement of the relative suitability of a coating system is previous experience with the peformance of that coating. Even first hand observation of the serviceableness of a coating can be misleading, however, when the new environment in which the paint is to serve is dissimilar, or if the composition of the paint has been modified. Knowledge of the chemical and physical

TABLE 3 — Summary of the Performance of Various Coatings in the Arabian Gulf Area

Coating System	Where Used	Typical Service Life	Minimum surface preparation required (1)	Applicator Skill Required
Zinc chromate with aluminum alkyd enamel top coats	Industrial atmospheres	8 to 10 years	Sa2	Moderate
	Marine atmospheres	3 + years	Sa2	Moderate
Catalitically cured epoxy	Industrial atmospheres	5 to 10 years	Sa2	High
	Marine atmospheres	5 to 10 years	Sa2½	High
	Immersion (tank and vessel coating)	10 + years	Sa3	High
Catalitically-cured coal tar epoxy	Immersion (tank and vessel coating)	10 + years	Sa2½	High
Epoxy splash zone compounds	Marine splash zone	5 to 10 years	Sa2	High
Hot applied coal tar enamels	Marine atmospheres not in direct sunlight	20 years	Sa2	Moderate
	Marine splash zone and marine atmospheres in direct sunlight	1 to 2 years	Sa2	Moderate

NOTE: — (1) Sa2 = Commercial blast, Sa2½ = near white metal blast, Sa3 = White metal blast

properties of a coating (acid and alkali resistance, heat resistance, abrasion resistance, etc.) is a selection aid, but should not be the sole basis for final selection. When coating expenditures run into the hundred of thousands of dollars, it pays to be sure about performance.[2,3]

An important factor in selecting coatings for the Arabian Gulf area is shelf life. High storage temperatures in the summer months, the long shipping times required, and the lack of proper storage facilities in remote areas make shelf life an important consideration. Coatings with shelf lives less than one year should not be considered unless they can be manufactured locally on an as needed basis.

When large volumes of a given type of coating are used by a company, it is essential to have stringent purchase specifications, and to insist on compliance with them. Technological requirements, such as condition in container, fineness of grind, and viscosity, as well as application properties such as minimum DFT per coat, curing time, pot life, and appearance are always required. In most cases, tests for hardness, adhesion, flexibility and chemical resistance should be specified. If the coating is intended for immersion service it should be immersion tested at temperatures at least as high as the maximum expected operating temperature. The test time should be long enough to clearly establish the suitability of the coating. Six months is reasonable. One and two week immersion tests prove nothing.

Summary and Conclusions

The service performance of several types of coatings in the Arabian Gulf area over the past 25 to 30 years is summarized in Table 3, and is discussed in the text of this paper. The most critical factor affecting the service performance of most of these coating systems is surface preparation, and the advantages of commercial abrasive blasting grits over wind blown desert sand are discussed. The advantages of conventional alkyd systems over catalysed systems, when the surface preparation is less than the minimum for the catalysed systems, are also discussed. The optimum selection of coatings for specific jobs involves a compromise between the technological superiority of a paint and its minimum application requirements. In general, conventional paint systems will outperform "exotic" paint systems when applied over wire brushed or brush blasted surfaces.

Several factors that should always be considered when painting on a large scale are:

1. The economic consequences of poor coating practices can be disastrous.

2. The number of coating materials available necessitates careful study if a proper material is to be selected.

3. Proper surface preparation is a major factor for economic coating in any environment.[4]

4. Difficult factors in the Arabian Gulf area, such as high temperatures and remote work locations, force a practical and common sense approach in selecting products that have long shelf lives and are easily applied over minimally prepared surfaces.

5. Miscellaneous factors may determine the success or failure of an expensive coating job. These include such things as skill of the workmen, proper in-

spection of the job in progress and at its end, and provision of proper equipment.

6. Continuous testing and evaluation of candidate systems and systems actually in service more than pays its way in providing a basis for proper coating selection.

The high cost of replacing metal that has been destroyed by corrosion makes it imperative that those who are responsible for the design and maintenance of structures and equipment be thoroughly familiar with: (1) the factors that contribute to the satisfactory performance of an applied coating; and (2) the factors that impair the integrity of the coating and facilitate the destruction of the metal under the paint film. Painting for protection in the Arabian Gulf should not be considered a necessary evil. It is a vital and necessary part of the overall effort to minimize corrosion.

Acknowledgements

Permission to present this paper was given by the Arabian American Oil Company. Mr. Mahboob Asghar was responsible for installing the paint panel test facilities, and for preparing the paint panels used to generate the data in Table 1. Mr. G.M. Bhatti typed the manuscript.

References

1. Sidney B. Levinson, *Selecting Paints*, Architectural & Engineering News, 1970.
2. NACE Basic Corrosion Course, Fourth Printing, December 1973.
3. Steel Structures Painting Council, **2,** Ninth Printing, 1976.
4. *Materials Performance*, **16,** No. 3, pp. 9-12, March, 1977.

Sulfonate Based Coatings: Their Chemical, Physical, and Performance Properties

L.S. Cech, J.W. Forsberg, W.A. Higgins*

Introduction

Sulfonate based coatings have been developed over the last twenty years, primarily to provide protection for body metal in automotive and truck vehicles. The inner metal surfaces of vehicles require high performance protective coatings to prevent corrosion due to: rock salt; calcium chloride; road dirt; mineral matter; and absorbed industrial pollutants in combination with water and oxygen. The sulfonate based coatings wet out on and adhere to surfaces which are not clean. They are flexible over a wide temperature range, either penetrate or seal off rolled or welded seams, and resist corrosive undercutting. They are thixotropic liquids; when applied they remain in place on vertical surfaces and the underside of horizontal surfaces. In many cases, they are applied in closed box sections, especially in unitized body construction. The solvent based and water based systems which will be discussed in detail resist being washed off by their own solvent vapors in the box sections. Vapors are not a factor if the materials are applied as hot melts.

We will illustrate the important chemical and formulating parameters which are responsible for the performance properties outlined. As these coatings have evolved, it appears that they may be useful in application areas other than vehicle rustproofing.

Sources of Sulfonates

The sulfonates which are the prime building block for this generic type of coating were used first in the lubricant additive industry. In the 1930s, the original detergent additives used in crankcase oils to prevent build up of deposits on the internal surfaces of internal combustion engines were alkaline earth salts of carboxy acids, such as naphthenic, phenyl stearic and chlorophenyl stearic acids. These materials removed surface deposits, but were corrosive to heavy metals in the bearings. Alkaline earth phenate esters such as magnesium lauryl salicylate were also useful detergents.

Starting in 1940, alkaline earth salts of petroleum sulfonic acids were used as detergents. These acids are by products of the white mineral oil manufacturing process. They are obtained by sulfonation and removal of alkylated aromatic hydrocarbons from refined lubricating oil. The paraffinic portion of the oil does not sulfonate, and is used as mineral oil after the aromatic fractions have been removed as sulfonic acids. The acids have the general structure:

$$RArSO_3H$$

where R is a side chain with about 20 carbon atoms, and Ar is an aromatic moiety.

The alkaline earth metal salts of these acids were effective detergents and did not attack bearing metal. Some of the salts were basic, that is, they had more than the stoichiometric quantity of base associated with the acid, for example:

$$RArSO_3 - Ba - OH$$

The basicity of these salts not only contributed to detergency, but also neutralized acid contaminants formed during engine operation.

In time, the advantages of basicity led to an evolution of sulfonate detergent technology in the 1950s. Overbased sulfonates were developed, in which the amount of oil soluble alkaline earth metal associated with a sulfonic acid molecule was increased 10 to 15 times the normal equivalency by a carbonation process. These oil soluble products provided high detergent action and a considerably higher alkaline reserve to control acidity formed *in situ*. These overbased sulfonates in oil solution are somewhat analogous to formulated water soluble household detergents in which significant amounts of inorganic salt builders are used in combination with water soluble alkali metal sulfonates to provide detergency and dispersion of soils.

The source of alkylate for the oil and water soluble systems described are also related. Although most of the early oil soluble sulfonates were derived from refined petroleum oil, many of the current oil soluble sulfonates are made by sulfonation of polydodecylbenzene bottoms. The distilled dodecylbenzene is sulfonated to produce water soluble alkali metal salts

---
*The Lubrizol Corporation, Cleveland, Ohio

for use in household detergents. The bottoms are primarily mixtures of mono and didodecylbenzene.

$$CH_3(CH_2)_{11} - \langle O \rangle \qquad [CH_3(CH_2)_{11}]_2 \langle O \rangle$$

They also contain some higher boiling hydrocarbons, which are formed by oligimerization of olefin.

In recent years, proprietary synthetic sulfonates have been produced which provide improved performance properties especially with respect to ultraviolet stability.

Description of Overbased Sulfonates

Overbased sulfonates have the general structure:

$$(RAr - SO_3)_2 - M \cdot x(MA)$$

where R = one or more alkyl side chains with from 12 to 30 carbon atoms; M = an alkaline earth cation such as Ca, Mg or Ba; A = an anion such as CO_3^{--}; x = 10 to 30

The unusual feature of these compositions is that amorphous inorganic alkaline earth salts (MA) are associated, or complexed with, alkaline earth sulfonate in such a way as to render them oil soluble. These products provide outstanding detergency, rust inhibition, and acid contaminant control in crankcase oils for internal combustion engines.

Basic Sulfonates For Coatings

In order to use basic sulfonates for coating applications, their physical properties had to be changed. When oil soluble overbased sulfonates (as described) are treated with combinations of certain polar liquids, the MA portion of the complex precipitates in the form of submicron sized crystals such as the calcite and vaterite forms of $CaCO_3$ and the witherite form of $BaCO_3$. An alternate procedure has been developed to produce other types of basic sulfonates containing crystalline particles, such as the brucite form of $Mg(OH)_2$. These crystals have average particle sizes in the 50 to 300 Å range. The crystals are associated with the alkaline earth sulfonate, and they remain suspended in organic solution even at high dilution. As the crystal forms in a given solution, the rheology changes from a Newtonian solution to a thixotropic liquid. The magnitude of the change depends on the concentration and type of crystals, as well as the size and size distribution of the crystals.

In addition to this beneficial rheological change, the crystalline form of the basic sulfonates provides improved corrosion resistance in metal coatings for several reasons. First, they mechanically reinforce the film so that it does not tend to erode or distort when it is exposed to sprayed liquids or splashed contaminants. Second, these basic complexes provide a large excess of reserve alkalinity. When coatings are exposed to moist environments, these inorganic compounds ionize and develop alkaline pH in the vicinity of the metal surface. The potential benefit of alkalinity in preventing corrosion of steel is well known. The selection of the proper inorganic compounds for use in coatings is important. It is desirable to provide an alkaline environment near the metal surface, but it is not desirable to build hydroscopic particles into a protective coating. The following table indicates the fundamental data which provide a basis for selection of the most desirable compounds.

Table I

Compound	Molecular Weight	Solubility Product Constant K_{sp}	Solubility in water g/100 ml	pH of Saturated Solution
$Mg(OH)_2$	58	10^{-11}	8×10^{-4}	10.4
$MgCO_3$	84	10^{-5}	3×10^{-2}	10.9
$Ca(OH)_2$	74	10^{-5}	1×10^{-1}	12.4
$CaCO_3$	100	10^{-8}	1×10^{-3}	10.1
$Ba(OH)_2$	171.4	10^{-2}	2	13.5
$BaCO_3$	197.4	10^{-8}	2×10^{-3}	10.1

As can be seen from the data, the carbonates of calcium and barium, and the hydroxide of magnesium, are the most desirable forms for use in coatings.

A third important reason for the effectiveness of the complexes is the dual nature of the organic sulfonate molecule.

$$\begin{array}{ccc} O & & O \\ \parallel & & \parallel \\ -S-O-M-O-S- \\ \parallel & & \parallel \\ O & & O \end{array}$$

The polar sulfonate group has an affinity for metal surfaces, and, thus, provides good adhesion, and the high molecular weight hydrocarbon portion of the molecule provides a physical barrier to moisture and atmospheric contaminants.

Formulation Modifiers

The heterogeneous basic sulfonate complexes (described) are resinous solids. To be used in coatings, they must be combined with plasticizers, resins, and polymers to obtain desired film properties. Typical auxiliary film forming ingredients include petroleum oils and synthetic oils of various viscosities, hydrocarbon resins and polymers, and waxes. The choice of ingredients will depend on the end application. In some cases, the resins and polymers may not be completely soluble in the film.

These separate organic filler phases can also modify flow and surface properties of the film. Low concentration of pigments may be used if required by the application.

A unique additive which has proved to be useful in certain sulfonate based coatings is a polymeric organic acid phosphate:

$$(RO)_a - P(O)(OH)_{3-a}$$

where R is a combination of oligomers containing hydroxyl groups.

It is important, at this point, to discuss the reason for using this type of functionality in some of these coatings. Inorganic phosphate coatings markedly improve the adhesion of coatings to metal surfaces which are exposed to corrosive environments. This beneficial effect is obtained on ferrous, zinc (galvanized), and aluminum surfaces. Coatings lose adhesion to unphosphated surfaces because of corrosion products which form at voids and edges, then grow and spread across a metal surface, undercutting the film. In many cases, phosphating of metal surfaces prior to application of a protective coating is not possible. It has been found that organic phosphates in coatings can, in some instances, produce some of the improvements attributable to phosphating of the metal surface.

The polymeric organic phosphate additive, when properly dispersed in certain sulfonate type coatings, contributes substantially to the corrosion inhibitive features of the coating. Although the exact mechanism is difficult to determine in this heterogeneous system, it appears that the organic phosphate maintains its acidity and does not react with the surrounding alkalinity because it is a separate dispersed phase. It can be shown, however, by an ammonium molybdate spot test, that it does react with the metal surface. The test data indicate that this association of the organic phosphate with the metal surface can improve the corrosion inhibitive action of sulfonate-based coatings.

Carboxy acid salts can also contribute to improved properties of sulfonate based coatings.

$$[R - C - (O) - O]_2 - M$$

where M is an alkaline earth cation.

These materials are used to modify the rheology and the physical properties of sulfonate based coatings. In general, salts of aliphatic acids will increase thixotropy while salts of aromatic acids will decrease thixotropy.

Alkanolamine salts of organic phosphates are used in water dispersions of sulfonate resins systems.

$$(RO)_a - P(O)(OH)_{3-a} \cdot N(R^1 - OH)_b H_{3-b}$$

where R is a poly alkoxy ether chain and R^1 contains two or three methylene groups.

When sulfonate type resins are dispersed or emulsified in water, it is necessary to use additives of this type to prevent flash rusting when the water based coating is applied. During the period when water is evaporating from the wet film, this relatively mild type of surface corrosion is possible.

Comparative Performance Data

To illustrate the performance contributions of the various types of basic sulfonates and some of the functional chemicals used with them, the compositions in Table 2 were evaluated in two types of exposure: weatherometer and salt fog.

Table 2 — Film Compositions

	A	B	C
1	Proprietary Synthetic Sulfonate (PSS)	Sulfonate from Detergent Alkylate (SDA)	Sulfonate from Petroleum Oil (SPO)
2	PSS + Organic Phosphate	SDA + Organic Phosphate	SPO + Organic Phosphate
3	PSS + Hydrocarbon Resin	SDA + Hydrocarbon Resin	SPO + Hydrocarbon Resin
4	PSS + Organic Phosphate + Hydrocarbon Resin	SDA + Organic Phosphate + Hydrocarbon Resin	SPO + Organic Phosphate + Hydrocarbon Resin

All systems contain petroleum oil as a plasticizer

These systems were all prepared as dispersions in aliphatic solvent. The one combination that was not included was organic phosphate plus hydrocarbon resin. It is not possible to make a stable dispersion of these two materials in solvent without thixotropic basic sulfonate being present. Wet films were drawn down on cold rolled steel panels to give solvent free film thicknesses of approximately 4 mils.

Weatherometer (ASTM G-23-69, Type D). Figures 1, 2, and 3 show the appearance of the films after various exposure times in the weatherometer. It is very apparent that if ultraviolet exposure is to be encountered in the application of the coating, a synthetic sulfonate must be used. The other sulfonates have been satisfactory for internal and underbody automotive applications. There are many types of hydrocarbons resins and polymers that can be used in these coatings. It is obvious that the particular resin used in

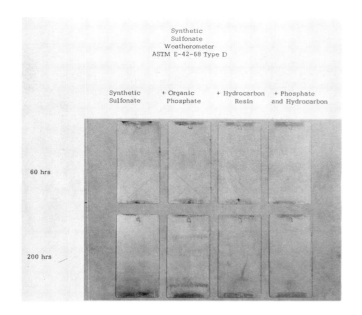

FIGURE 1 — Film compositions A 1, 2, 3, and 4 from Table 2 after 60 and 200 hours weatherometer exposure

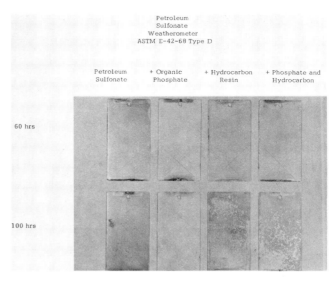

FIGURE 3 — Film compositions C 1, 2, 3, and 4 from Table 2 after 60 and 100 hours weatherometer exposure

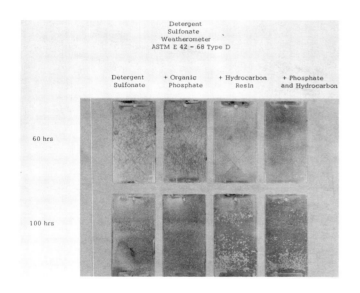

FIGURE 2 — Film compositions B 1, 2, 3, and 4 from Table 2 after 60 and 100 hours weatherometer exposure

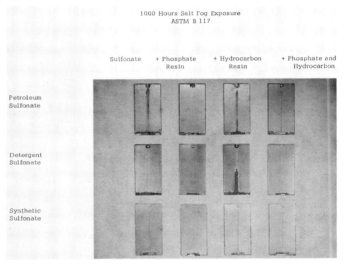

FIGURE 4 — Film compositions A, B and C from Table 2 after 1000 hours of salt fog exposure

this test series would not be satisfactory if ultraviolet exposure were to be encountered.

Salt Fog (ASTM B 117). The test panels shown in Figure 4 indicate the differences in sulfonates with respect to corrosion resistance and film integrity. The beneficial effects of the phosphate resin on both corrosion resistance and film integrity are apparent. This particular hydrocarbon resin can adversely effect corrosion resistance. Resins and polymers are used to increase film integrity and hardness. Overall, the synthetic sulfonate systems provide the best corrosion resistance.

Performance Properties of Proprietary Systems

Solvent based sulfonate type coatings have been in use for many years. Tables 3 and 4 indicate typical physical and performance properties of these materials.

This composition meets MIL-C-0083933(A)MR, the Post Office Specification VB-65-1, and the solvent system conforms to Rule 101 of the South Coast Air Quality Management District.

More recently, a *hot melt sulfonate type coating* has been developed for use on critical automotive components such as frames, suspension members, springs, and axles. Tables 5 and 6 indicate the typical physical and performance characteristics of this hot melt applied coating.

Table 3 — Solvent Based

Typical Properties

Pounds per U.S. gallon at 60 F.	8.20
Specific gravity at 60 F.	0.985
Flash point, P.M.C.C.,[1] F Minute	100
Typical	120
Color	Light Tan
Consistency	Thixotropic Fluid
Nonvolatiles, % Weight	58
% Volume	47

[1] Pensky-Martins Closed Cup

Table 4 — Solvent Based

Laboratory Performance Data

Sprayed Film Thickness	
Wet, mils	8 to 10
Dry, mils	3.5 to 4.5
Corrosion Resistance on Cold Rolled Steel	
ASTM-B 117 Salt Fog, hours	1,000
Undercutting at scribe	1/32
Salt Water Immersion	14 days +
High Temperature Flow Resistance	
300 F to 800 F	No flow
Low Temperature Flexibility	
No cracking, peeling or chipping	– 10 F or lower
Minimum Drying Time to Resist Wash Off By Water	
77 F	1 hour
40 F	1 hour

Table 5 — Hot Melt

Typical Properties

Color	Black
Flash point	400 F (204 C) minute
Fire point	400 F (204 C) minute
Autoignition Temperature	765 F (407 C)
Weight per gallon	
1. Solid at 77 F (25 C)	9.5 lbs (SG = 1.14)
2. Liquid at 250 F (121 C)	8.6 lbs (SG = 1.03)
Solids Content	99% minimum
Brookfiled viscosity at 250 F (121 C)	200 to 300 cps
Stability	
1. At room temperature	Indenfinite
2. At 225 to 250 F (107 to 121 C)	Constant slow agitation

Table 6 — Hot Melt

Laboratory Performance Data

Salt Spray, scribed	200 hours+ with excellent rating and no loss of adhesion
Gravelometer at – 20 F (– 29 C)	Indentations to visible metal, but no chipping
Gravelometer as above, plus 336 hours salt spray	Excellent, little or no rust
High Temperature Resistance 4 hours at 190 F (88 C)	No sag, flow or dripping
QUV (100 hours) plus Scribed Salt Spray (33 hours)	Good — Excellent
Detergent Wash plus Scribed Salt Spray (336 hours)	Good — Excellent
Water Immersion, 7 days at 100 F (38 C)	
1. Distilled Water	No Effect
2. 5% Salt Water	No Effect

Table 7 — Water Based

Typical Properties

Pounds per U.S. Gallon at 77 F	8.77
Specific Gravity at 77 F	1.054
Flash Point, P.M.C.C	greater than 170 F [1]
Color	Tan
Consistency	Thixotropic Fluid
Nonvolatiles, %Weight	37.0
% Volume	32.7
pH	8.43
Freeze thaw stability (5 cycles)	Stable

[1] Maximum temperature obtainable with this equipment because evaporation of water causes foaming.

A unique feature of this material is its carefully balanced rheology. Even though the material is fluid at application temperatures, it is possible to dipcoat very large objects such as automobile frames (up to 19 feet in length) and obtain a uniform film thickness from top to bottom. This is because of thixotropy in the liquid state, the temperature (230 to 250 F) of the liquid when an object is dipcoated, and the solidification point (180 to 190 F) of the composition. This material conforms to the application and performance specifications of several major automotive manufacturers.

Several factors, including availability, emission considerations, and increasing costs, have made it desirable to replace hydrocabon solvents with water for corrosion control coatings. In recent months an effective *water based sulfonate type coating* has been introduced which provides the corrsion inhibiting characteristics obtained with solvent based coating systems.

Tables 7 and 8 indicate the typical physical and performance characteristics of the water based systems.

There are two unusual features of this water based coating, in addition to its ability to prevent corrosion. First, it can be applied in very humid atmospheres and, before the coating has become completely water free, it can be exposed to rain and water without being washed off. Second, after three years of actual service exposure, it has been found that this coating maintains its plasticity to a much greater degree than do the solvent based systems.

Table 8 — Water Based

Laboratory Performance Data

Sprayed Film Thickness	
Wet, mils	12 to 16
Dry, mils	4 to 5
Corrosion Resistance on Cold Rolled Steel	
ASTM-B 117 Salt Fog, hours	greater than 1000
Undercutting at scribe	less than 1/32
High Temperature Flow Resistance	
300 F to 800 F	No flow
Low Temperature Flexibility	
No cracking, peeling or chipping	−10 F or lower
Minimum Drying Time to Resist Water	
Wash off 77 F	3 hours

Conclusion

1. Basic sulfonate coatings provide outstanding corrosion protection for metal surfaces because they are stable dispersions of several types of chemical functionality.

   A. They provide an alkaline environment at the metal interface when moisture is present.
   B. A passivating layer is formed on the metal surface as a result of a reaction with the dispersed organic phosphate phase.
   C. Submicron sized particles provide a beneficial thixotropic rheology during application and a permanent physical reinforcement in the applied film.

2. These coatings can be formulated and applied from solvent, as solventless hot melts, and as water dispersions. The latter two features are of great significance now, and in the future, because of ecological considerations, and because of the shortage of hydrocarbons.

3. The sulfonate coatings developed to date are generally not as hard as conventional paint systems. Improved formulating techniques involving higher molecular weight materials can provide sulfonate based coatings with improved hardness, abrasion resistance, and film strength properties.

Water Displacing, Corrosion Preventive Compounds

*Charles R. Hegedus**

Introduction

Performing corrosion control procedures in a marine environment is a difficult task. Salt spray and moisture on a substrate to be coated can lead to inadequate adhesion, voids, and other defects in even the best coatings. These defects lead to additional corrosion and maintenance problems. United States naval personnel have encountered such obstacles when attempting touch up procedures on both land and carrier based aircraft. Areas where paint has cracked or chipped leave bare metal exposed, consequently inviting corrosion. Subsequent paint touch up is difficult because of high humidity and salt spray on or near the substrate. The Navy is attempting to solve this problem by utilizing water displacing, corrosion preventive compounds. These compounds displace water upon application and perform as corrosion preventives during usage.

Three compounds which have been developed at the Naval Air Development Center (NADC) are:

1. AML-350, an ultra thin, corrosion preventive, water displacing compound.

2. AMLGUARD, a corrosion preventive water displacing compound which is applied to a dry film thickness of 25.4 to 50.8 μm (1 to 2 mils).

3. A pigmented, water displacing corrosion preventive paint which dries to a flexible and durable coating with good adhesion.

This paper discusses a mechanism for the displacement of water droplets on a metal surface that would be applicable to these three products. The specific uses for products are discussed.

It must be noted that when discussing water displacement in this paper, bulk water droplets are being considered, not water molecules attached to the surface by secondary chemical bonds.

Background

Displacement of water by organic compounds is a well researched subject. Much of this work has been reported by Zisman and coworkers at the U.S. Naval Research Laboratory (NRL).[1-9] This early work was performed to develop cleaning and preservation materials and techniques for salvaged naval equipment.[4-8] In this effort, the removal of gross amounts, or puddles, of water was a significant problem. The research at NRL lead to the evaluation of many materials as water displacing agents.[4]

As a result of the NRL research, a mechanism for the displacement of water puddles by organic compounds has been defined.[9] The means by which water is displaced by this mechanism are as follows: the displacing agent is dropped onto a puddle of water; being slightly soluble in water (usually between 1 and 10% by weight), the agent penetrates to the surface of the solid; by means of preferential adsorption, the agent adheres to the surface, thus displacing the water layer from the metal substrate. This phenomenon is illustrated in Figure 1. The alcohols, especially 1-butanol, have been found to facilitate this mechanism because of their amphipathic nature. The results and conclusions obtained from the research were based on the requirement that layers or puddles of water had to be displaced from a surface. This requirement resulted from efforts to salvage flooded equip-

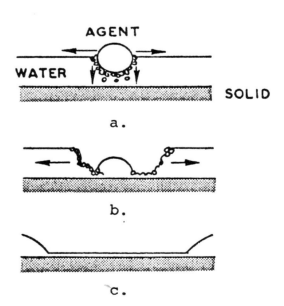

FIGURE 1 — Mechanism of water displacement: a) displacing agent applied to water surface, mixture with water begins; b) agent reaches the surface while pushing water aside; c) preferential adsorption of the agent over water allows water to be displaced from the surface.

---

*Naval Air Development Center, Warminster, Pennsylvania

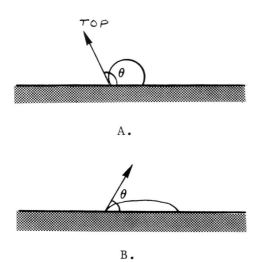

FIGURE 2 — A) a liquid drop on a solid surface with a contact angle, Θ greater than 90°. The liquid does not wet the surface. B) a liquid which has semiwet the surface, Θ between 0 and 90°.

ment which often has gross amounts of entrapped water that must be displaced. The compounds and coatings discussed in this paper, however, were developed to displace water from the exterior surfaces of aircraft. Surface preparation consists of light sanding and removal of dirt and sanding dust. It is assumed that only water droplets from the atmosphere will be present on or near this surface at the time of coating application. As will be discussed, the requirements for materials protecting exterior surfaces of aircraft are different than those for materials which displace puddles or layers of water.

Discussion

The phenomenon of water displacement occurs because of several chemical and physical interactions between the water, the water displacing compound, and the metal substrate. The chemical and physical interactions are effects of surface tension, miscibility, and secondary (bonding) forces.

The mechanism of water displacement requires the applied compound to: 1) spread over the surface upon application, entirely wetting the substrate (surface tension effects); 2) be immiscible with water to ensure no water entrapment in the compound (water displacing agent miscibility criterion); 3) have higher affinity for the substrate than water in order for the coating to creep under the water droplets, displacing them from the substrate (adsorption effects).

Surface Tension Effects. A liquid placed on a metal surface will generally be observed as a drop having a definite contact angle, Θ. This angle can be found theoretically by Young's equation:[10]

$$\cos \Theta = (\gamma_S - \gamma_{SL})/\gamma_L \quad (1)$$

where γ_S = the surface tension (surface energy) of the solid at the solid/vapor interface

γ_{SL} = the surface tension (surface energy) at the solid/liquid interface

and γ_L = the surface tension of the liquid

The liquid is said not to wet the surface if the contact angle is greater than or equal to 90°; that is, the liquid will form a drop on the surface. Conversely, if the liquid forms a contact angle of 0°, it is said to completely wet the surface. In many cases, a liquid will spread or "semiwet" the surface, with Θ ranging between 0° and 90°. Figure 2 illustrates these situations.

The total differential in surface free energy, G, of a liquid on a solid surface at constant temperature and pressure is given by:

$$dG = (\partial G/\partial A_S)dA_S + (\partial G/\partial A_{SL})dA_{SL} + (\partial G/\partial A_L)dA_L \quad (2)$$

where A = surface area
S = solid
L = liquid

at equilibrium

$$dA_L = -dA_S = dA_{SL} \quad (3)$$

$$\frac{\partial G}{\partial A_S} = \gamma_S \quad (4)$$

and

$$\frac{\partial G}{\partial A_{SL}} = \gamma_{SL} \quad (5)$$

$$\frac{\partial G}{\partial A_L} = \gamma_{SL} \quad (6)$$

The free energy change of a liquid spreading on a surface is given by $-dG/dA_L$ (solid surface area), which is defined as the spreading coefficient, $S_{L/S}$, of a liquid (L) on a solid (S). Substituting Equations 3, 4, 5, and 6 into Equation 2, and rearranging yields:

$$S_{L/S} = -\left(\frac{\partial G}{\partial A_L}\right) = \gamma_S - \gamma_L - \gamma_{LS} \quad (7)$$

The criterion for spreading is that $S_{L/S}$ be positive. Qualitatively, this is best accomplished by having a liquid with a low surface tension.[11]

Therefore, to satisfy the first criterion for water displacement, that is, spreading of the agent, this agent must possess a relatively low surface tension. This can be accomplished easily by incorporating organic solvents which have low surface tension into the water displacing compounds.

Table 1 lists the surface tension of the compounds discussed in this paper with that of water. As expected by these values, the water displacing materials spread on the steel surface with a near 0° contact angle. Water alone did not spread freely, but was found to have a contact angle of 85° on the steel sub-

Table 1 — Surface Tensions of Water Displacing Compounds and Water

	γ (at 21°C) (dynes per cm)
Water	72.8
AML-350	29.0
AMLGUARD	25.5
Water Displacing Paint (WDP)	27.3

Steel: AISI 1010 steel, surface ground to 10 to 20 microns and polished.

strate used. Thus, the water displacing agents satisfy the first requirement for water displacement; that is, they spread over the substrate upon application.

Miscibility Criterion. To displace a water droplet from a metal surface, a coating must creep under the water droplet. To ensure that no water remains on or near the metal surface, it is necessary that the coating and water be immiscible. If they were miscible, mixing could ensue after application, entrapping water. This water could lead to voids and/or adhesion loss of the paint to the substrate. The entrapped water would also be a potential corrosive environment to the metal.

To ensure immiscibility of the displacing agents and water, solvents and organic additives, which will prevent mixing, are incorporated in the agents. Common solvents used in these materials are trichlorotrifluoroethane and mineral spirits. When solvents which are highly water soluble are substituted, water displacing efficiency is decreased.

Adsorption Effects. When polar organic compounds such as alkali and alkaline earth metal salts of sulfonated oils are dissolved in organic solvents and applied to a metal surface, they can adsorb onto the surface. If water is present on the surface in the form of a droplet, the displacing agent, having a higher affinity for the metal, creeps under the water, and thus displaces it. The final requirement for the displacement of water, preferential adsorption of the agent, can, therefore, be obtained by incorporating easily adsorbed materials in the displacing agent.

The compounds discussed in this paper contain petroleum sulfonates which aid in the adsorption of the applied compound onto the metal surface. These sulfonates also aid in the protection of the substrate against corrosion. The corrosion inhibition properties of sulfonates has been reported by Baker, et al.[16] They act by forming a physically adsorbed layer at the metallic surface. This type of inhibition occurs because the adsorbed film isolates the metal surface from corrosive environments. Although the penetration of small amounts of water molecules is possible, adsorbed inhibitors restrict their movement, making direct contact of the water molecules with the metal more difficult.

In summation, the materials investigated in this work displace water droplets by: 1) spreading over the substrate, completely wetting it; 2) being immiscible with water to avoid water retention; 3) preferentially adsorbing, thus penetrating under the water droplets.

This mechanism is slightly different than that proposed by Zisman and colleagues. The compounds discussed are virtually immiscible with water, whereas the previous mechanism utilized solvents slightly soluble in water. Water solubility is necessary in displacing puddles of water because the means by which the displacing agent reaches the substrate is by diffusion of the solvent through the water layer. When the agent reaches the solid surface, it is preferentially adsorbed.

In contrast to this, the series of photographs in Figures 3, 4, and 5, illustrate the displacement of a water droplet by AML-350. The results illustrated in these figures are similar to those observed when AMLGUARD or the water displacing paint is used.

In Figure 3, the periphery of a water droplet (light ellipse) on a steel panel (dark background), is highlighted. AML-350 is introduced onto the steel panel next to the water droplet. As the AML-350 contacts the water, a small deformation of the droplet occurs. This is observed in frame 2 of Figure 3, where the arrow illustrates the deformation of the drop. Following the path which will decrease the total free energy of the system, the AML-350 travels around the droplet. This is observed as the increased deformation about the periphery of the droplet. The difference between each frame in Figure 3 represents an elapse time of 2.5 milliseconds.

Figure 4 is another illustration of the effect of AML-350 on water droplets. In Figure 4A, AML-350 is introduced onto a steel surface near a water droplet. In Figure 4B, the displacing agent has contacted the water droplet, and diffusion of the AML-350 under the water has begun. In Figure 4C, the displacing agent has crept further under the droplet. And in Figure 4D, approximately five minutes after initial contact, the AML-350 has completely displaced the water droplet from the steel surface. If allowed to remain in this condition for several days, this 1010 steel panel will show

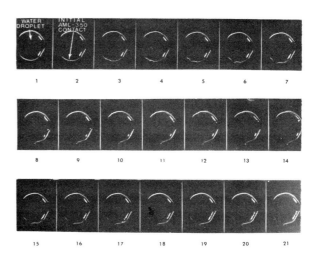

FIGURE 3 — Deformation of water droplet induced by AML-350.

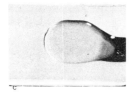

FIGURE 4 — Displacement of water droplet from steel panel by AML-350.

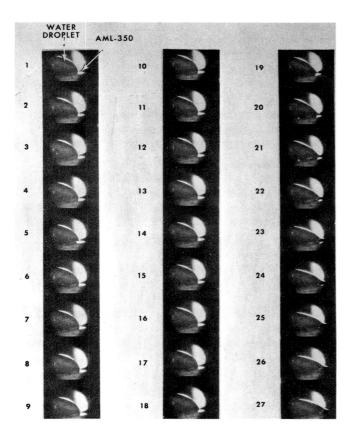

FIGURE 5 — Collapse of water droplet leading edge under 25X magnification.

no evidence of corrosion, whereas a panel not treated with AML-350 would show corrosion products in a period of hours.

Figure 5 is a series of photographs filmed at 400 frames per second under 25X magnification. A water droplet is observed on the left side of the photographs as a quarter of an ellipse. AML-350 is observed as the liquid approaching the drop from the right. The displacing agent contacts the droplet (Frame 18) and creeps under the leading edge. The interfacial surface tension between the water and the substrate is decreased and the contact angle becomes smaller, until the water droplet eventually collapses. At this point, the periphery of the water droplet is displaced. The remaining water is displaced by the lateral diffusion of AML-350 between the steel and the water droplet.

End Use of These Products

AML-350. This product is a corrosion preventive, water displacing compound which is to be applied to a film thickness of 2 to 5 μm. It is composed of a petroleum sulfonate and a mineral spirits type solvent. The combination of these materials allows the coating to spread over metal and creep under water droplets on a metal surface, thus displacing the water. Eventually the solvent evaporates and the petroleum sulfonate remains, forming a soft, thin, oily film. This film isolates the metal from the environment, thus acting as a temporary passive corrosion preventive. The petroleum sulfonate also assists in protection against corrosion by performing as a corrosion inhibitor.[16] The mechanism of this inhibition is the same as described for adsorbed polar type inhibitors.

AML-350 is intended for use on internal metallic parts and electrical connectors. It has widespread use on airborne electronic equipment. It is, generally, not intended for use on external areas because of its soft condition; however, under adverse weather conditions, or when aircraft are liberally water soaked, it may be applied as a temporary protective measure.

AMLGUARD. This coating is a water displacing corrosion preventive compound also. However, this material contains polymeric resins which, upon application and cure, form a dry, hard film. Generally, the coating is applied to a dry film thickness of 25.4 to 50.8 μm (1 to 2 mils).

AMLGUARD is a combination of organic solvents, silicone and silicone alkyd resins, barium petroleum sulfonate, and several other additives. It also displaces water by spreading over the metal and creeping under the water droplets. Drying occurs via solvent evaporation, leaving a solid film. Although AMLGUARD dries to the touch in 18 hours, it continues to cure. Upon aging 1 to 3 months, the film cures to a hard, flexible finish. Corrosion protection is provided by the physical barrier of the coating, and also by barium petroleum sulfonate and alkyl ammonium organic phosphate performing as corrosion inhibitors.

This material is intended for temporary use on external aircraft parts where it offers excellent corrosion protection. It has also been recommended for use on leading edges of aircraft wings and helicopter blades, and on exhaust and gun blast areas, as well as many other aircraft components where erosion resistance and corrosion protection is necessary. AMLGUARD has been used successfully on mild steel, stainless steel, aluminum, magnesium, zinc, cadmium, copper and brass.[17,18]

Water Displacing Paint. This material is a pigmented coating which will displace water, dry, and subsequently afford corrosion protection. It is composed of a petroleum sulfonate, silicone alkyd resin, organic solvents, pigments, and other organic additives. The mechanism by which this material displaces water is the same as that of the aforementioned compounds. This pigmented coating dries to a hard, flexible finish which protects the substrate from corrosion by: 1) the physical barrier of the coating; 2) corrosion inhibiting pigments; that is, molybdates, and chromates.

This water displacing paint is designed as a touch up paint on exterior surfaces of aircraft where original paint has cracked or chipped, and total repainting is not feasible. Such a situation is confronted on operational aircraft deployed on board aircraft carriers where paint touch up is necessary and must be completed quickly and efficiently. This paint was designed to be applied during deployment and to last indefinitely until total repainting of the aircraft is necessary.

Experimental Results

To test water displacing ability of a compond, a 5.1 by 10.2 cm (2 by 4 inch) AISI 1010 steel panel is placed at a 30° angle from the horizontal.

Synthetic sea water[1] is then sprayed over the entire panel to form water droplets. Using a pipette, one milliliter of the test compound is applied along the upper edge of the steel panel and allowed to flow down the panel. After one minute, another milliliter of test material is applied as before. The panel is then placed above distilled water in a closed desiccator at 21 C. After a period of 16 hours, the panel is observed for corrosion products. These corrosion products indicate the synthetic sea water was not totally displaced.

Figure 6 is a photograph of the steel panel sprayed with tinted synthetic sea water. Figures 7 and 8 illustrate the effects of AMLGUARD and the water displacing paint 5 minutes after the second application of the test material. Figure 9 shows the results of AML-350, AMLGUARD, the water displacing paint, and a control (only synthetic sea water applied) after water displacement test and 16 hours in the desiccator.

Figures 10 and 11 illustrate AMLGUARD's performance in this test. In Figure 10 the bottom half of the steel panel has been treated with AMLGUARD. The water droplets are immediately being displaced. Figure 11 illustrates the results 5 minutes after the ap-

---

[1] Synthetic sea water solution: 50 grams of sodium chloride, 22 grams of magnesium chloride, 3.2 grams of calcium chloride, and 8.0 grams of sodium sulfate in 1.0 liter of distilled water.

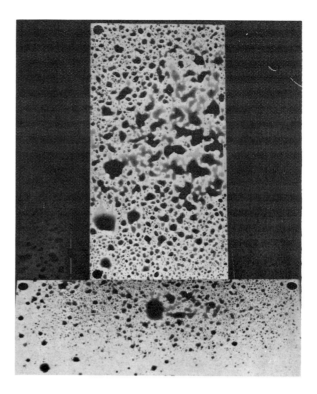

FIGURE 6 — Synthetic sea water applied to a bare steel panel resting 30° from the horizontal.

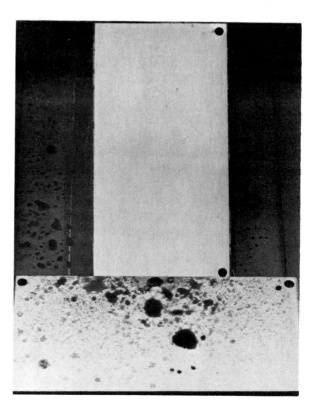

FIGURE 7 — Steel panel after AMLguard (clear coating) has displaced synthetic sea water from metal surface.

FIGURE 8 — Steel panel after water displacing paint (pigmented white) has displaced synthetic sea water from the steel surface.

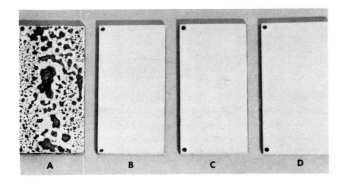

FIGURE 9 — Steel panels 24 hours after synthetic sea water application: A) no treatment; B) AML-350 treated; C) AMLGUARD treated; D) water displacing paint treated.

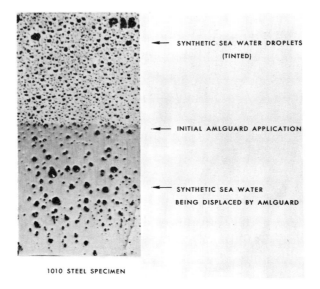

FIGURE 10 — Synthetic sea water displacement test.

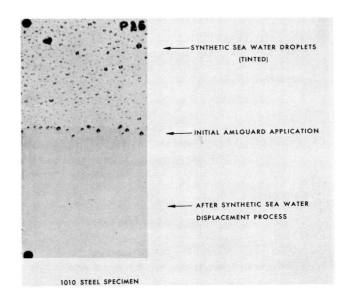

FIGURE 11 — Synthetic sea water displacement test.

plication of AMLGUARD — complete displacement of water.

Figure 12 shows a steel panel coated with an epoxy primer and a polyurethane top coat. An "X" has been masked and left unprotected. The upper half of the panel was then treated with AMLGUARD, the bottom half was untreated. After curing for seven days, the panel was placed in a 5% NaCl spray chamber for 7 days. Figure 12 illustrates the AMLGUARD treated area has no corrosion, while the untreated area has corroded severely.

Figure 13A shows a steel panel coated with the water displacing paint and placed in a 5% NaCl chamber for 500 hours with no corrosion products or coating defects on the main body of the panel. Slight rusting is noticeable on the upper right hand corner; this was due to a defective wax seal at the edge of the panel. Figure 13B also shows an aluminum panel[2] coated with the water displacing paint with an "x" scribed in the coating through to the aluminum substrate. After 1000 hours in a 5% NaCl spray chamber,

---

[2] 2024 bare aluminum treated with a chemical conversion coating meeting Mil. Spec. MIL-C-5541, Class 1A requirements. The "X" scribe went through this coating into the aluminum.

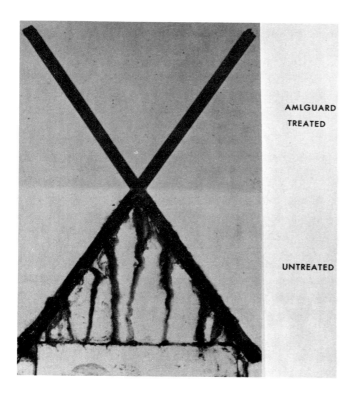

FIGURE 12 — Exposure to 5% salt spray cabinet for 7 days. AMLGUARD tinted for discernibility on aircraft surfaces.

FIGURE 14 — Sulfurous acid/synthetic sea water spray test in progress.

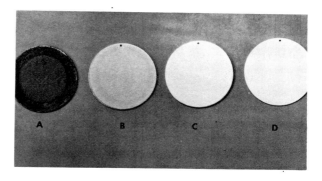

FIGURE 15 — Sulfurous acid/synthetic sea water spray test specimens.

FIGURE 13 — A) 1010 steel and; B) 2024 aluminum coated with water displacing paint and subjected to 500 and 1000 5% NaCl fog, respectively.

no uplifting, blistering, or loss of adhesion occurred. Slight corrosion occurred in the scribe after 500 hours exposure, and is considered minor.

Figure 14 is an illustration of a sulfurous acid/synthetic sea water spray corrosion test in progress. This test is defined in Mil. Spec. MIL-C-81309. It consists of a turntable which holds mild steel disks coated with the test compound. These disks are positioned horizontally, facing down as the turntable rotates over a sulfurous acid/synthetic sea water spray. Figure 15 illustrates test specimens subject to this test. Specimen A is uncoated 1010 steel used as a control after being subjected to the test for one hour. Specimens, B, C, and D are coated with AML-350, AMLGUARD, and the water displacing paint. Specimen B underwent 4 hours exposure, and specimens C and D underwent 10 hours exposure, each with no detrimental effects to the coating or substrate.

These coatings possess the ability to displace water from a metal substrate, and also provide protection against corrosion after application. They also display specific properties necessary to perform their intended function. For example, AML-350 is a soft film; hence, it is only temporary because it was intended to be used as a quick means for displacing water and preventing corrosion on electronic equipment and exterior areas of aircraft. AMLGUARD is clear, but dyed blue so that it will not detract from the appearance of the aircraft but will be noticeable. WDP is pigmented and durable, so that reapplication will not be necessary during deployment.

Conclusions

Water displacing compounds perform by spreading over the substrate. By means of preferential adsorption, the displacing agent creeps under water droplets and displaces them from the substrate. Because of their water displacing ability and corrosion protection properties, these compounds afford improved corrosion protection to a number of metals. These compounds can be used for specific needs, such as pigmented coatings and corrosion preventive compounds.

References

1. H.R. Baker, C.R. Singleterry, and W.A. Zisman, Factors Affecting the Surface — Chemical Displacement of Bulk Water from Solid Surfaces, Naval Research Laboratory Report 6368, Feb. 1968.
2. H.R. Baker, and W.A. Zisman, Water-Displacing Fluids and their Application to Reconditioning and Protecting Equipment, Naval Research Laboratory Report C-3364, Oct. 1948.
3. H.W. Fox, E.F. Hare, and W.A. Zisman, J. Phy. Chem. 59, p. 1097 (1955).
4. H.R. Baker, P.B. Leach, C.R. Singleterry, and W.A. Zisman, Surface Chemical Methods of Displacing Water and/or Oils and Salvaging Flooded Equipment, Part 1 — Practical Applications, NRL Report 5606, Feb. 1961.
5. H.R. Baker, P.B. Leach, and C.R. Singleterry, Surface Chemical Methods of Displacing Water and/or Oils and Salvaging Flooded Equipment, Part 2 — Field Experience in Recovering Equipment Damaged by Fire Aboard USS CONSTELLATION and Equipment Subjected to Salt-Spray Acceptance Test, NRL Report 5680, Sep. 1961.
6. H.R. Baker, and P.B. Leach, Surface Chemical Methods of Displacing Water and/or Oils and Salvaging Flooded Equipment, Part 3 — Field Experience in Recovering Equipment and Fuselage of HH 52A Helicopter after Submersion at Sea, NRL Report 6158, Oct. 1964.
7. HR. Baker, and P.B. Leach, Surface Chemical Methods of Displacing Water and/or Oils and Salvaging Flooded Equipment, Part 4 — Aggressive Cleaner Formulations for Use on Corroded Equipment, NRL Report 6291, Jun. 1965.
8. HR. Baker and P.B. Leach, Surface Chemical Methods of Displacing Water and/or Oils and Salvaging Flooded Equipment, Part 5 — Field Experience in Removing Sea-Water Salt Residues, Sand, Dust, and Soluble Corrosive Products from AN/FPS-16 (XN-1) Missile-and Satellite-Tracking Radar, NRL Report 6334, Oct. 1965.
9. R.N. Bolster, Removal of Fluid Contaminents by Surface Chemical Displacement, Surface Contamination, 1, Ed. K.L. Mittal, 1979.
10. T. Young, Miscellaneous Works, 1, G. Peacock, ed., Murray, London, 1955.
11. A.W. Adamson, Physical Chemistry of Surfaces, John Wiley and Sons, New York, 1976.
12. R.B. Waterhouse and J.H. Schulman, J. Oil and Col. Chem. Assoc., 38, p. 646, 1955.
13. E.L. Cook and N. Hackerman, J. Phys. Chem., 56, p. 524, 1952.
14. S.G. Daniel, Trans. Far. Soc., 47, p. 1345, 1951.
15. E.G. Greenhill, Trans. Far. Soc., 45, No. 319, part 7, p. 625, 1949.
16. H.R. Baker, C.R. Singleterry, and E.M. Solomon, Ind. Eng. Chem., 46, No. 5, p. 1035, 1954.
17. Avionic Cleaning and Corrosion Prevention/Control, NAVAIR 16-1-540 Manual.
18. G.J. Pilla, The Development of AMLGUARD, A Clear, Water Displacing, Corrosion Preventive Compound, Naval Air Development Center Report No. NADC-78220-60, May 1979.
19. G.J. Pilla and J.J. DeLuccia, AMLGUARD — A Corrosion Preventive Compound for Aerospace, Met. Prog., 117, No. 6, p. 57, 1980.

Corrosion Control Under Thermal Insulation and Fireproofing

J.F. Delahunt*

In recent years, there have been an increasing number of reports concerning corrosion occuring on carbon steel structures and equipment that are either thermally insulated for energy conservation or coated with concrete for protection from fire. Field investigations, as well as laboratory investigative programs, have been carried out to evaluate the cause of corrosion and to determine appropriate means to mitigate it. The cumulative result of these various programs is presented within this paper, and it includes discussions concerning:
- Examples of corrosion in refineries, petrochemical plants, and pipelines occuring on insulated or fireproofed structures and equipment.
- Descriptions of potential corrosion mechanisms.
- Corrosion mitigation systems used to prevent attack of such equipment.

The seriousness and the aggressiveness of corrosive attack beneath insulation cannot be underestimated. It is, by definition, insidious in nature because it cannot be detected by visual examination or, in most cases, nondestructive testing systems. Presently, equipment and structures where corrosion is thought to be occurring can only be evaluated by removing the fireproofing or thermal insulation system. Corrosion has occurred beneath all major types of insulating, including mineral wool, fiberglass, foam glass, calcium silicate, organic cellular insulation (such as phenolic and polyurethane materials), and both poured in place concrete and hydraulically applied gunite.

Corrosion of Insulated/Fireproofed Equipment Widespread

The following are brief descriptions of corrosion problems which have occurred worldwide: first beneath fireproofing; second, under thermal insulation. Generally, where we mention fireproofing, we are speaking of concrete and gunite (both cementitious based coatings are applied in sufficient thickness to obtain two hour fire protection as specified in ASTM E-119 fire tests). Regarding thermally insulated equipment, attention will be directed to both hot and cold insulation, as well as thermally cycled insulated equipment.

*Exxon Research & Engineering Co., Florham Park, New Jersey

The first section of the discussion will be centered on examples depicting corrosion of carbon steel facilities beneath fireproofing.

Corrosion Beneath Concrete Fireproofing

The first two figures show examples of severe corrosion at an oil refinery located on the island of Aruba, in The Netherlands Antilles in the Caribbean Sea. The environment at this location is moderately (dry as far as total rainfall is concerned), but with fairly high velocity tradewinds which subject the refinery to much airborne moisture laden with sea salts. Generally, at night there is substantial deposition of ocean salts and moisture which evaporates during the day from radiant sun energy. The cycle is repeated until the sea salts are washed from surfaces by rainfall.

Figure 1 shows a 1 x 1½ foot hole caused by corrosion in a vessel skirt that was protected from fire with a 1½ to 2" of hydraulically applied concrete. Figure 2, again in the same location, shows a corroded steel beam supporting an atmospheric crude oil topping

FIGURE 1 — Corrosion beneath concrete fireproofing occurring on a vessel skirt. Refinery located on an island in the Caribbean Sea. Hole measure 1½ x 1 feet.

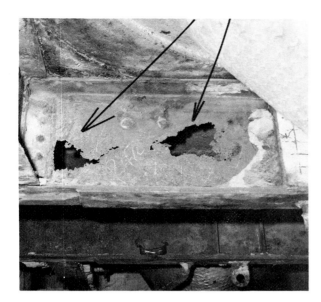

FIGURE 2 — Same location as Figure 1. Corroded steel support beam, again illustrating the aggressiveness of corrosion under concrete fireproofing.

FIGURE 3 — A photograph of a fireproofed beam in a process unit of a refinery located near Southampton, United Kingdom. Note spalling caused by expansion of rust scale and corroded edges of beams.

unit. Large holes, completely penetrating the web, were caused by corrosion occurring under the concrete and are clearly visible. It was reported that after 16 years from the date of construction of this unit, that fireproof structural steel had lost an average of 50% of its design thickness. The onset and severity of corrosion was accelerated, if not catalyzed by, too high a water/cement ratio, and the use of sea water to prepare the concrete jacketing formulation. Both of these factors will be discussed more fully in the subsequent sections of this paper dealing with corrosion mechanisms.

Figure 3 shows a supporting beam, at an atmospheric pipestill, at a refinery located near the Southampton coast in Great Britain. In this location, the climate is clearly different from that described previously. In general, the average temperatures are lower with less airborne chlorides, but the amount of precipitation is increased. As can be seen, there is much cracking and spalling of concrete, as well as corrosion of steel, occurring within this plant.

Corrosion under fireproofing has become a major concern at many other locations. For example, at a chemical plant located in Linden, New Jersey (at a solvents production facility), all concrete fireproofing had to be removed from structural steel, and large quantities of steel renewed because of accelerated attack. For comparison, New Jersey has about the same quantity of precipitation as the refinery located in Southampton, but is located in a more industrialized environment with colder average temperatures. Recent visits to facilities located at Singapore, Okinawa, and Saudi Arabia have indicated similar performance.

Thermal Insulation

There are a number of recent examples that can be called upon to illustrate the seriousness of corrosion that can occur on carbon steel thermally insulated. A few of these are summarized briefly in the next few paragraphs.

Figures 4 and 5 illustrate the extent and degree of corrosion that has occurred on four large cold (25 F) tanks, again located in southern England. These tanks store a variety of products including butane, propane and butadiene. The thermal insulation

FIGURE 4 — Overall view of several large refrigerated tanks located in the United Kingdom. Note rust streaking of galvanized steel jackets.

FIGURE 5 — Roof of tanks shown in Figure 4. Note deteriorated insulation (frothed in place polyurethane) and voluminous amount of rust scale.

FIGURE 6 — Polyurethane insulation (foamed in place) scraped from roof of heated crude oil tankage located in Europort in The Netherlands.

employed was an integral poured in place polyurethane foam (PUF)/galvanized steel jacket design. The jacket was constructed initially and offset from the tank shell by the use of trapezoidally shaped spacers of proper depth. The galvanized jacket was used as the vapor jacket to protect the PUF from water penetration. However, deep pitting, as well as general corrosion, has occurred on these 12 year old 80 × 120 foot high storage tanks; primarily, on the roof and the first two shell courses. As shown in Figure 4, the dark streaking is rust cascading down the shell from the freely corroding jacketing on the roof. Figure 5 shows the tank roof with polyurethane insulation removed so that a large quantity of corrosion scale is clearly visible.

Figures 6 and 7 show deteriorated polyurethane foam insulation being removed from heated crude oil floating roof tanks at a large oil terminal in Europe. In Figure 6, the polyurethane foam has been mechanically scraped from the roof and bagged in polyethylene sacks. Figure 7 shows the degree of scale formation occurring beneath the foam. Actually, in at least one area, there was crude oil on the roof, indicating complete penetration of the floating deck.

Metal loss can be the least of the problems caused by corrosion occurring beneath insulation. For example, at another refinery, an 8 inch diameter carbon steel heavy fuel oil pipeline reportedly operating at 250 F was externally insulated with calcium silicate block and protected with metal weather jacket. Corrosion occurred on this pipeline, causing a hydrocarbon leak. Ignition of oil escaping from the corroded pipeline led to an extremely large fire which, before it was extinguished, resulted in many hundreds of thousands of dollars in damages to process equipment. There are extensive investigations underway, at this refinery and others, to locate and identify all piping and equipment that is thermally insulated but not adequately protected against corrosion.

FIGURE 7 — Same tanks as noted before, but showing amount of rust and scale formation beneath polyurethane foam. One hole caused by external corrosion was discovered in the roof.

Four Corrosion Mechanisms Identified

There are four potential corrosion mechanisms that have been identified, which may cause aggravated corrosion beneath fireproofing or insulation. These are: (1) Formation and then destruction of a passive film formed on carbon steel; (2) formation of acidic water; (3) stress corrosion cracking of austenitic stainless steel; and (4) quality of design and application.

Each of these potential mechanisms is described in subsequent sections of this paper. The one common factor to each of these mechanisms is that water must be present. That in itself signifies failure of the system. For example, the ability of insulation to transfer energy increases 50 fold in the presence of water. Additionally, concrete containing water may provide less than 10% of its original fire protective properties.

Corrosion Beneath Fireproofing Promoted by Chlorides

Ideally, an intact cover of properly formulated concrete fireproofing will prevent corrosion of the underlying steel because of the passivating effect of the (OH^-) ions. Cement will react with water to produce (OH^-) ions from calcium hydroxide contained in the concrete mix. It has been shown by others that alkaline water solutions at pH 11 or greater almost completely inhibit the corrosion process. However, corrosion can occur if along with (or entrained by) water there are airborne or waterborne contaminants such as SO_2, SO_3, and especially chlorides. This mechanism has been developed extensively in the literature dealing with corrosion of reinforcing steel in highway and civil engineering works. Such penetration can occur if the concrete has a high permeability (resulting from a high w/c ratio), or if the concrete develops cracks by shrinkage which occur during normal curing or during service as a result of thermal fluctuations, mechanical damage, or at junctures with steel equipment. The chloride ion can readily penetrate the passive oxide film on the steel surface and initiate severe corrosion pitting. A threshold concentration of chloride ions is necessary to break the passive film, which results in general and pitting corrosion. The chloride content is estimated by others to be about 700 ppm at pH 12 in the water adjacent to the steel. In process plants, sources of chlorides include normal content in the air, washing equipment with sea water, and chlorides present in the cement, such as calcium chloride, which may be added as a curing accelerator, or from following poor construction practices and using salt water to prepare the mix.

Once the breakdown of the passive film occurs by pitting and general surface corrosion, then the subsequent buildup of corrosion products between the steel and the concrete creates a high stress and leads to cracking and spalling of the concrete, further accelerating corrosion. In addition, the electrolyte is trapped by the concrete, and the corrosion process, unlike atmospheric corrosion, is continuous.

Corrosion Beneath Insulation Follows Two Patterns

Corrosion beneath thermal insulation is believed to follow two patterns:
- The first, similar to concrete, occurs when water penetrates the insulation and leaches alkaline components to form basic passive noncorrosive solutions which are then destroyed by airborne contaminants.
- The second occurs when water penetrates organic cellular insulation but, because of its composition, forms acidic corrosive solutions.

In laboratory testing, it was found that in commercial insulations such as fiberglass, mineral wool, cellular glass, and ceramic fiber, during leaching with distilled water, the pH of the solution was alkaline and solutions were noncorrosive since no chlorides were present. It was determined that pH values ran from pH 7 to 11. Corrosion is almost completely stifled because of passivation of the steel surface, such as described with concrete fireproofing.

It is assumed that if such surface passivation is disturbed or broken by chlorides or other airborne contaminants, corrosion beneath thermal insulation will, like concrete, be initiated and become more severe with increasing temperature.

Organic Cellular Foams Can Cause Accelerated Attack

At the present time, organic cellular foams (polyurethane and phenolic) are used frequently in low and moderate temperature environments. Cellular structure is different, and both are different from cellular glass, an inorganic foam, which is used extensively for many of the same applications as the organic foams. For comparison, these are shown in Figures 8, 9, and 10. Figure 8 is a scanning electron microscope photograph at 50X of its well defined skeleton structure, as well as the thin plastic membranes which encapsulate the insulating Freon gas.

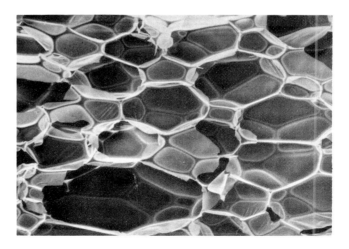

FIGURE 8 — SEM of new 2 lb/ft³ polyurethane foam at 37.5X. Note plastic skeleton, and then plastic membrane sealing the cells.

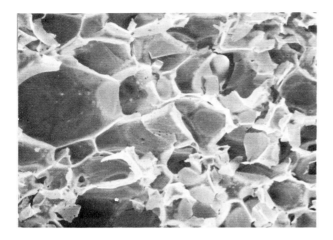

FIGURE 9 — SEM of phenolic plastic insulating foam at 37.5X. Note demolition of cellular structure compared to that shown in Figure 8.

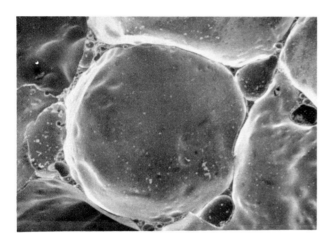

FIGURE 10 — SEM of cellular structure of foam glass at 50X for comparison purposes. The glass cells are much larger and more massive than the polyurethane shown in Figure 8. Density of foam glass is 8 lb/ft^3.

Figure 9, again at 50X, is phenolic foam, which illustrates the complete demolition of the cellular structure, which is the result of the manufacturing process for phenolic based foams. Figure 10 is foam glass. Here, at the same magnification, the glass cellular structure is shown to be much more massive than either the polyurethane or phenolic foam. The structure, in combination with the material properties of glass, results in a foam which is extremely inert and less permeable to water or water vapor, compared to organic foam insulations.

When polyurethane foam insulation, formulated with fire retardant chemicals (which are brominated or chlorinated compounds), are investigated in distilled water leaching tests, they, along with phenolic foams, are found to form aggressive acidic solutions. pH associated with solutions are frequently measured at levels of 2 to 3. Results of such tests are shown in

TABLE 1 — Room Temperature Results From Phenolic Foam and Polyurethane Foam Leaching Tests

	Phenolic Foam	PUF
pH free water	3.45	6.1
pH water in foam	(1)	(1)
water pickup	13g	0.1g

(1) insufficient water absorbed by the foam after 5 days to allow pH measurement.

TABLE 2 — Boiling Water Results From Phenolic Foam and Polyurethane Foam Leaching Tests

	Phenolic Foam	PUF
pH free water	2.37	8.4
pH water in foam	2.25	4.3
water pickup	154g	60g

Tables 1 and 2. Laboratory corrosion tests have shown corrosion rates of 15 to 20 mpy. Of the two, phenolic foams are the most corrosive.

Stress Corrosion Cracking Occurs Beneath Insulation

The third corrosion mechanism of concern beneath thermal insulation is stress corrosion cracking of stainless steel equipment. This has been experienced to a limited degree by Exxon affiliates, and reported by others. For example, external stress corrosion cracking was experienced on a stainless steel vessel in a chemical plant under a fiberglass blanket. It has also reportedly occurred on piping under asbestos insulation. However, in both cases, the insulation was found to contain little, if any, chlorides, and it is assumed that the chlorides again penetrated the insulation system along with water from outside sources. To illustrate, various chemical insulations were leached with distilled water. The results are shown in Table 3, showing hologen content with electrical conductivity. Only fire retarded polyurethane demonstrated measurable quantities of Cl$^-$ and Br$^-$.

TABLE 3 — Conductivity of Leached Solutions Thermal Insulation

Type	Conductivity at Temperature (micromhos)			Halogens (ppm)	
	RT	120 F	210 F	Cl$^-$	Br$^-$
Polyurethane (FP)	30	45	100	20	30
Polyurethane	40	50	400	—	—
Calcium silicate	200	350	450	—	—
Mineral wool	75	700	—	—	—
Cellular glass	40	60	300	—	—
Fiberglass	220	850	1200	—	—
Ceramic blanket	25	40	100	—	—

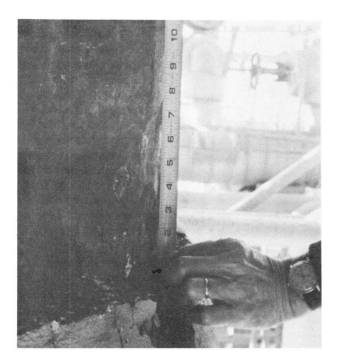

FIGURE 11 — Photograph of corrosion occurring at stiffening ring on vacuum tower. It is presumed water entered at ring and caused the corrosion shown. It was hidden by the calcium silicate solution.

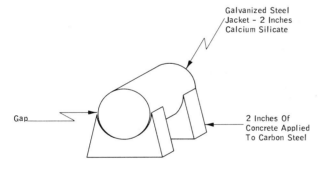

FIGURE 12 — Vessel initially insulated, and then supported saddles fireproofed with concrete. Gaps at tangent line permitted ingress of atmospheric water and excessive corrosion of support steel.

Design/Application Can Promote Corrosion

Special care must be taken in the mechanical design and application of thermal insulation and, also, concrete fireproofing systems. Design can promote corrosion by permitting water to enter the system directly, or by capillary action, soaking insulation and then corroding underlying equipment. Two examples are shown. The first (Figure 11), is a photograph of the corroded shell of a reaction tower operating at negative pressure. The reinforcing ring extended beyond hot thermal insulation, and rainwater creeped up the shell. The photograph shows significant shell metal loss. Properly, the reinforcing ring should have been enclosed within the insulation system or, alternatively, protected with a raincap.

The next example (Figure 12) shows the fireproofed support structure of a thermally insulated horizontal drum within the refinery. The support structure was heavily corroded after 10 years, and all concrete had to be removed. If the vessel supports had been fireproofed first, and then insulated, atmospheric water would have been excluded from the system.

Methods to Control Corrosion Can Be Specified

There are specific measures that can be employed to minimize and completely eliminate corrosion beneath fireproofing and insulation. These are described for individual situations.

Concrete Fireproofing

The first step in providing concrete fireproofing that will be long lasting and minimize corrosion of underlying steel is to use a water/cement ratio of 0.50 or less. This ratio will result in a strong, dense water-tight coating. All steelwork should be coated with suitable metal primers (epoxy, phenolic, etc.). Sophisticated systems should only be used if required by the situation. However, if bond of the concrete to steel is required for strength, only epoxy based coating systems should be used. In addition, the next figure shows a sketch of the termination point of concrete applied to a column. A rain shelter or "top hat" as shown in Figure 13 can be constructed. These are now used at certain Exxon affiliates located in the United States. If top hats, as shown in this figure, cannot be justified, then the concrete/steel interface should be caulked with a butyl, silicone or acrylic based caulking compound. These are relatively inexpensive and easily available; however, they do dry out with time, and lose their flexibility so that continuing maintenance is needed.

In those environments where these procedures are not deemed sufficient, the concrete fireproofing itself can be coated. In this respect, the types of coatings used most successfully on concrete to minimize

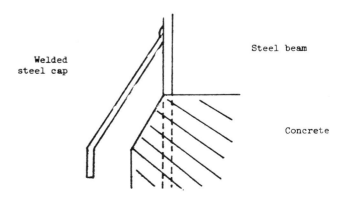

FIGURE 13 — A "top hat" commonly used at refineries and chemical plants to shed atmospheric water at columns where concrete fireproofing terminates.

water diffusion are water based vinyl copolymers or acrylic emulsions. Such topcoating has proven its worth in recent examinations of fire damaged structures, where little corrosion of wire reinforcement or structural steel was found beneath concrete fireproofing, which had been painted in the past.

Thermal Insulation

For above ground equipment, it is recommended that if the equipment is to be thermally insulated, and it operates continuously or intermittently between 25 F and 250 F, then it should be painted. This specification can vary at the higher end, from as low as 160 F to as high as 350-400 F. Normal maintenance primer or more sophisticated coating systems are again satisfactory. However, in no event should inorganic zinc rich coatings be used. In addition, to protect the steel, the insulation must be weather proofed externally; in the event that it is hot equipment, a metal jacket is sufficient, compared to cold insulation where a suitable nonpermeable vapor barrier coating must be used. In either case, the coating or the jacket must be integral, intact, and must not allow atmospheric water or water vapor penetration. It is important to note that more frequently, organizations are turning away from metal jackets, because of corrosion, and because of durability, and again are relying upon mastic coatings with, at times, fabric reinforcement.

To reiterate, if the equipment is cold, then the coating must be vapor tight; in other words, it must not allow water vapor to enter the system. This is most frequently accomplished with an elastomeric vapor barrier based upon butyl or hypalon rubber. Suitable reinforcement is required to obtain film build and film strength.

For thermally insulated underground pipe, it is recommended that the pipe be coated before insulation is applied, and then apply a coating over the insulation. However, this is such a rigorous environment that the coating systems used in this instance must be designed for much heavier duty. It is recommended that, for this application, the pipe itself be coated with either a coal tar epoxy, or with an extruded polyethylene jacket. Then it is recommended that a polyethylene jacket be applied over the insulation. However, the most critical part of this defensive coating system is the protection of the field joints. It is for this reason that extreme inspection measures are needed in the field to ensure high quality application at each pipe joint.

If the underground and the primary barrier are broached, it is possible for ground water to migrate through the insulation and attack bare areas of pipe or welded joints, at remote regions from the break in the primary barrier. Therefore, although external cathodic protection is often used with these systems, it is not effective. Therefore, to be effective, it is recommended that the anode be mounted in the insulation beneath the primary coating barrier, and above the secondary coating applied to the pipe surface.

Stress Corrosion Cracking Control

Stress corrosion cracking has been an infrequent problem because of leachable chlorides contained in the insulation. This may be almost completely overcome by using insulation with an inhibitor. However, two points to remember are:

• First, with insulations such as polyurethane, it is not possible to add inhibitors; fire retarded polyurethane foams for use over austenitic stainless steels are not recommended.

• Second, most cases of stress corrosion cracking are not caused by chlorides contained in the insulation. The chlorides migrate to the insulated steel surface with water, and their source is either the atmosphere or sea water. Therefore, more emphasis must be placed on the external barrier design and application.

For our part, we do not recommend coating stainless steels. In the event of a fire, catastrophic embrittlement can result because of zinc, titanium, or other metallic contaminants.

Corrosion Caused by Drying Paint Films

J.H. White, Z. Kielmanson, and P. Letai*

The process of drying oil paints has been investigated for a long time. Indeed, two hundred years ago Joseph Priestley, with his experiments on mice and candles, showed that linseed oil paint, when drying, removed the life and combustion supporting element we call oxygen from the air.

It has also been known since the early days of this century[1,2] that oil paints, while drying, are liable to evolve corrosive vapors, and it has been shown[3] that these vapors consist of the lower aliphatic acids, especially formic acid, formed directly or by oxidation of aldehydes. Despite some specific publications,[4,5] this aspect of corrosion is largely ignored in the general literature and is hardly mentioned in the better known corrosion handbooks, with the implication that damage resulting from this process is infrequent.

Instances of such corrosion had, however, come to our attention. The nature of the process is such that corrosion damage will be noticeable only in a closed space in which the corrosive vapors can accumulate. The phenomenon, therefore, occurs when freshly painted articles are enclosed in a more or less airtight package, or where the container itself has been painted. The use of such a container often implies that the packaged item is relatively costly, and we considered that the economic significance of the process merited further investigation. The study was designed to answer two questions: how far typical, locally manufactured paints were liable to cause such corrosion; and the drying time necessary to eliminate the danger. It was conducted along two lines: analysis of the evolved vapors using a gas chromatograph; and corrosion tests which were carried out on zinc, a metal known to be readily attacked by the lower aliphatic acids.

The materials included in the investigation were of two types: drying oils known from the literature to evolve the corrosive vapors and local paint products expected to evolve such vapors; and paints of types largely free from drying oils, and considered unlikely to evolve such vapors while drying. Details of the materials used are given in Appendix I.

The quantities of corrosive vapors expected to be emitted were small, and no procedure was available for the chromatographic detection of formic acid. Therefore, it seemed important to adopt an experimental procedure that would use as large a quantity of drying oil (or other material) as was possible, while using a relatively small container, to obtain a vapor concentration high enough to be measured with reasonable accuracy.

Accordingly, test tubes with an enclosed volume of 80 to 100 mls were selected as the experimental vessel, and it was planned to use rubber latex closures through which the enclosed gases could be sampled with a gas syringe. Since it was found that the available closures absorbed acetic acid vapor, test tubes were used with a ground glass stopper carrying a tap and a rubber closure fitted over the tap outlet. The length of the stopper and its outlet was kept short enough to allow the syringe needle to be inserted through the opened tap, and to reach the free vapor space of the test tube. For experiments carried out to test corrosivity, similar ground glass stoppers carrying a tap were used, but with two hooks fused side by side to the bottom of the stopper, so that zinc panels could be suspended in the vapor space.

Preliminary experiments showed that the use of 5 to 10 g of oil paint allowed the material to be spread over the tube wall, by rotating the tube, without contaminating the ground glass surface, although a uniform film could not always be obtained. With the slowest drying materials, no flow occurred after 48 hours. A rig was constructed to hold 12 test tubes on the circumference of a 15 cm diameter drum, rotated about its horizontal axis at a speed of 5 rpm, and this apparatus was used throughout the experimental program. The area of the coated surface was approximately 100 cm^2.

Experiments were then carried out with most of the drying oil materials from the first series, with the vinyl finish included for reference purposes. Approximately 5 g of material were introduced to the bottom of a test tube that was then mounted on the apparatus described. The open test tube was rotated at room temperature, removed from the apparatus, closed with the stopper, and held at room temperature for one month before sampling (initial experiments had shown no detectable concentration of acid vapors after a few days).

Analyses for acetic and for propionic acids were done, using a Hewlett Packard Model 5700 A gas

---

*Paint Research Association, Haifa, Israel
P. Letai is now at Mekorot, the Israel Water Company, Ltd., Tel Aviv, Israel

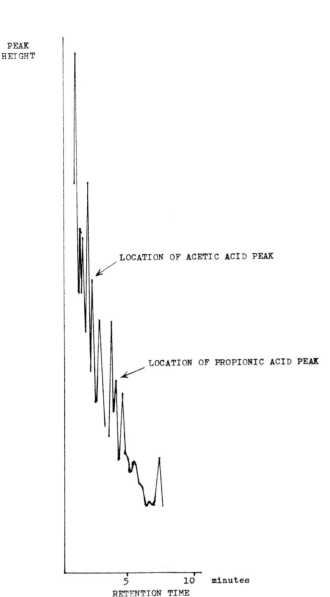

FIGURE 1 — SP 1200 Column: interference by solvent vapors.

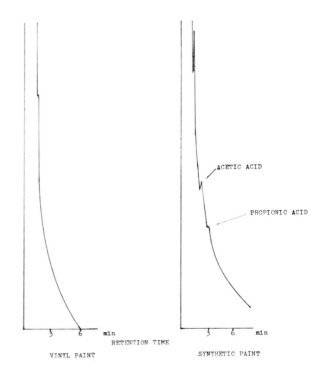

FIGURE 2 — Chromatogram showing traces of acids from synthetic paint.

chromatograph with Type 18711 thermal conductivity detector. A glass column packed with Carbowax 20M + H_3PO_4 was used, and details of the procedure are given in Appendix II. Supelco SP 1200 with 1% H_3PO_4 on 80/100 Chromasorb AV is reported in the literature,[6] to give better separation of the aliphatic acids, but it was found that residual solvent vapor interfered with the estimation of acetic and propionic acids when this column filling was used (Figure 1).

Since it was hoped to follow the development of the acid vapor as the coating material dried, the analytical procedure was to remove 5 mls of gas using a Hamilton gas tight syringe. However, in our first experiments, we ran into a difficulty we had not anticipated when the procedure was planned, due to the absorption of oxygen and the creation of a partial vacuum in the test tube. The effective quantity of gas thus removed depended on the pressure in the test tube. The pressure varied from material to material and prevented accurate comparison of the results from different experiments. Occasionally, the pressure was low enough to prevent the operation of the syringe. The difficulty was overcome by initially injecting clean air with the syringe, until the pressure in the test tube was approximately atmospheric. The volume of air added was measured, and the 5 ml sample was removed after a short interval to allow mixing of the enclosed gases. Priestley's experiments came to our attention at a later stage, but the difficulty should have been envisaged when the work was planned.

The initial results obtained one month and six weeks after coating the tubes were disappointing. Using the gas chromatograph system at its maximum sensitivity, peaks for acetic and propionic acids were sometimes, but not always, obtained, and were hardly ever large enough to allow an estimate of the vapor concentration (Figure 2). The sensitivity was such that 1 mm² of peak area was equivalent to about 0.01 μl of either of the acids.

A second series of tests was carried out using the maximum quantity of material that would allow a uniform coating to be obtained (from about 1¼ g of the vinyl finish to 5 g for the drying oil/alkyd primer) and, in addition to the conditions used in the first series, tests were carried out with test tubes that were not closed until 6 days after they had been coated. Analyses were carried out two weeks after coating of the first set of tubes, and one month after coating for those left open for the first six days. Again, only small peaks were obtained with the first set, and acetic and

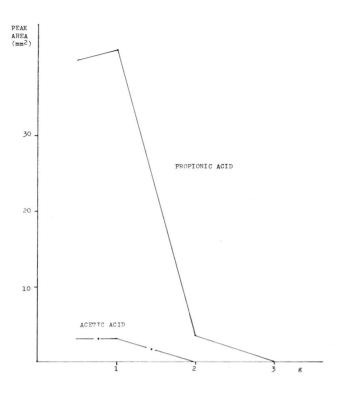

FIGURE 3 — Acid vapors from tung oil: dependence on coating weight.

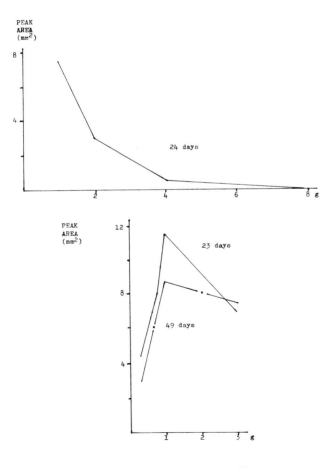

FIGURE 4 — Synthetic enamel: propionic acid concentration and coating weights.

propionic acids could hardly be detected at all in the tubes initially left open. Analysis carried out three weeks later showed no significant changes in the acid concentration.

Corrosion tests, parallel to the analytical experiments were carried out with the alkyd and phenolic paints. Two clean zinc panels, approximately 75 × 15 × 0.25 mm and weighing about 2¼ g, were suspended in each test tube. Two sets of panels were exposed with each coating material, one in the same condition as the analytical tests, and the second with the addition of 50 μl distilled water, calculated to give a saturated atmosphere, immediately before the tube was closed. Zinc panels were removed from the test tube, and fresh zinc panels inserted successively for periods from one to fifteen days, over periods up to three months. Unexpectedly, on the basis of past experience, no significant corrosion of the zinc occurred.

When the first of these results was obtained, it was thought that the extent of acid vapor production had been overestimated, and that the continuation of the research would require a more sensitive analytical procedure. However, as the results accumulated, it became apparent that the whole phenomenon was more complicated than had been thought. Although most of the experiments gave results so low that they could hardly be represented quantitatively, those with tung oil gave somewhat larger quantities of acetic, and especially of propionic, acid. An unexpected relationship gradually emerged between the quantity of acid produced and the quantity of tung oil used, with a big decrease in the acid concentration as the quantity of tung oil increased above the minimum amount used (Figure 3). The results obtained with the synthetic enamel gave similar indications (Figure 4).

At this stage, a flame ionisation detector (Hewlett Packard Type 18710 A) was obtained and put into use, allowing vapor concentrations to be estimated more readily by comparing peak heights, with sensitivities of about 1 cm for 0.02 μl acetic acid and 0.01 μl propionic acid. At the same time, the sample volume was reduced to 1 ml (instead of 5 ml), reducing the effect of withdrawing samples from the vapor space. The experiments with the synthetic enamel were repeated using this equipment, the weight of paint used ranging from about ¼ to 3 g. Again, the quantity of acid formed reached a maximum with an intermediate quantity of paint, though at different quantities for acetic acid, about 1 g, and propionic acid, about ¾ g (Figure 5). Experiments with the drying oil/alkyd primer, and with the phenolic finish, showed the same effect (Figures 6, 7). With both these paints, though, after a decrease in acid formation when the quantity of paint was about 1 g, the acid concentration rose again when 2 g or more of the paint was used. These results were obtained approximately 5 weeks after the tubes had been painted, and further analyses carried out 9 days later showed, in general, an increase in acid formation, except for the tubes with the smallest quantities of paint initially.

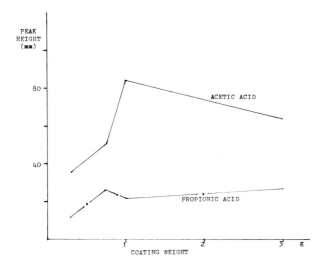

FIGURE 5 — Acetic and propionic acids from synthetic enamel.

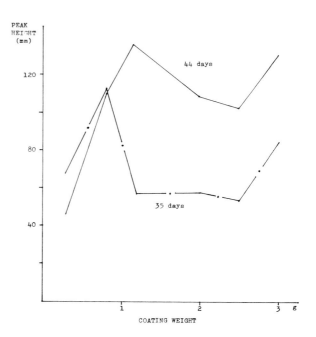

FIGURE 7 — Acetic acid from phenolic finish.

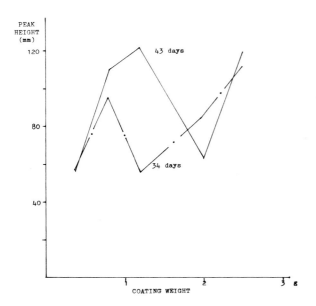

FIGURE 6 — Acetic acid from drying oil/alkyd primer.

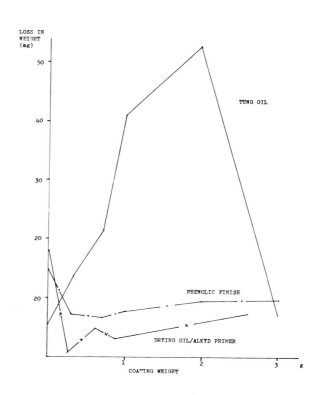

FIGURE 8 — Effect of coating weight on corrosion of zinc.

Parallel corrosion tests were carried out in test tubes coated with tung oil, the drying oil/alkyd primer, and the phenolic finish, using the same coating weights, from about ¼ to about 3 g. Duplicate panels were exposed for 14 days, and although the results for the test tubes containing only tung oil or paint were inconclusive, the corrosion results for panels exposed in test tubes to which water had been added (Figure 8) were analogous to the analytical results: tung oil giving a large decrease of corrosion with coating weight above 2 g; the primer and the phenolic finish showing similar effect, but to a lesser extent, and at a lower coating weight.

Owing to the failure to understand the importance of the weight of coating material used, the results of earlier experiments carried out with paints that did not contain drying oils, and were not expected to be corrosive, could not be considered to be significant. However, at this stage, it was considered possible to test different paints, and determine whether or not they were liable to lead to a corrosion risk, due to the production of aliphatic acid vapors linked with the drying process. As the production of acetic and propionic acids had been found to be dependent in a nonregular fashion on the weights of material used, each paint was tested at two sample weights, to give dry film weights of approximately 0.5 and 1.5 g. Test tubes were coated as described, previously thinning the paint (if necessary) to give a reasonable uniform coating, and closing the test tubes 48 hours after the paint had been introduced. The closed test tubes were held at room temperature for three weeks, and the enclosed gases sampled and analyzed for acetic and propionic acids. Operating conditions for the gas chromatograph, using the flame ionisation detector, are given in Part B of Appendix II, and were designed to give a peak height of about 50 mm for 0.1 mg of acetic acid. The results obtained for a number of the paints listed in Appendix I are given in Table 1.

We conclude that no significant corrosion risk exists unless the total acid concentration in at least one of the two test tubes is greater than 20 μl/ml. On this basis, of the paints listed in Table 1, only the quick drying finish presents such a risk.

This conclusion requires some modification, although in a direction that is not relevant to the main topic of this paper. It has been mentioned that, in the earliest corrosion tests, the vinyl finish (10 in Appendix I) was included for reference purposes. However, it has not been mentioned previously that this paint was the only material to cause significant corrosion of zinc in these early experiments, with weight losses (with added water) approaching 200 mg; experiments with other materials had given weight losses of about 5 mg, which was a typical result for blank experiments with uncoated test tubes. It was found that this corrosion had been caused by hydrochloric acid, a known phenomenon with vinyl paints, but not expected in the conditions of these experiments. As a matter of interest, a series of experiments was carried out repeating the tests with the vinyl finish, but including the vinyl zinc chromate primer and two chlorinated rubber paints (9, 13 and 14 in Appendix I). The experiments were carried out in two stages. The first pair of panels was placed in the test tube 2 days after application of the coating and exposed for 15 days. The second pair was exposed for a period of 18 days, two days after the first pair was removed. The results given in Table 2 are the mean loss of weight for each pair of panels after removal of corrosion product using a saturated solution of ammonium acetate.

TABLE 2 — Corrosion of zinc by vapor from vinyl and chlorinated rubber paints

Paint No.	Type of paint	Weight of paint used (g)	First Specimens (mg)	Second Specimens (mg)
9	Vinyl primer	3.98	76	75
10	Vinyl finish	3.75	91	115
13	Chlorinated rubber finish	3.77	2.5	24
14	Chlorinated rubber finish	4.44	3.0	7.5

These results show clearly the corrosive nature of the vinyl materials, under the exerimental conditions used, which did not involve exposure to direct sunlight. The chlorinated rubber paints appear to break down in a similar fashion after initial delay.

We were not able to include in our investigation any work on the reaction mechanism leading to the production of the corrosive aliphatic acids, but in considering the type of process involved, a certain amount of additional information is available. We have mentioned that difficulty was experienced in extracting 5 ml gas samples from the test tubes in early experiments because of reduced pressure caused by oxygen absorption, and that we overcame this problem by introducing air to bring the gases in the test tube more or less to atmospheric pressure. The quantity of air used was measured to allow a suitable correction to be made to the concentration of acid measured by the gas chromatograph. This informa-

TABLE 1 — Tests of paints using standard procedures

Paint No.	Type of Paint	Dry film weight (g)	Peak height (mm) Acetic acid	Peak height (mm) Propionic acid	Estimated vapor concentration (μg/ml) Acetic	Estimated vapor concentration (μg/ml) Propionic
5	Zinc chromate synthetic primer	0.6	—	—	0	0
		1.15	—	—	0	0
7	Quick drying synthetic finish	0.6	11.5	15.5	20	15
		0.15	15	17	30	15
11	Strontium chromate epoxy primer	0.5	—	—	0	0
		0.95	—	—	0	0
12	Epoxy polyamide finish	0.85	—	—	0	0
		1.3	—	—	0	0
15	Polyurethane finish	0.55	1.5	1.5	3	2
		1.0	4	2.5	8	3

0 = none detected

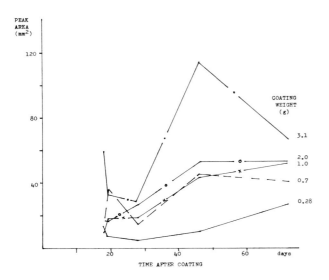

FIGURE 9 — Production of propionic acid from tung oil: dependence on time and coating weight.

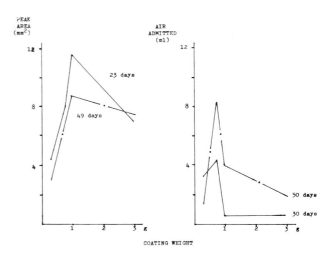

FIGURE 11 — Synthetic enamel: production of propionic acid and absorption of oxygen.

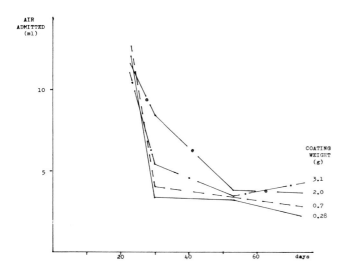

FIGURE 10 — Absorption of oxygen by tung oil: dependence on time and coating weight.

tion was largely useless, as our first experiments gave negligible quantities of acetic and propionic acids. However, as we have shown, the tung oil coatings gave larger quantities of acid vapors, so that even the earlier results for this material were significant. While we learned that the quantity of acid produced depended in a somewhat complex manner on the quantity of tung oil used, it also became apparent that the quantity of oxygen absorbed varied in a somewhat similar manner, reaching a minimum for an intermediate weight of tung oil, and with a tendency for high acid production to be linked with low oxygen absorption. This relationship can be seen in Figures 9 and 10, showing the results obtained from a series of experiments with different weights of tung oil, carried out as described; by sampling 5 ml of gas at intervals, after bringing the pressure in the test tube to atmospheric by addition of a measured quantity of air. Similar results were obtained with the synthetic enamel (Figure 11).

Taking all of our results into consideration, it becomes clear that the production of corrosive acid vapors does not inevitably accompany the drying process of oils and the paints produced from them. The two processes appear to be competitive, with the drying process preferred, so that the production of free acid vapors occurs only when a sufficient excess of oxygen is available. The production of acid, and a reduced absorption of oxygen in the drying process, are apparently connected, and the fact that a synthetic resin paint containing a chromate pigment did not produce acid vapors may be significant.

Avoidance of the corrosion risk by ensuring adequate drying of a corrosive paint does not seem to be a reliable solution to the problem. We have found that the production of acid vapors continued under our experimental conditions for periods of more than six weeks. However, it now seems clear that corrosion will occur only when metal and paint are confined together in conditions defined by fairly narrow limits. A small sealed volume will not contain sufficient oxygen for acid vapor production, while ventilation of a container will prevent the build up of a corrosive concentration of acid vapors. Only in intermediate conditions will the corrosion risk be significant, and this is presumably the reason that the incidence of this type of corrosion is low and that the phenomenon is not better known. Unless it is clear that there is no danger of acid vapors accumulating in an enclosed space, only paints known to dry without involving this risk should be used.

References

1. F. Fritz, *Z. angew. chem.* **28** (1), p. 272 (1915).
2. W.H.J. Vernon, *Trans. Faraday Soc.* **19,** p. 839 (1924).
3. P.D. Donovan and T.M. Moyneham, *Corrosion Sci.* **5,** p. 803 (1965).
4. V.E. Rance and H.E. Cole, Corrosion of Metals by Vapours from Organic Materials, H.M.S.O. (1958).
5. Defence Guide for Prevention of Corrosion of Cadmium and Zinc Plating by Vapours from Organic Materials, H.M.S.O. (1959).
6. J. Hrivnak, L. Sojak, E. Beska and J. Janak, *J. Chrom.* **68** (1), p. 55 (1972).

Appendix I

The Materials Investigated

Series 1. Drying oils and materials manufactured using drying oils.

1. Linseed oil, iodine number 180, with added driers 0.05% Co and 0.45% Pb, as naphthenates.
2. Tung oil.
3. A safflower alkyd resin (044/630), 67% in white spirits, containing lead, manganese and cobalt naphthenate driers.
4. A commercial drying oil/alkyd resin primer, to Israel Standard IS 539 (10442).
5. A phenol modified zinc chromate primer to MIL-P-8585 (2212).
6. An alkyd synthetic enamel ("Etan," 10713).
7. A synthetic quick drying gloss finish (2213).
8. A phenolic finish ("No. 309," 10581).

Series 2. Paints free from drying oils

9. A zinc chromate pigmented vinyl anticorrosive primer ("Chromovil brown," 10134).
10. A vinyl finishing paint ("Chromovil Finish," 10139).
11. A strontium chromate epoxy primer to MIL-P-23377 (Production batch 06957, 2214).
12. An epoxy polyamide semigloss finish to MIL-E-22750 (Production batch 07626, accelerator No. 07379; 2215).

Appendix I (Cont'd.)

13. A white chlorinated rubber gloss finish (10360).
14. "DKS Emaille," a chlorinated rubber paint supplied by the Imerit Co., Switzerland (10151).
15. A white polyurethane finish to MIL-C-83286, a product of U.S. Paint, Lacquer & Chemical Co. (2216).

Apart from the two last materials, the resin and paints were products of the constituent companies of Tambour Askar Paints Ltd., Acre, Israel.

Appendix II

Procedures for Gas Chromatogaph Analyses

A. Using Thermal Conductivity Detector
Column filling: Carbowax 20M + H_3PO_4
Column dimensions: internal diameter 4 mm, length 183 cm (6 feet)
Column temperature: 140 C
Carrier gas: helium at 60 p.s.i.; 50ml/minute
Injection and detector temperatures: 200 C
Detector sensitivity: 5 Attenuation: 1
Recorder sensitivity: 1mV full scale
Chart speed: 0.2 inch/minute
Volume of gas injected: 5 ml

B. Using Flame Ionisation Detector
Column filling: Carbowax 20M + H_3PO_4
Column dimensions: internal diameter 4mm, length 183 cm (6 feet)
Column temperature: 120 C
Carrier gas: Helium at 60 p.s.i., 60 ml/minute
Injection and detector temperatures: 200 C
Air at 60 p.s.i., 240 ml/minute; hydrogen at 30 p.s.i., 60 ml/minute
Range: 10 Attenuation: 4
Recorder sensitivity: 1mV full scale
Chart speed: 0.2 inch/minute
Volume of gas injected: 1 ml

An Anomalous Effect of Limited Drying Time on Performance of a Vinyl Coating

J.H. White and W. Rothschild*

Introduction

The Sea of Galilee is Israel's largest reservoir, and supplies water to the National Water Carrier, a distribution system to the more arid south. The first stage of the Carrier, through hilly country, comprises open concrete channels linked by steel pipes protected internally by a paint coating. There are three sections of pipe; the first from the pumping station on the lake shore, and two syphons traversing valleys, a total of about four kilometers. The pipe diameter is about 3 meters.

When installed, the internal surface was protected by a two component sytem: about 100 μm of a zinc chromate pigmented vinyl paint applied over a wash primer, followed by about 320 μm of an asphalt type material. After some 15 years, damage to the metal was insignificant, but the asphalt coating showed embrittlement and loss of adhesion to the vinyl undercoating, and there was some blistering of the vinyl coating. Some measure of repair was considered desirable, and tests were carried out on a number of types of coating, including epoxy and chlorinated rubber; however, none were found to be significantly better than the coating originally used. A modification of this coating, though, with a vinyl intermediate coat to improve adhesion to the asphalt material, seemed to be of some benefit.

However, the vinyl asphalt system (compared with the best of the epoxy systems), is at a disadvantage, owing to the larger number of coats, 6 or 7, and the lengthier application period. This disadvantage was not relevant when the carrier was built, but for a repair system could involve a heavy sacrifice. The preferred time for painting is early autumn, when the rains begin and demands from the Water Carrier for irrigation are low. At this time, if the water level is still sufficiently high, water is pumped to increase the available capacity of the lake. Failure to pump sufficient water, at this stage, means that later on it may be necessary to allow water to flow down the Jordan valley to the Dead Sea, with a consequent loss to the economy, and an extra week's drying time could well prove to be an unacceptable penalty for an otherwise satisfactory paint system.

We investigated, therefore, on behalf of Mekorot, the Israel Water Co., Ltd., the effect of varying coating thickness and between coat intervals on the drying properties and protective properties of zinc chromate vinyl coatings, and the unexpected results we obtained may be of general interest.

The protective properties were evaluated using two tests which had been shown together, in earlier work, to correlate with the behavior of test panels exposed on the wall of the Carrier itself. These were: 1) salt spray, as described in ASTM B117-73 and parallel specifications; 2) two-thirds immersion in warm distilled water (45 ± 3 C). Each test was carried out for periods of two and three months. This procedure had been adopted as a laboratory screening test for selecting materials for test *in situ* and was applied to mild steel panels with a total dry coating thickness of 180 μm.

Two series of vinyl coated panels were prepared, the first on chromate treated aluminum, and the second on mild steel with an approximately 10 μm coating of the two pack wash primer specified for use with the system. The aluminum panels were used for drying time tests, and the steel panels for corrosion tests as well. The zinc chromate vinyl paint was supplied in two colors, green and brown, and different production batches were used in each series. Particulars of these paints are given in Appendix 1.

Drying properties of the Vinyl Films

When it came to studying the drying process of the vinyl coatings, it was found that different methods gave different pictures, and in the search for the required information five different procedures were investigated: 1) loss of weight; 2) the "mechanical thumb" described in Part C3 of British Standard BS 3900; 3) scratch hardness; 4) the toothed wheel of the "Erichsen" Model 338 Universal drying time recorder; 5) the I.C.I. Munk micro indentation apparatus, using an 0.8 mm spherical indenter under a 2 g load. The load was applied for two minutes, and measurements continued for another two minutes. The instrument records movements over a limited range, up to 6 μm.

The drying process was followed on vinyl films of different thicknesses, up to about 200 μm, and on two

---
*Paint Research Association, Haifa, Israel

TABLE 1 — Two and three-coat systems

System number	Between coat intervals (hours) First/second coats	Second/third coats	Total dry film thickness (μm)
A1	6	18	130 to 180
A2	18	6	180 to 210
A3	6	18	250 to 300
A4	6	18	260 to 290
B1	24	—	approx. 180
B2	48	—	approx. 180
B3	24	24	approx. 180
B4	48	48	approx. 180

In each system the thicknesses of the individual coats were approximately equal.

and three coat films with different between coat intervals, as shown in Table 1. In system A4, all coats were of the brown zinc chromate vinyl, but in the other systems alternate coats of brown and green were used.

The mechanical thumb and the Erichsen notched wheel were found to give inadequate indications of the progress of the drying process. Coatings up to 60 μm thick withstood the mechanical thumb after 6 hours, and all the coatings up to 240 μm thick after 18 hours. Coatings only 25 μm thick still showed the marks of the Erichsen wheel after two days, and no tests were carried out on coatings more than 130 μm thick with this instrument.

The scratch test hardness measurements were somewhat more useful, giving a certain amount of information on changes in film properties during the first few days, depending on the film thickness. However, the test was not capable of giving much information on thicker coatings and, as with the mechanical finger and Erichsen wheel tests, was not used on two and three coat systems. For these systems the measurement of weight loss gave a continuous record of the drying process, and the indentometer measurements (which were carried out on all except the two thickest systems listed on Table 1), gave a parallel indication of changes in the physical properties of the paint films once they had hardened sufficiently to bring the penetration below the 6 μm upper limit of the recorder. Changes in film weight were measureable over longer periods than expected, and measurements were continued for 60 days, with the indentometer showing appreciable changes in film properties for the whole of this period.

A difficulty arose, due to the absence of a controlled atmosphere room in which the paint could be allowed to dry without any limitation on the solvent evaporation process. The effect of temperature variation (the mean temperature over the period of the experiment was about 23 C) was not expected to be significant, but parallel experiments with the asphalt topcoat showed some oscillations in weight which may have been caused by humidity changes. Accordingly, indentometer measurements were made on specimens drying in the air, and parallel specimens previously kept for 24 hours in a desiccator, to stabilise the moisture content of the film. There were no significant differences between the results for the two sets of measurements, so that the indentometer could be used without interfering with the drying process by keeping the specimens in a desiccator. Indentometer measurements were carried out at 25 C.

The Corrosion Tests

The corrosion tests were designed to examine not only the effect of using thicker and thinner coats with different between coat intervals, but also the effect of the total drying time on the protective properties of the system. Laboratory tests of thick coatings of high quality paints present a problem, as they may be expected to withstand reasonable test conditions for long periods. Two possibilities exist for coping with this problem: increasing the severity of the test conditions and decreasing the coating thickness. The first has its own dangers, as departing too far from actual exposure conditions by changing the corrosive medium, increasing the temperature unduly, or introducing exaggerated conditions of erosion or other physical stress, can cause a type of breakdown completely different from what occurs in practice, and give completely misleading results. Too great a reduction of film thickness, especially for a multicoat system, is liable to make the results too susceptible to accidental faults in the test panels, and may also affect the breakdown mode. For these reasons, it was considered undesirable to alter the laboratory test conditions that already had been found to correlate with exposure *in situ,* nor was it practical to increase the test period beyond the three months that had been found necessary in previous tests. As a result, it was decided to introduce a reference standard in the test programs, designed to lead to relatively rapid breakdown of the coating, and provide a basis for comparison for the coatings in which we were interested. The final drying time recommended by the manufacturer for a system consisting of 2 coats of the zinc chromate vinyl paint, followed by 3 to 4 coats of the asphalt paint (the coats being applied at 1 to 3 day intervals), was 14 days, and the test program was designed to cover the range from 3 to 14 days. To this program we added a final drying time of only 24 hours, which we expected to lead to a relatively rapid breakdown of the coatings and to provide the required base line for evaluating the other panels. The other drying times decided on were 3, 7, and 14 days, so that the corrosion test program, for the four systems B1 to B4 listed in Table 1 (each in duplicate), involved 32 panels. After 2 and 3 months exposure, the coating adhesion was tested by the cross cut method, cutting 1 mm and 2 mm squares, and using masking tape, and the appearance of the coating evaluated.

Evaluation of the Results

As the tests progressed, it became apparent that the relatively rapid breakdown, expected for the

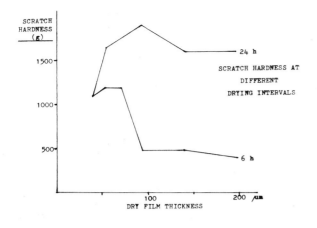

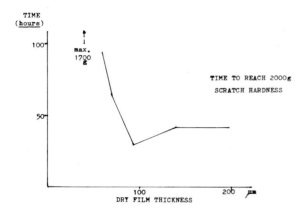

FIGURE 1 — Scratch hardness related to film thickness and drying time.

panels immersed the day following application of the final coat of paint, was not occurring. Accordingly, when it came to the final evaluation of the panels, it was a problem to find a system of scoring that allowed the effects of the different drying time parameters and the number of coats to be evaluated. Where visible breakdown occurred, it was almost entirely in the form of blistering, with almost no rust on the metal surface; the two factors mostly involved were the blistering and loss of adhesion shown by the cross cut test. Both of these could be evaluated more or less objectively: blistering by the scales of size and density laid down in ASTM Specification D 714-56, and adhesion as a percentage. The problem arose for reasons only too common in assessing paint tests — the increase in the size of blisters and their density did not always run in parallel, and as other systems involving the asphalt paint were tested at the same time, there were also instances of loss of intercoat adhesion which had to be taken into consideration.

The problem was dealt with by a series of assessments carried out by three experienced chemists, in a Monte Carlo type of operation. These assessments were designed to give overall orders of merit as regards blistering and adhesion, as well as an overall merit mark. At each stage, three independent assessments were made, the objective of each stage being sufficiently limited to give close agreement between all three assessors. The procedure led to separate 0 to 10 scales for blister size and density, and for an expanded scale from 0 to 15 for adhesion, with the one degree of corrosion found described as slight, and rated as 5. The blistering was rated in greater detail than in the usual ASTM notation, including the odd numbers up to 9 to denote size, and additional grades FF, MF, MD, and DD to denote blister density. The final scale adopted is shown in Appendix II, with ratings for the various defects being added together to give the breakdown rating: a perfect panel is rated 0, and a panel with severe blistering, slight surface corrosion, and a total loss of adhesion is rated 40. Even this somewhat complicated procedure failed to deal adequately with all panels, and to avoid disagreements in the final assessments, a number of rules, as noted on the table, had to be introduced.

1. Where duplicate panels gave different assessments, or the assessors gave different values to the same panel, mean ratings were used.

2. Where blistering occurred only at the edges of a panel, one quarter of the ratings for size and density was included in the total.

Two further rulings applied only to the additional systems with asphalt paint: loss of intercoat adhesion was rated at three-quarters of the appropriate rating, and a few panels removed before the end of the test because of complete breakdown had 10 points added to their total ratings (in no instance was more than one of a pair of panels removed before the end of these tests).

This method of assessment was reached only after a lengthy process of considering other procedures for weighting the results, but achieved its objective of giving almost complete agreement between the three assessors on the order of merit of the different systems in the two corrosion tests.

Results

The results relevant to the drying process are presented graphically in Figures 1 to 6. The effect of thickness on scratch test hardness is complex, since a thicker film is more liable to withstand the 2000 g maximum load, but is slower drying, and the results are accordingly presented in two different forms in Figure 1.

Figure 2 shows the weight losses over two months for systems A1 to A4, each consisting of three coats of vinyl paint, differing in between coat drying times and total thickness. Figures 3 to 6 compare the changes in weight and indentometer measurements for all the systems except A3 and A4, which were thicker than the others and remained too soft to be tested with the indentometer. The data from Figure 2 are repeated to allow the drying process to be followed more readily. The data for sytems B1 to B4 cover only 14 days drying, as this was the maximum period before corrosion tests were started.

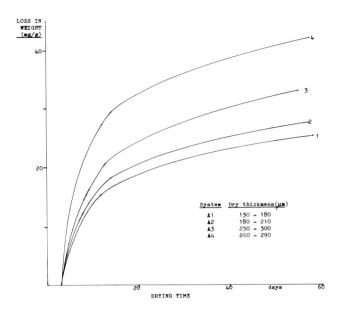

FIGURE 2 — Three coat systems: loss of weight on drying.

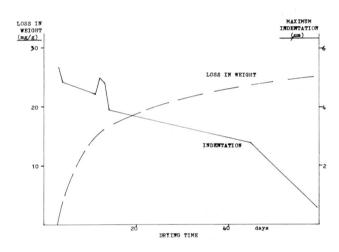

FIGURE 3 — Drying of system A1 (3 coats, 130 to 180 μm, 6 + 18 h).

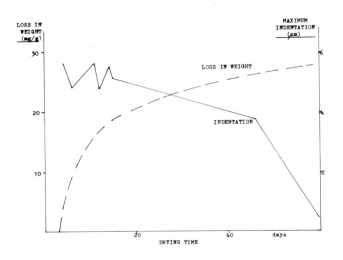

FIGURE 4 — Drying of system A2 (3 coats, 180 to 210 μm, 18 + 6 h).

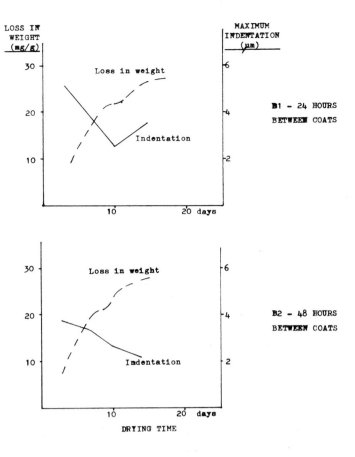

FIGURE 5 — Drying of systems B1, B2 (2 coats, 180 μm).

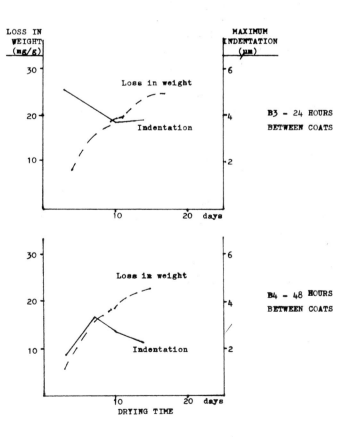

FIGURE 6 — Drying of systems B3, B4 (3 coats, 180 μm).

175

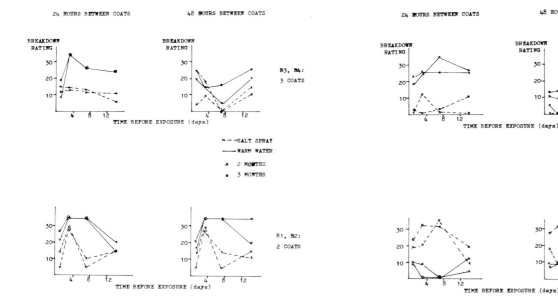

FIGURE 7 — Corrosion resistance of vinyl coatings (180 μm).

FIGURE 8 — Corrosion resistance of vinyl/asphalt systems (180 μm).

In accordance with the purpose of the investigation, the corrosion results have been presented in Figure 7 by plotting the 0 to 40 point breakdown rating against the final drying time of the system. The results for the 2 and 3 month salt spray and water immersion tests are presented on the same graph, but different graphs are used for the same system with 24 hour and 48 hour between coat intervals, mainly to reduce the confusion of overlapping curves.

Discussion

The reason we have seen fit to present this work is the unexpectedly good results obtained for the systems tested 24 hours after painting was completed. It is clear that at this stage, despite the initial rapid loss of solvent, the paint coatings are far from dry, but their protective properties were satisfactory. Even more unexpectedly, films dried for 3 days before testing gave less satisfactory results, and this effect sometimes continued to 7 days. It is doubtful whether too much reliance should be put on individual results, but the general tendency is clear.

Equally unexpected was the lengthy period over which the drying process extended. After the first week or so, the changes in weight were small, but the indentometer measurements showed that they were accompanied by significant changes in the properties of the film.

The unexpected corrosion results demand some explanation and while we have no evidence on which we can base a reasonable theory, we put forward the suggestion that the films exposed to water (or salt spray) at a relatively early stage of the drying process continue to dry in water as in air, by diffusion of the solvent into the water, analogously to its diffusion into the air in normal drying conditions. However, if the drying process has progressed somewhat further, the surface of the film may become sealed, or less readily wetted, so that this process is largely inhibited and the solvent remains trapped in the film to a greater extent.

Some support for this theory comes from a consideration of the solvent make up of the vinyl paint: together with aromatic hydrocarbons, it contains alcohols, methyl isobutyl ketone, cellosolve acetate and a little nitropropane. Reference has been made to experiments that included, together with the vinyl material, the asphaltic top coat. This material has a xylene white spirit solvent, and the vinyl solvent is, therefore, a more polar material. It is significant, perhaps, that the systems including the asphaltic top coat exhibited the unexpected protective qualities after 24 hours drying to a much lesser extent, and that the effect was completely absent when 48 hours drying between coats was allowed, instead of 24 hours. Figure 8 shows the results for two systems, each of two vinyl coats and an asphalt top coat, with one system (C2) including a vinyl intermediate instead of a second coat of the zinc chromate vinyl.

From the practical point of view, we should not like to go further than approve the renewal of the flow through the Water Carrier as an emergency measure much earlier than recommended; for instance, in the case of an unexpected drought. However, our results are perhaps of some general interest, and may be relevant, for example, to the apparently successful practice of allowing short drying periods for antifouling paints before removing ships from dry dock.

Appendix I

Particulars of the Zinc Chromate Vinyl Paint Samples

Color of Paint	Green		Brown	
Sample Number	3023	3432	3024	3433
Nonvolatile content (% paint w/w)	29.5	32.0	32.3	34.0
Binder (% paint w/w)	19.5	21.5	20	21
Density of paint (kg/l)	1.03	1.02	1.04	1.04
Density of nonvolatiles (kg/l)	1.57	1.60	1.64	1.70

Appendix II

The Panel Scoring Scale for Breakdown Rating

A separate score is given for each type of breakdown, measured as follows:

Adhesion. percentage adhesion (%)
Corrosion. extent of rusting of metal. The single grade recorded was described as "slight".
Blister size. numbered from 1 to 9
Blister density. F, M, D and combinations of these letters (the last two designations to ASTM D 714)

The score for each type of breakdown is given in the last column of the following table:

Adhesion	Corrosion	Blister size	Blister density	Score
100	—	—	—	0
95 to 99		9	FF	1
90 to 94		8	F	2
		7	MF	3
80 to 89			M	4
	slight	6		5
70 to 79		5	MD	6
				7
50 to 69		4	D	8
		3		9
40 to 49		2	DD	10
30 to 39				11
				12
				13
20 to 29				14
0 to 19				15

The final Breakdown Rating was obtained by addition of the scores for each type of breakdown.

Notes

When duplicate panels gave different scores, or when an assessor was unable to decide between scores, mean values were used.

When blisters occurred only at the edges of the panel, one quarter of the score was counted.

When adhesion tests showed a loss of adhesion between coats (vinyl/asphalt systems) ¾ of the score was counted.

When one of a pair of panels was removed early because of severe failure, 10 points were added to the score.

Evaluation of Linings for SO$_2$ Scrubber Service

*Dean M. Berger, Robert J. Trewella, Carl J. Wummer**

Introduction

In 1970 Congress enacted the Clean Air Act. This legislation promulgated a vigorous attack on air pollution by all segments of industry. Emissions by industrial installations, such as electric power plants, paper mills, steel mills, and other fossil fuel facilities, have been carefully monitored. Gas scrubbing systems have been designed, mainly to control SO$_2$ emission, and it has been shown that controlling SO$_2$ presents a severe problem for corrosion protection of all surfaces within the system.

Some of the first units built used carbon steel lined with various types of coating materials, in an attempt to prevent corrosion. Each unit differs from another, and there are practically no ground rules as to why some areas remain intact while others fail rapidly. The severity of the problem appears to center around the temperature of the gas being emitted. Temperatures above 210 F do not appear to be as severe as those units operating with gas emissions below the condensing temperature of water. The use of reheat to increase the temperature is a costly operation; therefore, materials must be found, or substitute better materials, from these gas condensates at operating temperatures of 130 to 190 F.

Billions of dollars have already been invested in various SO$_2$ scrubber systems. Numerous papers have been presented, and a wide variety of materials have been used and tested for this service.[1-8] Some of the major problems facing the power industry have caused severe economical strain, usually related to outages, down time and labor costs. Materials have been a serious problem in the corrosive environment within the SO$_2$ scrubber system. The most severe corrosive areas are usually downstream of the scrubber, outward toward the breeching and flues. Wherever condensation occurs, such areas are under a constant rain of acidic water confined within the system.[9] Limestone is used to neutralize the emitted SO$_2$ in many scrubbers.

Usually the water and lime slurry creates a pH inside the venturi area, ranging from as low as 4 to a high of 11, with standard operating conditions of about pH 6 to 7. As the gases flow out the system, water evaporates until the acidity of the droplets and the condensate reaches a pH of 1.0 to 1.5. Inside the flue, the side walls condense the gas, and water runs down, causing the low pH condition.

Not much information has been published concerning the contents of these gas condensates which are found in coal fired units. A program was initiated to collect the gas condensate at various heights of the flue, and run an analysis of this condensate. Not all elements were identified, however acidity, Cl, SO$_2$, and NO$_2$ were monitored. Table I contains the analysis of some[20] of these condensate samplings which were taken from various elevations of the flue and from the drain water. Total organic carbon was highest during start up. Organics were not further identified.

Chlorides, nitrates, and sulfates varied considerably. Chlorides were highest in the drain samples, and sulfates reached a level of 12,000 Mg/L. The pH varied from 1.5 to 5.0.

In analyzing the type of stack lining failure which has occurred, there is a definite progression of corrosion. A spray applied polyester lining showed early signs of breakdown. A chemical analysis of this product revealed that over 100 ppm of water soluble halogens were present, a situation which speeds up corrosive attack on steel substrates. In SO$_2$ scrubbers, coatings often are blistered after less than one year of service. Low pH and elevated temperatures cause premature corrosion to occur.

Protective coatings act as a barrier. They prevent moisture, oxygen, and other chemicals from attacking the steel substrate. Most coating films, however, permit chemicals, moisture vapor, and oxygen to permeate and attack the steel. This phenomena is accentuated at temperatures between 150 and 210 F. Gaseous penetration not only occurs through pinholes and other micropores, but also through the general coating film. The movement of penetrants through the coating film is fostered principally by osmotic and electroendesmotic pressures, and the constant thermally induced movements and vibrations of the coating film molecules.

First Stages of Underfilm Corrosion — Spot Rusting and/or Blistering

Early stages of corrosion often are left unattended. They have been described many times as rust spotting on the coating. The standards for determining

---

*Gilbert/Commonwealth, Reading, Pennsylvania
Printed by permission of Gilbert/Commonwealth, Gilbert Associates, Inc.

Table 1 — Analysis of Flue Gas Condensate and Drain Water

		1.	2.	3.	4.	5.	6.	7.	8.	9.	10.
Acidity	mg/l CaCO$_3$	335	2990	836	1232	2192	342	2560	292	2480	314
Chloride	mg/l Cl$^-$	9	98	10	28	433	5	151	6	160	7
Nitrates	mg/l NO$_3$ N	4.70	81.3	9.66	16.3	50.6	3.00	81.3	3.00	83.5	2.14
pH		2.7	1.7	2.0	1.9	1.8	2.4	1.8	2.4	1.7	2.3
Solids, Total	mg/l @ 105 C	457	6780	1128	1963	9710	456	6783	402	6735	378
Solids, Susp.	mg/l @ 105 C	43.5	148	41.5	114	274	39.0	392	14.5	219	27.3
Solids, Vol. Susp.	mg/l @ 550 C	18.0	54.0	18.5	34.0	42.0	16.5	56.0	7.0	32.0	12.0
Sulfate	mg/l SO$_4$	389	5509	913	1564	7706	411	5712	341	5814	372
TOC Total Org. Carbon	mg/l	14	29	12	12	12	11	16	6	17	10
Odor		None	Present	None	Present	Present	Present	Present	Present	Present	Present
		11.	12.	13.	14.	15.	16.	17.	18.	19.	20.
Acidity	mg/l CaCO$_3$	1500	1756	1772	816	1816	372	517	428	2068	102
Chloride	mg/l Cl$^-$	24	238	236	27	12	8	8	20	34	102
Nitrates	mg/l NO$_3$ N	28.0	37.4	37.4	13.6	28.8	3.39	1.94	4.58	22.5	1.87
pH		1.6	1.8	1.8	1.9	1.6	2.1	2.0	2.2	2.2	5.8
Solids, Total	mg/l @ 105 C	2778	6540	6526	2615	3086	473	758	2690	5482	1676
Solids, Suspended	mg/l @ 105 C	233	529	510	277	501	50.3	95.3	180.7	122	73.0
Solids, Vol. Susp.	mg/l @ 550 C	44	44	43	53	101	14.5	18.7	36.7	25.3	10.3
Sulfate	mg/l SO$_4$	2265	4952	4884	11969	2248	434	679	2003	3617	668
TOC	mg/l	25	15	17	32	51	13	13	29	200	9
Odor		Present	None	Present	Present	Present	None	None	Present	None	No Identifiable Odor

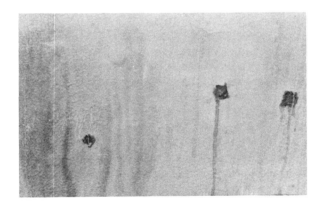

FIGURE 1 — Early signs of corrosion blisters in polyester lining of SO$_2$ scrubber flue are the size of half dollars. Water, pH of 2.5 is entrapped. This 40 mil lining began to show corrosive attack of the steel after only 6 months service.

and evaluating degree of rust spotting are found in Steel Structures Painting Council Vis-2 or ASTM D 610-68. One rust spot found in one square foot may provide a 9+ rating, but 3 or 4 rust spots drop the rating to 8. The fact that the rust spots go unattended allows for further corrosion to take place.

Early stages of corrosion can be recognized as blistering (Figure 1). Frequently, blistering occurs without external evidence of rusting or corrosion. The mechanism of blistering is attributed to osmotic attack, or a dilation of the coating film at the interface with the steel under the influence of moisture. Water and chemical gases pass through the film, dissolving ionic material either from the film or the substrate, which causes osmotic pressure greater than that of the external face of the coating. This establishes a solute concentration gradient, with water building up at these sites, until the film eventually blisters. Visual blistering standards are found as ASTM D 714-56.

Blistering is also dependent upon electrochemical reactions. Water diffuses through a coating by an electroendesmotic gradient. After corrosion has started, moisture is pulled through the coating by an electrical potential gradient between the corroding area and the protected areas that are in electrical contact. Therefore, osmosis starts the blistering, and as soon as corrosion begins, electroendesmotic reactions accelerate the corrosion process. By adding heat and acidic chemicals, the breakdown occurs most rapidly. At temperatures of 150 to 200 F, the chemical reaction is accelerated. Thus, steel will literally dissolve in the chemical environment at these temperatures. Moisture is ever present, and often condenses on the surface behind the blister. This condensation offers a solute for gaseous penetrants to dissolve. In acidic environments, the water pH behind the blister can drop as low as 1.0 to 2.0; therefore, the steel is subjected to severe attack.

Second Stage of Underfilm Corrosion — General Rusting

After one rust spot has been observed, or a few blisters are found, the condition advances to general rust spotting. The second stage of corrosion can be

FIGURE 2 — In the second stage of corrosion, black iron oxide or iron sulfide appears through weakened lining of SO_2 scrubber flue.

FIGURE 3 — In the third stage of corrosion, the polyester lining has lost adhesion. Spot corrosion or rusting appears beneath the coating.

described as general rusting. This is most frequently seen as red iron rust, Fe_2O_3. However, within SO_2 scrubber systems, which do not have sufficient oxygen during normal operation, the rusting is usually black FeO. This Fe_3O_4 (Figure 2) corrosion product converts to Fe_2O_3 eventually as the unit is shut down and more oxygen is available. The second stage of rusting may, therefore, be more accurately described as a number 5 or number 3 rusting condition according to SSPC Vis-2.

Third Stage of Corrosion — Coating Disbondment

The next advanced stage of corrosion is the total disbondment of the coating, and direct exposure of the steel to the environment occurs (Figure 3). No longer is the coating protecting the steel substrate; therefore, corrosion can now occur at an uninhibited rate. The acidic conditions within the SO_2 scrubber system quickly corrode the steel at these elevated temperatures, especially with plenty of moisture present. Disbondment occurs because of chemical attack on the substrate. One might think of disbondment as one large blister.

Fourth Stage of Corrosion — Pitting

Once the coating has been removed and no longer acts as a protective barrier, the steel is left to be directly attacked. This attack most frequently is not uniform, but is localized into many electrolytic cells. Each cell, with an anode and cathode, an electrolyte (moisture and acid), and the steel surface provides the perfect mechanism for pitting to occur, the concentration cell.

Pitting develops when the anodic (corroding) area is small in relation to the cathodic area. These concentration cells develop where oxygen or conductive electrolyte concentrations in water differ. Mill scale, for example, is cathodic to steel; therefore, areas surrounding mill scale would be eaten away and pitted first. In a short time, the pitting will undercut the mill scale, and flaking occurs. Pitting will cause structural failure from localized weakening effects while there is still considerable sound metal remaining (Figure 4).

Fifth Stage of Corrosion — Flaking of Steel and Development of Holes

As the corrosion cell becomes more active, the rust and pitting becomes more advanced. Deep pits in steel may eventually penetrate completely to cause holes to occur. Such penetrations generally result in structural loss and replacement of the steel at considerable expense to the owner. Within the corrosion cell, pitting has occurred to such a degree that undercutting, flaking, and delamination of the steel is noted. Most of these pits have a conical configuration.

After the small hole develops, the electrolyte can now seek other fresh surfaces on the reverse side, enabling corrosion to occur on both surfaces front and back (Figure 5).

FIGURE 4 — The fourth stage of corrosion shows severe pitting and loss of steel. The polyester lining has completely lost adhesion. Red rust is noted.

FIGURE 5 — In the fifth stage of corrosion, considerable loss of steel and delamination has occurred. A small hole completely through liner permits the corrosive media to attack both front and back surfaces of the steel liner.

FIGURE 6 — Final stage of corrosion, hole about 1 x 2 feet through the ¼" thick A—242 type steel. A plate will be welded over this area and new lining applied.

Final Stage of Corrosion — Complete Loss of Steel

As the chemicals present condense on the unprotected substrate, corrosion occurs at its most rapid and aggressive rate. Large gaping holes are found, causing considerable structural damage (Figure 6). These holes are enlarged rapidly, because the electrolyte is ever present on both front and back surfaces of the steel. Costly repairs are necessary. Replacement of steel and welding of large plates to the surface are necessary. Complete inspection is required to determine functionality of the structure.

The necessity for some test program to evaluate materials for flue liners became apparent when the industry was surveyed. Many lining failures have occurred throughout the country. None of the coating and materials people guarantee the performance of their lining for more than one or two years service. Since few of these lining suppliers had any test data or meaningful field experience, the Atlas Test Cell program was initiated. A-36 steel panels or A-242 Mayari R type steel ¼ x 7 x 8 inches were used.

FIGURE 7 — The Atlas Test Cell. Two coated panels under test.

FIGURE 8 — A fluorelastomer is in excellent condition after 180 day test. Cross section view 25X magnification.

In each case, application of the coating material was done according to the coating manufacturers specific instructions. Their representatives were present at the time these panels were prepared. Some of the panels were coated at the job site, others in the laboratory.

The "Atlas Test Cell" (Figure 7) is an opened faced glass cell, described in National Association of Corrosion Engineers (NACE) TM-01-74, "Laboratory Methods for Evaluation of Protective Coatings Used as Lining Materials in Immersion Service." Each cell was half filled with gas condensate, collected from a large number of daily samples taken at various times from the chimney of an operating power station. The concern for the variety of chemicals found in the gas

Table 2 — Atlas Test Cell Data for SO₂ Condensate Evaluation of Lining Materials for Chimney Flues

Coating Material	mils Film Thickness	Visual Observations Liquid Phase Days				Visual Observations Vapor Phase Days				Microscopic Observations Liquid Phase Days				Microscopic Observations Vapor Phase Days				Performance Rating
		30	60	90	180	30	60	90	180	30	60	90	180	30	60	90	180	
Fluoropolymer	40			2D	2D				8F			S,R	S,R			E	E	7
Fluoropolymer	40			E				E				E				E		10
Fluoropolymer	40			E				E				E				E		10
Fluoropolymer	40			E				E				E				E		10
Chloro Polyester (Spray)*[1]	40			2MD	2D			2M	2MD			S	R,S				S,R	0
Bromo Polyester (Spray*[1]	40			E	cc			E	cc			cc	R,S				S,R	0
Chloro Polyester (Trowel)	80			E	8D (Surface)			E	8D (Surface)				S,R				SR	8
Bromo Polyester	80			E	8D (Surface)			E	8D (Surface)				S,R				S	8
Polyester 4 Trowel	120				E,cc				E				S,R				S	8
Polyester 1 Trowel	120				cc				4-6MD (Surface)				E				E	10
Epoxy/Polyamide Spray	30			4M	4MD,cc				4MD				S,R				S,R	4
Epoxy/Polyamide Trowel	125			6M	6D,cc				6M				S,R				S,R	2
Epoxy/Amine	50	cc, D-2			D-1							R,S cc				R,S		0
Epoxy/Amine	50	D-2			D-1							R cc				R,S		0
Vinyl/Ester (metallized)	40		2F	2F,cc				2F,cc				S				S		6
Vinyl/Ester (metallized)	40			2F,cc				2F,cc				S				S		6
Vinyl/Ester	40			4F				4F				S,R						5
Coal Tar Epoxy	20	D-8			D-8							R,S				R,S		0
Coal Tar Epoxy	20	cc			D-6							R,S				R,S		0
Asphaltum	2	4MD		4MD												R		0
Asphaltum	2	F8		4MD												R		0
Cement, Chem. Resistant	1-1/2 inches		E				A				S,SR				A			9
Cement, Chem. Resistant	1-1/2 inches		E				A								A			9
Modified Inorganic	20		E	E				ST				SR			ST	ST		9
Modified Inorganic	20		E	E				ST				SR			ST	ST		9
Asphalt Urethane	125		4MD				4MD					S,R	S,R					5
Foam Glass Block (closed cell foam)	2 inches			E				E				E				E		10

*Control system which failed after 1 to 2 years service.
C.A. Corrosive Attack on surface
Microscopic examinations, panel cut in half
C.C. Chemical attack of coating

2,4,6,8   ASTM D 714 Blistering
F,M,MD,D
A = Ablated
E = Excellent
S = Separation from substrate
R = Rust
SR = Sl Rust
ST = Stain

condensate stimulated this test program. The main purpose of the Atlas Test Cell program was to evaluate coating materials for this service. A realistic secondary purpose was to develop a laboratory procedure which would be an effective method to evaluate all materials for this service, so that coating materials could be screened by the Atlas Test Cell technique. Recommendations could, therefore, be based upon results found in the laboratory.

The liquid gas condensate contains chloride, sulfate, nitrate, and organic carbon in quantities significant enough to damage severely organic coatings and steel surfaces. It is believed that this synergistic test is more effective than merely

Table 3 — Partial Analysis of Flue Gas Condensate

	Original Test Solution Total mg	Atlas Cells after test 180 days Total mg	
		(A)	(B)
pH	1.9	2.3	4.9
Chloride	63	96.8	118.4
Sulphate	2393	2305	1855
Total Org. Carbon	60.8	8.3	206
Fe	58.0	110.9	95.3

evaluating for a few specific chemicals. The combinations found in the collected samples are damaging to steel, and seem to penetrate nearly all coatings below a 40 mil thickness.

Discussion

After 30, 60, and 90 days, the cells were opened and observations were made. The visual condition of each panel was noted. The cells were reassembled, using the identical media. No replacement solution was added. This procedure was changed in later work. The media can be replaced each 30 days if desired, allowing for any losses or changes in pH which might occur due to reaction within the cell. In any case, some tests ended after 180 days with the same solution with no volume addition. No water was added to maintain the liquid level. In fact, the liquid level dropped, due to slight losses during the test. Some tests were discontinued after 30 or 60 days. All cells were adjusted to a constant temperature of 165 F. This temperature was chosen because it is close to the average gas temperature within the chimney, where the condensate gas was collected. Alternate testing was done on some systems at 195 F.

Some systems failed after 30 days testing. After the 180 day period, the remaining cells were opened. The liquid level retained in each cell was measured, and the solution was analyzed as to the chemical composition. Each panel was photographed, and visually examined, for blistering and rust immediately after dismantling. ASTM D 714-56 (Reapproved 1974) "Evaluating Degree of Blistering of Paints," was used to report blister size and frequency. Rusting was reported as severe, moderate, or slight, before and after microscopic examination of the steel substrate. The visual observations made on each of the panels were therefore, substantiated by the chemical analysis of their respective solutions.

Microscopic Examination

After testing, the panels were cut in half vertically, with a band saw, and a cut edge was polished. Photomicrographs were taken, showing the condition of the interface of the coating and steel exposed to both the liquid and vapor phases. A small area of coating was removed to expose the steel substrate. Photomicrographs were taken again to illustrate the condition of each area of the substrate. Many of the coatings appeared visually to be intact; however, at the steel interface rusting had started.

Table 2 provides data regarding lining materials which have been tested in the Atlas Test Cell.

Typical of Atlas Test Cell procedures, the media was analyzed before and after testing. Table 3 indicates the changes which occurred from the original test solution to that found at the end of testing.

Cell A increased in chloride, indicating water soluble chlorides were extracted from the coating. The sulphate remained the same, but total organic carbon was consumed, either onto the coating or reacted with the contents or other extractants. The Fe concentration doubled, indicating attack on the steel panel. Cell B had a significant increase in pH which accounts for the drop in sulphate. TOC increased severely, indicating a loss of lining material into the media. Fe also increased, indicating attack on the steel substrate. The chloride content nearly doubled, indicating the lining material contained significant water extractable chloride.

One technique used to identify attack on the steel substrate was to cut the panels in half and polish one edge. These edges were carefully examined under 25X magnification. Separation or delamination was determined easily. Rusting was observed in some cases. Usually, rusting was found after the coating was removed. Generally, the findings of coating deterioration were confirmed after analysis of the gas condensate used as the Atlas Test Cell media. A rise in iron content was evident in several cells, indicating corrosive attack of the steel substrate.

The first sample panel of the fluoropolymer blistered, but no rusting was evident until the blister was opened and a cross section made by cutting the panel in half. Some slight rusting appeared, and the obvious separation of the film was noted. Rusting did not occur on the subsequent 4 panels; therefore, it was concluded that this spray applied fluoroelastomer provides satisfactory service (Figure 8). This good record has been born out under actual service conditions in the flues of two power stations for over a 3 year period. One additional station in England has been in service 6 years without failure.

The spray applied chlorinated polyester and brominated polyester coatings were considered to be control panels. These linings failed badly under actual service conditions. Representative panels were tested in the Atlas Cell, and these coating materials failed, showing severe blistering and rusting after 180 days exposure (Figure 9). Large voids in the film (25 to 30 mils) were noted.

The heavier, trowel applied chlorinated and brominated polyesters (80 mils) were better. They lasted 180 days in the test cell but, by cross section

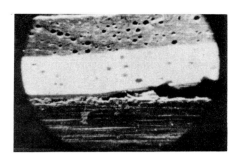

FIGURE 9 — A brominated spray applied polyester failed by delamination after 180 days test. Cross section view 25X magnification.

FIGURE 10 — This trowel applied 125 mil polyester system shows delamination after 180 days test. 25X magnification also shows large void in the film.

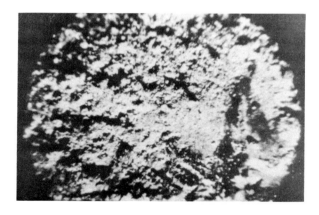

FIGURE 10A — Considerable rusting is evident on the steel surface beneath the polyester coating shown in Figure 10. 25X magnification.

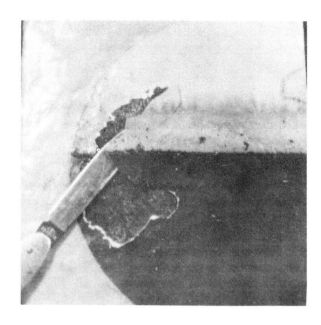

FIGURE 11 — A 40 mil epoxy/amine system failed by delamination in 30 days.

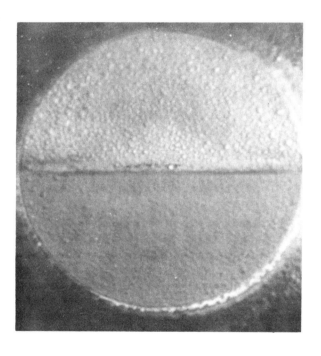

FIGURE 12 — An amine cured coal tar epoxy is severely blistered after 30 days test. Upper portion is vapor phase. Lower portion is liquid phase.

examination, and by mechanically removing the coating, some rusting was evident on the steel surface (Figure 10). These materials have provided good protection in the ductwork of an operating unit. They would not be recommended for use in chimney flues. Large voids (50 mils) were noted in the film.

Heavy, 120 mil, trowel applied polyesters showed good results in the Atlas Cell Tests, with only one panel exhibiting slight separation and rust after cross sectioning. This coating has performed well in ductwork of several operating units.

The epoxy/polyamide and epoxy/amine systems failed badly by showing separation and rusting (Figure 11). These systems would not be recommended for service.

The spray applied vinylester systems failed by exhibiting blister formation and separation when applied over metallized aluminum. When applied over steel, the coating showed few blisters and rusting of the surface after cross sectioning.

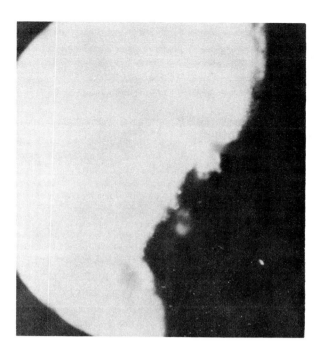

FIGURE 13 — The surface of this acid resistant mortar has been ablated away, loosing about 50 mils of material. Magnification 25X.

Coal tar epoxy, asphaltum, and an asphalt urethane blistered severely. They all showed rusting of the steel after only 30 days testing (Figure 12). One asphalt rubber product could not be tested in the Atlas Cell because the 165 F temperature caused the material to flow, allowing leakage to occur. These products also burned severely in a fire resistance test.

Several chemical resistant cements were evaluated. These products are spray applied at 1½ to 2 inches thick. One product was too porous, and could not be mounted in the Atlas Cell because leakage occurred through the pores of the material. Tests inside the flue, under actual operating conditions, showed severe erosion and ablation of the material, with rusting of the steel surface underneath the coating. The chemically resistant cement tested in the cell showed a loss of material (about 50 mils) after 180 days (Figure 13). It had good overall resistance. During the test, the texture of the surface changes from a hard material to a soft, easily damaged material.

A modified inorganic coating showed excellent resistance to the corrosive media. A superficial stain was noted on the surface in the vapor phase. Examination of the steel substrate did not show rusting after cross section examination. The coating was subjected to a mandrel bend, and only then could some minor rusting be observed. The gasket outline of the cell was noted.

A closed cell foam glass block system was found to be completely intact, after 180 days testing. This system has performed well within the flue of one operating power station for 2 years. The foam glass blocks are damaged easily by any physical impact; therefore, great care must be taken upon installation of this material.

A flexible furan was evaluated. This material is extremely flammable and therefore not recommended for chimney flues.

Rubber lining materials were not evaluated in this program. Some specialty steels are currently under test. A program has been initiated to analyze the flue gas condensates of various operating SO_2 scrubber units to establish the different types of corrosive environments found therein.

Summary

The Atlas Test Cell is an effective means to evaluate coating materials used to protect steel surfaces for SO_2 scrubber service. The analytical and test data correlate closely with field experience and field test data.

References

1. McDowell, D.W., Sheppard, W.L., Material and Corrosion Problems in Gas Scrubbing Systems, NACE Paper 166 CORROSION/77.
2. Smock, R., Sulfur Scrubbers come for Age, Electric Light and Power, Sept. 1977.
3. Mockridge, P.C., McDowell, D.W., Materials and Corrosion Problems in a Fly Ash Scrubbing System, *Materials Performance,* Apr. 1977.
4. Burda, P.A., Corrosion Protection of Wet Scrubbers, *Power Engineering,* Aug. 1975.
5. McDaniel, C.F., Wet Scrubber Coating Experience, Utility Wet Scrubber Conference, Feb. 1977.
6. Cement Lining Protect Scrubber, *The Wet Scrubber Newsletter,* Feb. 1977 and May 1977.
7. Hall, G.R., Corrosion Resistant Linings: A Materials Engineering Approach, Liberty Bell, 1976.
8. Singleton, W.T., Vinyl Ester Resins for the Air Pollution Industry, *Modern Paint and Coatings,* March 1978.
9. Berger, D.M., The Six Stages in the Corrosion of Coated Steel, *Chem. Eng.,* Aug. 18, 1978, pp. 121, 122.

The Impact of Environmental Restraints on Corrosion Resistant Coatings

*Robert N. Washburne**

Introduction

The protective coatings markets are undergoing changes in emphasis as to definition of new coating types. In the past, several market segments, notably the trade sales segment, have experienced a slow but steady change from conventional solvent based coatings to latex based coatings. Using the trade sales coating market as an example, the reasons for this change are well known and well understood: longer protection of the substrate; longer durability in terms of aesthetic appearance; greater ease of application and clean up; low odor; and others. This change of emphasis on coating type has been possible only because the latex binder manufacturers have been successful in developing latex polymers with inherently high levels of protective and aesthetic performance.

Beginning in the late 1960s, a new influence has been encountered which will continue to affect the direction of the coatings industry. I will call this influence "environmental concern." This encompasses considerations of pollution of air, water, soil, all of the natural resources of our planet. Formalization of this concern began with enactment of Rule 66 in California, the outlawing of lead and chromates in coatings for the do it yourself markets, the proposal of CARB rules which will limit solvent emissions in a wide variety of air drying coatings, and, currently, the close scrutiny of CARB type restrictions by both state and federal agencies, with the objective of adopting such restrictions on what could be a national basis. There is good agreement on the positive value of such restraints as a philosophy; there is less agreement as to the current practicality of putting this philosophy into universal practice.

All of these things have had a profound effect on the direction and emphasis of research efforts on development of new binders for coatings. We find included in the list of definitive properties/characteristics the requirements "uses little or no solvent," "is highly water miscible," "can be made in emulsion form." The development of new chemistry now also comes under the focus of potential applicability to low pollution systems. In some cases, the possibility that a newly developed or known type of polymer chemistry may fit the nonpolluting goal is sufficient to justify additional research dollars to carry it to a decision point.

I would like to review some newer developments in the area of corrosion resistant coatings binders which are the result, in significant part, of the imposition of the "nonpolluting" qualification on research and development.

Emulsion Binders

Emulsions as binders for corrosion resistant paints are not new. The first candidate acrylic emulsion binder for this application was developed over 15 years ago, and several technically improved emulsion vehicles have been developed since. These emulsion binders possessed, in common, a substantial degree of inherent corrosion resistance, which was easily demonstrable in both accelerated laboratory testing and actual exterior performance testing. Many of the paints formulated with these acrylic emulsions also contained reactive pigment such as strontium chromate and lead silico chromate. These pigments further increased the level of corrosion resistance of the coating by helping to overcome the physical "errors" in the films: thin spots, pinholes, film damage, etc. The long term performance of these latex paints was, in many cases, better than that of conventional solvent based alkyd systems (which also employed the same reactive pigments for the same reasons) because of their retention of film flexibility with time, and better resistance to weathering. There were also shortcomings unique to the emulsion paints in the early performance properties of flash rusting and early rusting.

Flash rusting is the formation of rust spots in the paint film during the film drying stage. Flash rusting is caused by the solubilization of some iron ions in the water phase of the latex paint, followed by rapid oxidation to ferric oxide. This phenomenon normally occurs over freshly blasted steel—the blasting operation exposes these reactive ions. Here, latex design was used to develop an acrylic emulsion vehicle which will inherently stop flash rusting.

---
*Rohm and Haas Company Research Laboratories, Spring House, Pennsylvania

I would like to digress for a moment to review the mechanism of film formation of an emulsion. This mechanism was postulated by Brown in 1956,[1] and many studies since that time have tended to support it as valid. A simplified schematic representation of the stages of film formation is as follows:

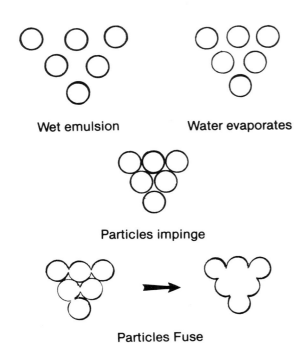

FIGURE 1 — Simplified schematic representation of the stages of film formation.

Film formation is a dynamic process which depends on evaporation of water, polymer hardness, presence of coalescent, and temperature.

The other unique shortcoming of an emulsion based paint for metal is early rusting. This phenomenon has been identified and studied only within the past several years. Early rusting is the appearance of rust spotting in the paint film after the film is dry to touch. Rusting can occur hours or days afterward, and as such is particularly insidious. Three conditions must be present for early rusting to occur: a thin film ($\cong 1.0$ to 1.4 mils); relatively low temperature ($\cong 50$ F); and high relative humidity ($\cong 75\%$ or greater). The presence of low temperature and high relative humidity retard the complete evaporation of water, and these conditions, in turn, retard complete film fusion at or near the following condition:

FIGURE 2 — Retardation of complete film formation at or near the pictured condition, as a result of low temperature and high humidity.

The microscopic voids between the dispersed particles allow the permeation of iron ions into the film. Subsequent oxidation to Fe_2O_3 creates the brownish spots. The contribution of low film thickness to early rusting is thought to be by reducing the statistical probability of these voids being blocked by a sufficient number of particulate layers.

Part of our ongoing research effort involves studies of the factors influencing film homogeneity. One discovery had led to the ability to produce an acrylic emulsion having extremely good film forming qualities.

In addition to the degree of emulsion film formation as a factor, it was determined that the compatibility of such paint ingredients as pigment dispersant and thickening agent with the polymeric binder was also an important aspect. Low compatibility among these materials can lead to disruptions in film continuity from both heterogeneity and high hydrophilicity, creating potential "holes" in the film. Substrate protection can be lost in these areas. A pigment dispersant and a nonionic thickener have been developed which are highly compatible with the emulsion polymer backbone. These compatible ingredients should minimize weak or hydrophilic areas in the paint film, thus increasing the ability to protect the metallic substrate.

Verification experiments were made in a practical paint formulation containing only a low level of a mildly reactive pigment, zinc oxide. Analogous paints were prepared with other conventional emulsion based maintenance vehicles, the binder being the only variable. Using the laboratory method developed by Grourke,[2] early rust resistance of these paints was determined. The inherent performance of the "engineered" system was far superior to the analogous "conventional" latex paints.

A similar comparison of resistance to salt spray and to high humidity was made. In addition to the latex paints, a commercial alkyd based red lead primer was also included. Test results are summarized in Table 1.

The performance superiority of the "engineered" acrylic emulsion and its attendant pigment dispersant and thickener is being substantiated by actual exterior exposure in both industrial plant locations and seashore environments. It should be noted that the formulations used fall well below the CARB maximum allowable solvent level, and are also free of lead and chromate pigments.

Water Based Epoxy Coatings

Epoxy based coatings are significant in industrial maintenance painting. They are used in areas where abrasion resistance, chemical resistance, and high film hardness are required. Generally, these are solvent based two package ambient curing systems. Good surface preparation is required for best performance, and the metallic or previously coated surfaces must be essentially free of all moisture for best

TABLE 1 — Comparative Effects of Salt Spray and High Humidity Testing

Test and Properties "Engineered" System	Primer[1]		
	Acrylic A in Formulation A[2]	Acrylic B in Formulation A	Commercial Alkyd Red Lead Primer[3]
Salt Spray Test[4]			
Blistering[5]			
1 week	9F	9D	10
2 weeks	9F	8D	9F
3 weeks	9M	7D	9F
Rust Bleeding			
1 week	None	Severe	None
2 weeks	Trace	Severe	None
3 weeks	Trace	Severe	None
High Humidity Test[6]			
Blistering[5]			
1 week	9F	9F	8M
2 weeks	9F	9F	7M
3 weeks	9F	9F	7M
Rust Bleeding			
1 week	None	Moderate	None
2 weeks	None	Moderate	None
3 weeks	None	Moderate	None

[1] Each primer was applied to a clean cold rolled steel panel by drawdown bar to give a dry film thickness of ≅1.5 mils. Films were air dried 2 weeks at 75 F/50% RH before testing.
[2] CARB VOC = 132 gm/liter-water
[3] Designed to meet Federal Specification TT-P-86e, Type II
[4] ASTM Method B-117.
[5] Blister size: 10 = no blistering, 2 = very large blisters
Blister density: F = few, M = medium, D = dense
[6] ASTM Method D-2247

results. The major reactive epoxy systems used in maintenance painting are the epoxy/amine, epoxy/polyamide, and epoxy ester types.

During the last several years, there has been increased activity in development of water based epoxy systems. The combination of CARB legislation in the state of Californa, with the interests of EPA and OSHA in examining the CARB rules for possible adoption on a federal basis, has helped to spur this interest. Currently, the Federal Highway Administration is starting an evaluation of low solvent containing coatings, many being water based, which may give equivalent (or satisfactory) performance over steel substrates, with an eye using such low pollution coatings on highway bridges, guardrails, and other metallic substrates.

There are several water based epoxy systems which have been developed recently. These generally are composed of a preemulsified epoxy component and a curing agent component generally a polymeric material containing amine groups, and acidified to keep the polymer in water solution. In about half the cases, the epoxy component carries the pigmentation; in the others, the polymeric amine component carries the pigmentation. Formulation constants of several systems are summarized in Table 2.

From this breakdown, it can be seen that most of the curing agent components are at acid pH to achieve solubilization in water. Most of the finished formulations carry greater than 50% curing agent by weight, implying that the curing agents may be high in molecular weight and/or low in reactive functionality.

TABLE 2 — Formulation Constants

Designation	A[1]	B[1]	C[1]	D[1]	E	Comm. Solvent Epoxy[1]
Epoxy Component						
Solids, % wt.	—	60.7	—	84.4	55.0	53.0
Form	Pigmented	Pigmented	Clear	Clear	Pigmented	Clear
Viscosity, cps (#4 × 12)	1500	31,000	500	450	500	1000
Curing Agent Component						
Solids, % wt.	39	34.5	—	42.6	45.2	65.0
Form	Pigmented	Clear	Pigmented	Pigmented	Clear	Pigmented
Viscosity, cps (#4 × 12)	2000	2000	2000	1250	2750	4500
pH	6.2	8.2	4.7	4.2	10.6	—
Epoxy Finish						
Solids: % wt.	—	50.1	—	47.3	53.1	60.2
% vol.	—	36.7	36.4	33.1	38.0	46.4
Epoxy/Curing Agent						
by wt.	—	43/57	—	46/54	65/35	66/34
by vol.	1/4	1/1	1/6	1/6.4	2.9/1	1/1
TiO_2/Binder	—	51/49	—	56/44	53/47	47/53
PVC	—	22.3	—	26.7	23.5	19.6
CARB No. (gm/l-H_2O)	—	259	—	249	232	508

[1] Calculated based on label analyses or on suggested formulations.

TABLE 3 — Paint and Early Film Properties

Designation	A	B	C	D	E	Comm. Solvent Epoxy
Viscosity (KU)						
Initial	133	>141	>141	>141	80	85
After 6 hr	>141	>141	>141	>141	102	89
Film (Room Temp)[1]						
Set to Cotton (hr)	2.5-3.0	3.5-4.0	2.5-3.0	2.5-3.0	6-7	1.5
Tack Free Time, Zapon, 500 gm (hr)	18-19	23-24	18-22	16-22	27-28	8-16
Flash Rusting (Rusty Hot Rolled Steel)	Heavy	None	Heavy	Heavy	None	None
Early Water Resist.						
4 hr. dry, 30 min. soak	Dissolved Washed off	Dissolved Washed off	Dissolved Washed off	Dissolved Partially dissolved	Insoluble Insoluble	Insoluble Insoluble
6 hr. dry, 30 min. fog						
Gloss, 2 wks. (60°/20°)						
Alodine Aluminum	97/88	92/69	84/51	63/17	98/92	97/75
Bonderite 1000	75/22	93/68	50/11	42/9	98/89	63/13
Tukon, 2 wks (KHN)	10.4	2.0	18.2	11.6	12.6	11.6

[1] 1.4-1.6 mils dry.

All systems carry only TiO_2 as pigmentation; undoubtedly, to obtain maximum gloss. Finally, where calculable, the water epoxy systems are close to the proposed CARB value of 250 g/L minus water. Note that this level is approximately half the solvent present in the commercial solvent based epoxy system.

Research projects over the last several years have resulted in methods of controlling acrylic polymer molecular weight, functionality distribution, and polymer solubility. It was decided to apply this know how in the development of a functional polymer for use in aqueous epoxy systems. First, it was decided that such a polymer should be of sufficient functionality to employ epoxy as the majority binder component. In this way, the inherent resistant properties of the epoxy should be maintained. Second, the reactive polymer should be miscible at alkaline pH. Our experiences with flash rusting and early rusting in emulsion paint systems dictated strongly that the water based epoxy coating should be alkaline during the

TABLE 4 — Film Development Properties[1]

Dried at 77 F/50% RH

Designation	A		B[3]		C		D		E		Comm. Solvent Epoxy	
	1 wk	4 wk	1 wk	4 wk	1 wk	4 wk	1 wk	4 wk	1 wk	4 wk	1 wk	4 wk
Tukon Hardness (KHN)	5.0	13.7	1.8	2.0	16.5	20.0	8.1	14.0	3.4	15.0	6.7	10.6
Gloss (60°/20°)	97/88	97/88	92/68	92/68	84/51	84/52	63/17	63/17	97/91	99/92	97/74	96/74
Pencil Hardness												
Before soak	H	H	F	F	4H	4H	4H	4H	3H	4H	2H	3H
10% NaOH, 30′	F	F	4B	4B	3H	2H	2H	2H	H	3H	F	2H
10% HCl, 30′	F	F	Lifted	Lifted	2H	2H	H	2H	H	3H	F	2H
10% HOAc, 30′	HB	HB	Lifted	Lifted	H	F	5B	H	6B+	2B	F	F
Toluene, 30′	6B+	6B+	6B+	Lifted	6B+	4B	6B+	6B+	6B+	4B	6B+	4B
Cleaning Soln.[2], 30′	6B+	6B+	6B+	6B+	6B+	6B+	6B+	6B+	6B+	6B+	6B+	6B+

[1] 1.5 mil dry film on Bonderite 1000.
[2] Butyl Cellosolve/28% NH$_4$OH/H$_2$O — 50/3/47.
[3] Samples may be atypical — cure is poor.

TABLE 5 — Solvent Resistance of Aged Epoxy Films

All films are at 1.5 mils dry,[1] aged 4 months at room temperature

Pencil Hardness

Designation	C	D	E	Comm. Solvent Epoxy
Solvents[2]				
Initial	5H	5H	5H	4H
Skydrol	5H	5H	5H	4H
MEK	6B+	6B+	F	6B+
Varsol	4H	5H	5H	2H
Butyl Cellosolve	4H	4H	4H	4H
Gasoline	5H	5H	5H	3H
Propanol	4H	4H	4H	3H
Acetone	6B+	6B	B	6B+
28% NH$_4$OH	4H	B	4H	4H
Lestoil	5H	5H	5H	4H

[1] Substrate — Bonderite 1000.
[2] Cheesecloth soak, 30′ under watchglass.

early film drying stages, to minimize these tendencies.

Third, the functional polymer should exhibit high compatibility with both epoxy resin and dispersed pigment, to maximize gloss capability. Finally, resistance properties of the polymer, when reacted with epoxy resin, should be maximum within the parameters of molecular weight, functionality, and compatibility already defined. After considerable evaluation, rebalancing of compositional and manufacturing parameters, reevaluation, and rebalancing, a polymer was developed which met most of the criteria. The result is system (E), shown in Table 2. The test formulation details are given in Appendix I.

The new acrylic/epoxy system was evaluated against several other commercial and semicommercial water systems for a variety of key film properties. These are listed in Table 3. The new acrylic/epoxy system (E) is showing very high gloss and excellent early water resistance relative to the other water based systems, even though tack free time is somewhat longer. Flash rust resistance is excellent.

The results of film development studies are summarized in Table 4. The results of studies on 4 month old films with a wider variety of solvents appear in Table 5. In general, the resistance of the new acrylic/epoxy system to solvents, acids and caustic, is equal or superior to that of the other water based epoxy systems tested, and is generally equal to the performance of the commercial solvent based epoxy/polyamide system. After 4 months aging, the resistance of the new acrylic/epoxy system to highly polar solvents, such as methyl ethyl ketone and acetone, is good, relative to that of the solvent epoxy system.

TABLE 6 — Corrosion Resistance Properties of Topcoats

Designation	A	B	C	D	E	Comm. Solvent Epoxy
Hot Water Resistance,[1,2]						
Bonderite 1000						
(1 wk @ 150°F)						
Blist./Rust	7D/Heavy	5D/Heavy	6D/Heavy	5D/Heavy	10/0	10/0
Underfilm Corrosion	Heavy	Heavy	Heavy	Heavy	None	None
Salt Spray Resistance,[1,2]						
CCRS, 336 hours						
Blist./Rust	5D/Heavy	1D/Heavy	9F/Heavy	7D/Heavy	9F/0	6F/Trace
Underfilm Corrosion	Heavy	Heavy	Heavy	Heavy	None	None
Undercutting/Rust at scribe	4/1	5/7	6/5	5/4	5[3]/7	7/4
High Humidity Resistance,[1,2]						
CCRS, 336 hours						
Blist./Rust	7D/Heavy	1D/Heavy	7D/Heavy	6D/Heavy	9F/0	8D/Trace
Underfilm Corrosion	Heavy	Heavy	Heavy	Heavy	None	Heavy
Undercutting/Rusting at scribe	5/5	1/6	9/8	6/6	9/9	9/9

[1] ≅1.5 mil films, dried 2 weeks at 77°F/50% RH before testing.
[2] ASTM type ratings, 10 = no failure, 1 = complete failure.
[3] Delamination on drying. No delamination over blasted steel.

TABLE 7 — Salt Spray Resistance Properties of Systems

System Designation[1]	I	II	III	IV	V	VI
React. pigment in primer	N.A.	N.A.	None	M-212[2]	Zinc Phosphate	N.A.
Topcoat	B	C	E	E	E	F
340 hours						
Blistering	4D	9F	9F	9F	9F	9F
Surface Rust	Heavy	0	0	0	0	0
Undercutting/Rust at scribe	7/5	9/8	9/8	9/8	9/8	9/8
2225 Hours						
Blistering	3VD	9F	9M	8M	8M	8F
Surface Rust	Heavy	Trace	0	0	0	0
Undercutting/Rust at scribe	5/4	7/7	7/7	7/7	7/7	8/8
3500 Hours						
Blistering	1D	9F	9M	8M	8M	8F
Surface Rust	Heavy	Trace	0	0	0	Trace
Undercutting/Rust at scribe	1/1	5/5	6/5	7/6	5/5	7/7

[1] 3 mil primer, 6 mils topcoat. Substrate: blasted rusty hot rolled steel.
[2] Molywhite® 212

The corrosion resistance characteristics of the several epoxy systems were also studied. Films of ∼1.5 mils dry were applied to clean cold rolled steel, and salt spray and high humidity tests were run. The results are tabulated in Table 6. Once again, the new acrylic/epoxy system gave performance superior to the other water based systems, and compared well with the performance of the solvent based epoxy. This is especially pertinent, since the paints tested are topcoat formulations not necessarily optimized for best corrosion resistance. A more practical comparison of some of the binders as systems (primer plus topcoat) over sandblasted hot rolled steel for salt spray resistance is summarized in Table 7.

The characterization effort with the new acrylic/epoxy system is continuing. In late 1979, a portion of a chemical storage tank in an industrial plant location in Philadelphia was painted. After nine months, the new acrylic/epoxy primer/topcoat system is virtually unchanged in terms of gloss, corrosion resistance, and general appearance. Our oldest panel series comparing the new system to some of the com-

TABLE 8 — Exterior Durability of Epoxy Coatings Systems

Exposure: Six months South 45° at Ocean City, N.J.

System Designation[1]	I	II	III	IV	V	VI
React. pigment in primer	N.A.	N.A.	None	Molywhite 212	Zinc Phosphate	N.A.
Topcoat	B	C	E	E	E	F
Gloss (60°/20°)						
Initial	99/71	99/73	99/72	100/79	99/73	84/48
6 mo.	4/1	62/11	61/16	64/15	63/15	64/1.2
Dirt (6 mo.)	9	9	9	9	9	9
Chalk (6 mo.)	3	10	7	7	7	4
Surface Rust (6 mo.)	10	10	10	10	10	10
Creepage at scribe (6 mo.)	10	10	10	10	10	10

[1] 3 mils primer + 3 mils topcoat. Substrate: blasted rusty hot rolled steel.

mercial and near commercial water epoxy systems and the commercial solvent epoxy system has been on exposure at Ocean City, N.J. for over six months. The data are given in Table 8.

The work reported here is another part of a continuing program to develop low or nonpolluting coating systems with performance properties similar to those of currently utilized solvent based systems.

High Solids Coatings

An obvious approach to making solvent based coatings which conform to the CARB solvent limitation is to use less solvent per unit volume of coating. This can be done to a minor degree with existing solvent coatings, but the practical limitation of the solids/viscosity relationship soon interferes by giving such high viscosities that application to the substrate is impossible. This situation can be improved by use of much lower molecular weight polymeric binders, which increase achievable solids content while maintaining viscosity satisfactory for application. To maintain coating properties, the crosslink density of the binder must be increased. Usually, a rebalancing of the binder composition is required to keep desirable properties. Care must be taken not to overly degrade such physical properties as toughness and flexibility, both of which tend to be compromised as molecular weight is reduced.

An alternative to that approach is the replacement of solvent with a "reactive diluent," a material which behaves as a solvent for the conventional binder in the wet state, but which, after application of the coating, reacts with itself or into the solid coating matrix, in either case becoming a part of the coating. A rough analogy to this is found with the use of "monomeric" melamine resins in thermosetting acrylic systems, instead of the polymeric melamines. One general result of such a change is that the coatings can be applied at higher solids content at a given viscosity. The reactive diluent approach just carries this further.

To be viable, a reactive diluent must possess several properties: 1) low molecular weight; 2) solvent for the polymeric ingredients; 3) reacts well with itself and/or with the polymer matrix; 4) maintains or improves the performance of the coating; 5) low volatility and reasonably low odor.

We have identified such a material, namely a special acrylic monomer of low volatility and with the ability to self initiate its own polymerization upon exposure to oxygen.

The ability to achieve good reactivity of vinyl groups in relatively thin films is difficult, due to the presence of oxygen from the air. The mechanism of oxygen inhibition is as follows:[3,4]

$$A \sim CH_2-\underset{\underset{CO_2R}{|}}{\overset{\overset{CH_3}{|}}{C}}\cdot + CH_2=\underset{\underset{CO_2R}{|}}{\overset{\overset{CH_3}{|}}{C}} \xrightarrow{fast} \sim CH_2-\underset{\underset{CO_2R}{|}}{\overset{\overset{CH_3}{|}}{C}}-CH_2-\underset{\underset{CO_2R}{|}}{\overset{\overset{CH_3}{|}}{C}}\cdot$$

$$B \sim CH_2-\underset{\underset{CO_2R}{|}}{\overset{\overset{CH_3}{|}}{C}}\cdot + O_2 \xrightarrow{very\ fast} \sim CH_2-\underset{\underset{CO_2R}{|}}{\overset{\overset{CH_3}{|}}{C}}-O-O\cdot$$

$$C \sim CH_2-\underset{\underset{CO_2R}{|}}{\overset{\overset{CH_3}{|}}{C}}-O-O\cdot + CH_2=\underset{\underset{CO_2R}{|}}{\overset{\overset{CH_3}{|}}{C}}$$

$$\xrightarrow{slow} \cdot\sim CH_2-\underset{\underset{CO_2R}{|}}{\overset{\overset{CH_3}{|}}{C}}-O-O-CH_2-\underset{\underset{CO_2R}{|}}{\overset{\overset{CH_3}{|}}{C}}\cdot$$

The rate of radical initiation must be greater than the rate of diffusion of oxygen into the film, to overcome oxygen inhibition and achieve cure. With typical radical initiation rates, cure is possible only in very hard or very thick films. The effects of competing oxygen inhibition reaction B can be reduced by introducing a competing oxygen reaction which has a neutral or enhancing effect on the desired reaction A.

A reactive diluent meeting most of the property requirements listed and possessing the ability to reduce the adverse effects of oxygen diffusion is the following:

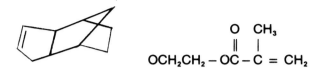

Dicyclopentenyloxyethyl methacrylate (DCPEMA)[5,6,7,8]

FIGURE 3 — DCPEMA is a reactive diluent which meets most of the property requirements listed. It also possesses the ability to reduce the adverse effects of oxygen diffusion.

The typical physical properties of this material are summarized in Table 9. Its high boiling point and solubility parameter of 8.6, giving it low volatility at even low baking schedules (180 F/30'), and good solvency for most common organic film formers is noteworthy. Of equal importance, is the determination of the toxicity characteristics of DCPEMA. These are listed in Table 9. These data suggest that its acute toxicity is lower than those of many commonly used organic solvents.

The candidate reactive diluent DCPEMA contains both a polymerizable methacrylate double bond and a group which can be a free radical source in the presence of oxygen and conventional metallic driers. In the presence of oxygen, the following reactions are postulated:

TABLE 9 — Typical Properties of DCPEMA

Physical Properties

Appearance	Clear Liquid
Viscosity	15-19 cps (25 C)
Boiling point	350 C (760 mm Hg)
Solubility parameter	8.6 $(cal/cm^3)^{1/2}$
Flash Point (Pensky-Martens CC)	>200 F
Cured film hardness (KHN)	15

Toxicity Properties

Acute oral (rats)	LD_{50} >5 g/kg, non-toxic
Acute dermal	LD_{50} >5 g/kg, non-toxic
Eye irritation (rabbits)	Non-irritating
Skin irritation (rabbits)	Draize = 1-2
Acute inhalation (rats)	Non-toxic
Ames mutagenic Test	Nonmutagenic
Skin sensitization (guinea pig)	Not a sensitizer

TABLE 10 — Effect of DCPEMA Modification on an Alkyd Paint

Alkyd[1]/DCPEMA Ratio	100/0	60/40
TiO_2/Binder	45/55	45/55
Volume Solids at 85-90 KU	55%	79%
VOC, gm/liter (CARB)	350	155
Set time, hours	1	3
Dry to touch, hours	4-5	6-7
Tack free time (Zapon), hours	30	7-8
After two weeks air dry		
Tukon Hardness (KHN)	0.6	1.0
Pencil Hardness	2B	HB
Impact Resistance, Direct (inch lbs)	>56 80-90	40 80-90
Gasoline Resistance (Pencil)	5B	B

[1] 30% phthallic, 60% linseed oil

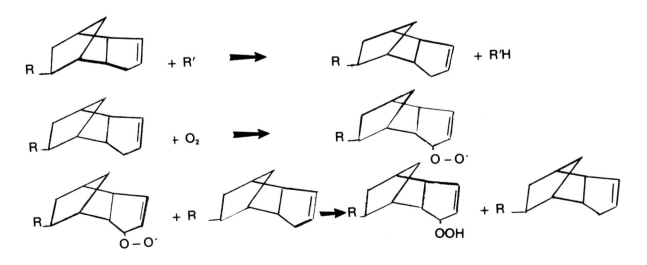

FIGURE 4 — Postulated reactions of the candidate reactive diluent DCPEMA in the presence of oxygen.

This side chain autooxidation of DCPEMA provides a mechanism for scavenging oxygen from thin films, thereby relieving the oxygen inhibition of the methacrylic polymerization. In thin soft films containing DCPEMA and cured at room temperature, oxygen inhibition can still be a problem. The presence of more autooxidatively active materials (drying oils, drying alkyds, etc.) in such systems is sometimes necessary to promote a reasonable rate of cure.

Another important property contributed by the side chain dicyclopentenyl group in DCPEMA is that of autoinitiation with metallic driers:

modified oils and alkyds; oil free unsaturated polyesters (maleates, fumarates); acrylated oils and resins; monomeric, oligomeric and poligomeric dienes.

This unsaturation in the resin permits copolymerization with the reactive diluent, rather than having only homopolymerization of the diluent taking place. This has been done with a brushing linseed oil alkyd formulation. DCPEMA was used as part of the solvent (replacing mineral thinner), such that the overall binder composition was 60 parts long linseed oil alkyd to 40 parts of DCPEMA. This modification yielded a

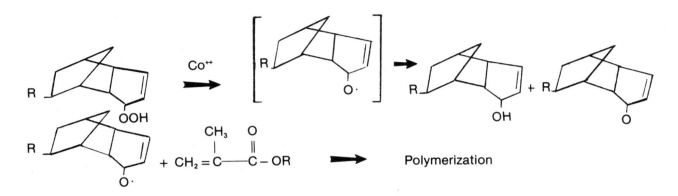

FIGURE 5 — Autoinitiation of DCPEMA with metallic driers. This property is contributed by the side chain.

In most cases, the rates of hydroperoxide formation and decomposition are rapid enough to maintain a good polymerization rate. The addition of more initiator has been found to be only mildly effective in increasing the cure rate. The generation of radicals by the autoinitiation reaction increases the reactivity of the total system.

Investigation of cured films containing DCPEMA shows the development of an insoluble (gel) fraction, which implies that a low level of crosslinking has taken place. The evidence is as yet circumstantial and the exact crosslinking mechanism is not yet understood. It is postulated that the crosslinking involves the dicyclopentenyl group in Figure 6.
This point is being investigated in more detail.

Earlier, we alluded to the use of DCPEMA with polymeric resins containing double bonds. These would include: drying oils; drying alkyds; urethane

paint at near 80% volume solids which has a CARB VOC of 155 g/L. Table 10 summarizes the property screening results obtained.

The drying characteristics of the DCPEMA modified alkyd paint show an interesting facet of its cure chemistry. The set time is extended relative to the unmodified control, which reflects the plasticization of the alkyd binder by the reactive diluent prior to curing. As polymerization begins, the dry to touch time is only slightly longer than that of the control. Finally, the tack free time is actually shorter with DCPEMA modification. The improved early through hardness development and gasoline resistance are further plusses for DCPEMA modification.

Currently, systems composed of hard non-crosslinking thermoplastic polymeric binders and DCPEMA are under investigation. These systems are potentially less complex, since homopolymerization

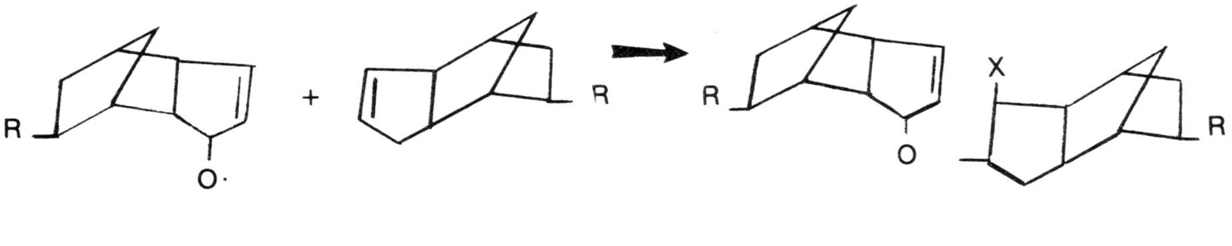

FIGURE 6 — The postulated crosslinking of cured films containing DCPEMA involves the dicyclopentenyl group.

of the reactive diluent is the only mode of cure. The homopolymerization product of DCPEMA is relatively hard and inflexible (Table 9), and attention must, therefore, be paid to the final hardness and mechanical properties of the cured coating. Thought is being given to use of reactive diluents yielding softer polymers as coreactants with DCPEMA. Resins having high toughness and flexibility characteristics will be prime candidates here.

Only three examples have been discussed of how environmental considerations have influenced research efforts in coatings. There are others. Allusion was made to high solids coatings using low molecular weight binders (oligomers, poligomers). Effort is being spent in this direction as well: reactivity with isocyanates is being studied as a means of crosslinking such systems. Marriage of the reactive diluents with oligomers could be promising as well.

Trademark References

Aroplaz	Ashland Chemical Company
Atlas	ICI Industries
Atomite	Thompson, Weinman and Company
Busan	Buckman Laboratories, Inc.
Dalpad	Dow Chemical Company
DMP-30	Rohm and Haas Company
Dow Corning	Dow Corning Corp.
Foamaster, NOPCO	Diamond Shamrock Corp.
Gen Epoxy	General Mills
Hydrite	Georgia Kaolin Company, Inc.
Imsil	Illinois Minerals Company
Kadox	New Jersey Zinc Company
Molywhite	Sherwin-Williams Company
Natrosol	Hercules, Inc.
Propasol, Carbitol	Union Carbide
Rhoplex, Skane, Tamol, Triton	Rohm and Haas Company
Texanol	Eastman Chemical Products, Inc.
Ti Pure	E.I. Du Pont de Nemours & Company.

Acknowledgement

The developments discussed here involve numerous coworkers in the Research Division. The author acknowledges the major contributions of Messrs. R. Flynn, R. Fulton, R. E. Harren, E. Lewandowski and Drs. W. D. Emmons, A. Mercurio, R. Novak, and P. Sperry.

References

1. Brown, G., *J. Polymer Sci.*, **22**, p. 423, (1956).
2. Grourke, M. J., *J. Coatings Tech.*, **49**, No. 632, p. 69, (1977).
3. Mayo and Miller, *J. am. Chem. Soc.* **80**, p. 2493, (1958).
4. Moglerich, *Russian Chem. Rev.*, pp. 48, 199, (1979).
5. W. D. Emmons, unpublished work.
6. U.S. Patent No. 4,097,677, June 27, 1978.
7. U.S. Patent No. 4,141,868, February 27, 1979.
8. U.S. Patent No. 4,145,503, March 20, 1979.

Appendix I

White Latex Primer

Material	Pounds	Gallons
Cowles Grind		
Water	48.5	5.82
Dispersant (Exptl. QR-681M)	20.0	2.29
Surfactant (Triton®CF-10)	2.0	0.20
Defoamer (Nopco®NDW)	1.7	0.22
Prime pigment (TiPure®R-960	150.0	4.58
Extender (Atomite®)	53.0	2.35
Reactive pigment (Kadox®515 - Zno)	12.0	0.26
Letdown		
Water	7.0	0.84
Ammonium hydroxide, 28%	7.0	1.00
Acrylic Latex (Rhoplex®MV-23)	667.0	74.70
Coalescent (Texanol®)	5.7	0.72
Ethylene glycol	28.0	3.00
Mildewcide (Skane®M-8)	2.1	0.25
Defoamer (Foamster®VL)	4.0	0.44
Thickener (Exptl. E-845)	25.0	2.87
	1,034.6	99.55

Appendix II

Latex Primers for High Early Rust Resistance

Material	Pounds	Gallons
Cowles Grind		
Water	77.0	9.24
Defoamer (Nopco® NDW)	7.7	0.23
Thickener (Natrosol® 250MHR, 100%)	0.7	0.04
Ammonium hydroxide, 28%	1.0	0.14
Methyl Carbitol®	50.0	5.79
Dispersant (Exptl QR-6B 1M)	22.0	2.48
Surfactant (Triton® CF-10)	2.7	0.31
Red Iron Oxide	50.0	1.22
Extender (Atomite®)	66.1	2.93
Reactive pigment (Kadox® 515-ZnO)	6.0	0.13
Reactive pigment (Molywhite® 212)	74.4	2.96
Letdown		
Acrylic latex (Rhoplex® MV-23)	497.3	56.69
Coalescent (Texanol®)	5.3	0.67
Long oil akyd (Aroplaz® 1271), drier treated	54.8	6.56
Defoamer (Foamaster® VL)	1.0	0.13
Thickener (Exptl. E-845)	35.0	4.02
Water	68.9	8.27
Sodium nitrite, 13.8% Aq.	7.2	0.86
	1,018.4	102.31

PVC 20%
Volume solids 35.4%

Material	Pounds	Gallons
Cowles Grind		
Water	114.0	13.68
Defoamer (Nopco® NDW)	1.7	0.23
Thickener (Natrosol® 250 MHR, 100%)	0.7	0.04
Ammonium hydroxide, 28%	1.0	0.14
Dispersant (Tamol® 850)	9.0	0.99
Surfactant (Triton® CF-10)	2.7	0.31
Red Iron Oxide	50.0	1.22
Extender (Atomite®)	148.0	6.56
Reactive pigment (Kadox® 515 - ZnO)	12.0	0.26
Reactive pigment (Busan® 11M-1)	100.0	3.65

Appendix II (Cont'd.)

Latex Primers for High Early Rust Resistance

Material	Pounds	Gallons
Letdown		
Acrylic latex (Rhoplex® MV-23)	468.0	53.35
Coalescent (Texanol®)	5.0	0.63
Long oil alkyd (Aroplaz® 1271), drier treated	51.6	6.18
Surfactant (Triton® X-100)	3.8	0.43
Ethylene glycol	25.0	2.69
Defoamer (Foamaster® VL)	1.0	0.13
Water	41.5	4.98
Thickener (Exptl. E-845)	33.0	3.80
Sodium nitrite, 13.8% Aq.	7.2	0.86
	1,075.2	100.13

PVC 30%
Volume solids 39%

Appendix III

White Water Based Epoxy Topcoat

Material	Pounds	Gallons
Component A		
Epoxy Emulsion (GenEpoxy® 370H55)	316.0	35.03
Surfactant (Triton® N-101)	3.5	0.40
Dispersant (Atlas® G-3300)	4.8	0.45
Defoamer (Foamaster® NS-1)	2.5	0.33
Thickener (Natrosol® 250HR, 100%)	1.5	0.09
Water	100.0	12.00
Prime pigment (TiPure® R-900)	308.0	9.36
Extender (Hydrite® UF)	50.0	2.29
Water	139.2	16.72
Component B		
Acrylic reactive polymer (Exptl QR-765M)	187.0	22.29
Solvent (Propasol® P)	5.5	0.74
Coalescent (Dalpad® A)	6.2	0.68
Accelerator (DMP® −30)	1.4	0.17
Amine (ZZL-0822®)	7.9	0.94
Slip Aid (Dow Corning® 11)	2.0	0.02
	1,137.3	101.51

PVC 29.5%
Volume solids 39.0%

Appendix IV

Red Water Based Epoxy Primer Formulations

Material	A Pounds	A Gallons	B Pounds	B Gallons
Component A				
Epoxy Emulsion (GenEpoxy® 370H55)	316.0	35.03	316.0	35.03
Surfactant (Triton® N-101)	3.5	0.40	3.5	0.40
Dispersant (Atlas® G-3300)	4.8	0.45	4.8	0.45
Defoamer (Foamaster® NS-1)	2.5	0.33	2.5	0.33
Thickener (Natrosol® 250HR, 100%)	2.0	0.11	2.0	0.11
Water	100.0	12.00	100.0	12.00
Red Iron Oxide	100.0	2.45	100.0	2.45
Extender (Imsil® A-25)	148.4	6.72	148.4	6.72
Reactive pigment (Molywhite® 212)	50.0	1.99	—	—
Reactive pigment (Zinc Phosphate)	—	—	50.0	1.91
Water	139.2	16.72	139.2	16.72
Component B				
Acrylic reactive polymer (Exptl. QR-765M)	187.0	22.29	187.0	22.29
Solvent (Propasol® P)	5.5	0.74	5.5	0.74
Coalescent (Dalpad® A)	6.2	0.68	6.2	0.68
Accelerator (DMP −30)	1.4	0.17	1.4	0.17
Amine (ZZL-0822)	7.9	0.94	7.9	0.94
Slip Aid (Dow Corning® 11)	2.0	0.02	2.0	0.02
	1,078.2	101.10	1,078.2	101.02

PVC 28.4% 28.6%
Volume Solids 38.6% 38.6%

Influence of Pigments on the Effectiveness of Anticorrosive Primers

*P. Kresse, V. Szadkowski, R.H. Odenthal**

Introduction

This paper summarizes some of the research work that was carried out by Dr. Peter Kresse and his associates in the Pigment Application Development Department of our parent company over the last ten years. In the past, most of our research work was related to the performance of the so called inactive or barrier pigments; such as, iron oxides in solvent based alkyd primer systems. In the recent past, intense research resulted in worldwide development of new active lead and chromate free corrosion inhibiting pigments. New work space regulations make the processing and application of lead and/or chromate containing pigments more and more difficult.

These topics will be discussed: 1) synthetic iron oxide pigments, some general properties relating to applications in primers; 2) mechanisms of water permeation and water vapor diffusion through pigmented films; 3) parameters in formulating primers with micronized iron oxide red pigments or other inactive pigments (pigment volume concentration and critical particle distance; degree of dispersion; wettability and degree of flocculation of pigment); 4) ferrites, a new class of active corrosion inhibiting pigments; chemical composition and physical data; mechanism of corrosion inhibition); 5) synergistic effects of blends of active and inactive corrosion inhibiting pigments.

Synthetic Iron Oxide Pigments: Some General Properties Relating to Applications in Primers

Iron oxide pigments, primarily the reds, represent the biggest product line, as far as inactive or barrier pigments for primer applications are concerned. In the past, substantial quantities of natural red iron oxides were used for primer applications, primarily for cost reasons. Today, the percentage of synthetic iron oxides being used in primers has increased and continues to increase. There are several reasons for this trend.

Better Consistency

Producing synthetic iron oxide pigments on a large scale ensures a better batch to batch uniformity. This uniformity not only relates to the color properties, but also to other characteristics, such as water soluble salts, pH, and trace contamination which might affect the performance of these pigments in primer systems.

Easier Processing

The development of micronized, easy dispersible, synthetic iron oxide pigments permits the use of high speed dispersers such as a Cowles dissolver for easier processing of these pigments.

Availability/Pricing

There are only a limited number of deposits where natural iron oxides can be mined. Traditionally, red iron oxides came from Spain and the Near East. Availability of some of these products has become a problem for political reasons. Also, importing natural oxides into the U.S. has become more and more difficult, due to rising freight costs. From the standpoint of tinting strength and processing costs, today synthetic iron oxide pigments are in most cases more economical.

Synthetic iron oxide red pigments are available in a wide range of colors. Although these products are chemically identical, having the composition $\alpha - Fe_2O_3$, the hue varies from a yellow to a blue cast. The shift in the color shade is caused by a change in the particle size. The color shade shifts from a yellowish red to a purple red, as the particle size increases from approximately 0.1 to 0.9 μm. Although the masstone color of a red iron oxide is of more interest in color matching work than for use in primer systems, it will be demonstrated that the particle size and the particle size distribution is directly related to the performance of an iron oxide red pigment in primer systems.

The oil or binder absorption also is related directly to the particle size and the color shade of an iron oxide red pigment. Since the oil or binder absorption determines the critical pigment volume concentration of a pigment in a given binder system, there is another relationship between the particle size of an iron oxide red pigment and the formulation of a primer. The selection of an iron oxide red pigment, from the standpoint of particle size, is an important parameter in formulating and optimizing primers.

---
*Mobay Chemical Corporation, Pittsburgh, Pennsylvania

The development and commercialization of easy dispersible iron oxide pigments started some 15 years ago. Jet milling is used to produce micronized red and yellow iron oxide pigments. Pressurized steam is used to mill the pigments. The velocity of the steam jet varies between 100 to 200 m/second. The grinding of the pigment is done primarily by direct impact of the particles. One major advantage of this technology is the fact that no moving mechanical parts are used, which ultimately reduces maintenance costs. Another advantage is the fact that the steam is recycled which, in turn, allows for energy conservation and savings.

The primary particle size distribution of crude material is determined by the calcining temperature during the manufacturing of the pigment. The micronization does not change the particle size distribution or the predominant particle size, compared with the regular grade. Therefore, most pigmentary characteristics, like specific surface and oil absorption, are identical for the regular grade and the micronized version. Since particle size distrubiton and predominant particle size affect the color shade, the masstone, as well as the tint color, are close. The micronization, however, results in a significant lowering of the mesh retention or amount of oversizes. The sieve residue drops from 0.05 to 0.001%. Lower sieve residue improves the performance of the micronized version in a primer system.

Contrary to what was discussed for red iron oxide pigments, jet milling of acicular yellow iron oxide pigments slightly changes the particle shape and size. Some of the needles are crushed during the jet milling. This change in the primary particle size has some influence on the technical data and color properties of the micronized version, in comparison to the regular grade.

Mechanisms of Water Permeation and Water Vapor Diffusion through Pigmented Films

A summary will be given of the correlation between water permeation, water vapor diffusion, and the degree of corrosion protection for coating systems pigmented with inactive pigments that serve as a barrier.

If one measures the water permeation and the water diffusion as a function of increasing pigment volume concentration, then one can see two different trends. First, the water vapor diffusion generally decreases in a linear way, as the pigment volume concentration increases, until the critical pigment volume concentration is reached. The reason for this is that the water vapor can only diffuse through the binder and not through the pigment. Second, one should expect that the water permeation decreases in the same way. In practice, however, one notices a sharp increase in the water permeation significantly below the critical pigment volume concentration of a given pigment/binder system.

We have measured the water permeation and the water diffusion as a function of the pigment volume concentration for an untreated and treated titanium dioxide pigment. In the case of the untreated titanium dioxide pigment, we notice a sharp increase in the water permeation significantly below the critical pigment volume concentration. If the treated titanium dioxide pigment is used, one can see a steady decrease in the water permeation and water diffusion until the critical pigment volume concentration of the system is reached. The differences in the behavior of the two systems, as well as the differences in the correlation between water permeation and water vapor diffusion and the pigment volume concentration, is caused by the varying amount of water that is absorbed on the pigment/binder interface.

We assume that no water absorption or water incorporation takes place at the pigment/binder interface. Therefore, the water permeation decreases as the pigment volume concentration increases. Once the critical pigment volume concentration is reached, the water permeation increases sharply. If it is assumed that a significant amount of water is accumulated at the pigment/binder interface, then an increase in the water permeation, significantly below the critical pigment volume concentration occurs, because a system of capillaries is formed through the film.

Parameters in Formulating Primers with Micronized Iron Oxide Red Pigments and Other Inactive Pigments

Pigment Volume Concentration and Critical Particle Distance

The formulation of primers with micronized synthetic iron oxide red pigments will now be discussed. Due to the relationship between water diffusion, water permeation, and pigment volume concentration one can expect that the particle distance in a film will significantly influence the degree of corrosion.

Salt spray tests of coatings that were pigmented with iron oxide reds of different particle size, ranging from 0.1 to 0.9 μm, but at the same pigment volume concentration of 32%, show the significant impact of the particle size on the corrosion inhibiting properites. The lighter pigment grades which have smaller particles, higher oil absorption, and therefore, a lower critical pigment volume concentration, perform less favorably than the coarser, more bluish shades. The degree of corrosion goes down as the particle size increases. In the case of the lighter, finer products, a system of capillaries was already formed at the chosen pigment volume concentration of 32%.

In another series of tests, the pigment volume concentration was only 6%. The same series of iron oxide red pigments was chosen. The degree of corrosion was constant, and not at all affected by the particle size of the iron oxide red pigment. In conclusion, this means that there is, obviously, a critical particle distance value for this system. Once the distance of the particles comes under this critical value, the formation of a capillary system across the film will occur, and sharply increase the degree of corrosion.

We have verified the correlation between the degree of corrosion and the critical particle distance in the dry film through several test series. We have prepared coating systems that were pigmented with iron oxide red pigments at pigment volume concentrations of 7 and 12%, below the critical pigment volume concentration. The particle distance was varied either by using iron oxide red pigments with different particle sizes or by blending these pigments. Panels were made up with the different pigment systems, and subsequently subjected to a salt spray test for a given period of time. The panels were then inspected, and the degree of corrosion was rated according to a rust scale that runs from 0 to 12. There was a reasonable correlation between the particle distance and the degree of corrosion, taking into account the many variables that are involved in carrying out such tests. The degree of corrosion increased sharply, as the particle distance is approximately 0.5 μm or less. The conclusion derived from this test series is that it is advantageous to formulate primers with somewhat coarser, that means more bluish, iron oxide red pigments at higher pigment volume concentration levels. If lighter reds with a smaller particle size are to be used, we recommend decreasing the pigment volume concentration which, in turn, increases the particle distance in the dry film. It is immaterial whether one single pigment grade or a combination of two different grades are used in formulating and optimizing primers.

We have compared the results from the salt spray tests with exposure tests that were carried out at the sea shore. The pigment volume concentration for this series was 8 points below the critical pigment volume concentration for each pigment grade. Again, the particle distance in the dry film was varied by using iron oxide red pigments with particle diameters from 0.1 to 0.9 μm. The degree of corrosion was determined vs the particle distance in the dry film. Again, as in the case of the salt spray test, a critical particle distance of approximately 0.5 μm was calculated. It is not possible to give one value for a critical particle distance for one type of pigment, such as iron oxide reds, which would be applicable in different binder systems. But it is obvious, from the test just described, that geometrical considerations are the primary concern in formulating primers with iron oxides or other inactive pigments.

Degree of Pigment Dispersion

Besides the particle size itself, the degree of dispersion of a pigment also affects the particle distance in a dry film. Pigments always contain a certain amount of aggregates and agglomerates. Pigment agglomerates normally can be dispersed and reduced to the primary particle size during the dispersion of the pigment in a binder system. Different dispersing equipment and different dispersing techniques might be required to break these pigment agglomerates. Pigment aggregates, however, are particles that have not been reduced to the desired primary particle size during manufacturing specifically during milling. Usually, these particles cannot be broken down during the pigment dispersion, and will remain in the finished paint or coating system.

If the size of these aggregates is in the range of the dry film thickness of a primer, the formation of a capillary system through the film along these oversized particles is most likely. It is obvious that such a capillary system would favor the water permeation through the film, and thus increase blistering and corrosion. In the earlier part of this presentation, micronized red iron oxide pigments were discussed, and it was pointed out that the mesh retention or the amount of oversizes is significantly lower for the micronized products, as compared with the unmicronized grades. Therefore, we can expect a much better performance from the micronized than from the unmicronized products.

A series of salt spray tests was carried out in which micronized red iron oxide was compared with regular versions of the same grade. Beside the pigment grade and pigment volume concentration levels, the dispersion technique was varied as a third parameter. Two sets of panels were coated with primers that were dispersed on ball mills for either 8 or 72 hours. Another set was coated with primers that were made on the Cowles dissolver. The dry film thickness of all the panels was in the range of 40 to 45 μm. The panels were subjected to the salt spray test for 570 hours. Two facts became obvious from the test.

First, the micronized product gives superior corrosion inhibiting properties at higher pigment volume concentration levels, compared with the unmicronized grade.

Second, the method of dispersion did not noticeably affect the performance of the primer in the salt spray test. This holds true only for the micronized pigment, as seen by the differences between the panels that were coated with primer made on the three roll mill, ball mill, or high speed disperser. This indicates the advantage of using a high speed disperser over other dispersing equipment for manufacturing a primer with micronized pigments. It is an energy saving and efficient method for producing primers. Based on our laboratory test results, it does not pay to use energy intensive dispersion equipment, or to disperse these micronized pigments for a long period of time.

Wettability and Degree of Flocculation of Pigments

Varying amounts of water can be accumulated on the pigment binder interface. This accumulation of water depends primarily on the wettability of a pigment in a specific binder system. The best wettability can be reached in such a system where a physical or physical/chemical bonding between the surface of the pigment particles and the resin molecules is possible. In the case of iron oxide red pigments, this can be a bonding between the OH groups on the pigment surface and functional groups of the polymer resin. In the case of a poor wetting of a pigment, in a certain resin system, the chance for the accumulation of water on the pigment binder interface and, ultimately, the formation of a capillary system through the film is most likely.

An example will now be given of how the pigment surface can influence the wettability and the corrosion inhibiting properties in a given binder system. We have compared two different titanium dioxide pigments with rutile structure in a salt spray test. One pigment is an untreated rutile grade and the other is a treated rutile grade. Both pigments are similar in particle size and were dispersed in the same manner. The untreated grade showed a poorer wettability and a high degree of corrosion at higher pigment volume concentrations after 240 hours of salt spray testing. The panels that were made with the treated titanium dioxide were still unaffected after 750 hours of salt spray testing.

The same holds true for iron oxide red pigments that are produced by different manufacturing processes. We have compared iron oxide red pigments of similar color shade, specific surface area, produced by either direct precipitation and drying or by calcination. In the particular resin system chosen, we determined a better performance of the precipitated product, compared with the calcined product. In comparing the corrosion inhibiting properties of chemically identical products that are produced by different manufacturing processes, it is important to select such products that are in the same particle size range, oil absorption and fineness. Otherwise, factors such as degree of dispersion and particle distance can also affect the test results, and ultimately lead to a misinterpretation.

Poor wettability of a pigment in a binder system can also cause a high degree of flocculation. Indications of this could be a significant color change or a drop in gloss. Flocculation can also affect the effectiveness of an inactive primer pigment in a positive way. If the pigment particles are perfectly dispersed in a binder system at a high pigment volume concentration level, the particle distance is relatively small. If the same pigment particles, however, are flocculated and form loose agglomerates in the dry film, the particle distance will actually increase, even at high pigment volume concentration levels. Therefore, we should expect an improvement in the corrosion inhibiting properties if a higher degree of flocculation exists.

To investigate this connection, we prepared two series using primers that were pigmented with titanium dioxide of the anatase structure. We purposely increased the level of the lead drier in one series from 0.3 to 1.3%, in order to induce a higher degree of flocculation. The water permeation was then measured on free films as a function of the pigment volume concentration. The water permeation was actually lower at higher pigment volume concentration levels for the system with the higher dryer level. The water permeation was lower in this sytem due to a higher degree of flocculation and greater particle distance compared with the unflocculated system. The degree of corrosion was determined as a function of the pigment volume concentration for the same coating system. Again, as in the case of the water permeation, the coating system with the increased drier level performed better than the unflocculated system.

Ferrites: A New Class of Active Corrosion Inhibiting Pigment

Chemical Composition and Physical Data

Chemically, the two products can be described as a zinc and calcium ferrite. Both products are produced by calcination of the basic oxide components. The ferrites have a spheroidal particle shape, and the particle size is close to a medium red iron oxide pigment. This similarity makes these products particularly suitable to compare and determine their mechanism of corrosion inhibition with inactive iron oxide pigments. Of specific interest to us was the question of how the pigment volume concentration affects the mechanism of corrosion inhibition.

The corrosion inhibiting properties of active pigments can be achieved by either one or several mechanisms summarized:

1. **Reducing the permeation of corrosion inducing chemicals, such as water and oxygen,** through the film. This can be achieved by both inactive and active pigments. Active pigments that form metal soaps with the binder can improve the mechanical stability of the film at high pigment volume concentration levels and, therefore, reduce the formation of capillaries through the film.

2. **Passivating the substrate.** The passivating function of chromate ions is the best known example for this type of corrosion inhibition.

3. **Preventing the saponification of the binder.** The uncontrollable formation of hydroxyl ions by the reaction of water with a ferrous substrate can accelerate the degradation of the binder. Certain active pigments can reduce this reaction by a controlled release of hydroxyl ions and the formation of metal soaps.

It is our conclusion from numerous investigations on inactive and active corrosion inhibiting pigments, as well as combinations of both, that the mechanism of corrosion inhibition is often a combination of several of these factors; the pigment volume concentration, in particular, is the determining factor as to which mechanism will prevail.

How the pigment volume concentration can actually reverse the mechanism of corrosion inhibition and the performance of pigments will now be demonstrated. Two series of primers were prepared, one with an inactive pigment, an iron oxide red, and the other with an active pigment, zinc chromate. The total pigment volume concentration was 8% for one set and 40% for the other set for both pigments. The upper portion of the panels are coated with the primer system only, whereas the lower portion contained an additional topcoat for complete hiding. The panels were exposed for 4 years at the seashore. At a pigment volume concentration of 8%, the iron oxide red does show a significantly better performance compared to the zinc chromate. The failure of the zinc chromate was caused by the poor hiding of this pigment compared with iron oxide red. This lower hiding power accelerated the UV degradation of the binder system, and the passivating effect of the chromate ions were ineffective. The iron oxide red provides better opacity at the low pigment volume concentration, and has a higher UV absorption, thus reducing the UV degradation of the binder system. At the higher pigment volume concentration of 40%, the results are as expected: the difference in the hiding power between the two pigments did not affect their performance. The better performance of the zinc chromate was caused by the passivating effect of the chromate ions.

Mechanism of Corrosion Inhibition

To determine the corrosion inhibiting mechanisms of these ferrites, these products were compared with a medium shade micronized red iron oxide. Pigmentation series up to the critical pigment volume concentration were prepared, and the coated panels were tested in a salt spray test for approximately 1700 hours. During this period of time, the panels were evaluated 13 times and the degree of corrosion was recorded for each panel. There are only minor differences in the performance between the inactive and the active pigments at low pigment volume concentrations. As we approach the critical pigment volume concentration of the system, however, the two active pigments performed much better. The active pigments probably act as barrier pigments at low pigment volume concentrations, whereas a chemical reaction takes place at higher pigment volume concentrations.

Next, the permeability of paint films pigmented with these ferrites was tested. Again, the same iron oxide red pigment used in the previous series was included in this series for reference and comparison. The water and water vapor permeation through the film was determined as a function of the pigment volume concentration. In both cases, no significant difference between either one of the pigments was noted. It is not possible to explain the differences in behavior between the ferrites and iron oxide red that were found in the salt spray test.

Both ferrites have alkali pH values of approximately 10 in an aqueous extract. To determine the formation of passivating hydroxyl ions in the presence of excess water, we prepared pigment volume concentration ladders with both ferrites and an iron oxide red pigment in a long oil alkyd system. The free paint films were then shaken for one week in distilled water, and the pH of the extracts were determined as a function of the pigment volume concentration. The pH of the aqueous extract from the coating containing the iron oxide red pigment stayed constant over the whole pigment volume concentration range. The pH values of the coatings pigmented with the ferrites approached the neutral value as the pigment volume concentration went up. Therefore, one can expect that hydroxyl ions contribute to the corrosion inhibiting action of these products. The free hydroxyl ions can either induce the formation of metal soaps or they can passivate the substrate. To determine the formation of zinc or calcium soaps, extraction tests were run on films that were pigmented with the zinc and calcium ferrite. The free films were immersed in distilled water for a certain period of time. The formation of zinc soaps was indicated through noticeable amounts of zinc that remained in the binder after the binder was separated from the pigment. The amount of calcium left in the film under the same testing conditions was considerably lower. Therefore, we assume that the calcium ferrite becomes effective by passivating the substrate with hydroxyl groups in binder systems such as alkyds.

Synergistic Effects of Blends of Active and Inactive Corrosion Inhibiting Pigments

Synergistic effects of blends of active and inactive primer pigments can be used to compare the effectiveness of various active pigments with different chemical compositions. A comparison was made of the ferrites with other well known active pigments. Iron oxide red and a titanium dioxide pigment of the anatase modification were chosen as good and poor inactive pigment. The blending ratio of the inactive pigment to the active pigment was 85:15, in the case of titanium dioxide, and 80:20 for iron oxide red. The panels were subjected to a salt spray test. As one would expect, the panels coated with the straight inactive pigments came out worse than any combination of inactive pigments with active pigments. However, we also found that the ferrites outperformed zinc phosphate and red lead. In the binder system chosen, and under our test conditions, both the calcium and zinc ferrite proved to be almost comparable in the corrosion inhibiting properties with zinc chromate.

Varying combinations of a micronized red iron oxide with the two ferrites were compared with zinc phosphate and zinc chromate in a salt spray test. After completion of the salt spray test, part of the coating was removed on each panel in order to check the substrate for rusting. We noted that the combinations zinc phosphate/iron oxide red, as well as zinc ferrite/iron oxide red, show rusting on the substrate. The calcium ferrite, AC 5071, did perform better than these two products, and was comparable to zinc chromate. The binder system chosen for this test was a long oil alkyd resin.

In the meantime, we have completed further comparative evaluations with both products in other binder systems such as chlorinated rubber, epoxy, and wash primers based on polyvinyl butyral. Based on these investigations, the calcium ferrite seems to be more effective in oleoresinous vehicles than the zinc ferrite. The best results were obtained at blending ratios of 60:40 parts by weight of the inactive pigment to the calcium ferrite. In other resin systems the zinc ferrite generally outperformed the calcium ferrite. For these systems, blending ratios of 80:20 and 70:30 for inactive to active pigments proved to be the best.

Study of the Use of Inhibitors in Coatings to Control Stress-Corrosion Cracking of Line Pipe Steel

E. W. Brooman, D. M. Lineman, W. E. Berry, R. R. Fessler*

Introduction

Inhibitors have been used for controlling a wide variety of corrosion problems.[1,2] They have been used in the liquid phase, solid phase, and vapor phase, and they have been included in primers, paints, sealants, and coatings to afford protection under a wide variety of circumstances.[3-6] When used in the solid phase, as part of coating systems, usually relatively large quantities of the inhibitor, in particulate form, have been felt to be necessary. Primers, for example, have been loaded with large quantities,[5] much of which will either remain immobilized or be lost to the environment. In theory, only a small quantity of inhibitor is required, if it is available where it is needed; for example, at the metal surface under holidays in the coating.

Recently there has been interest in using inhibitors to control stress corrosion cracking (SCC). Studies have been conducted to prevent SCC of aluminum alloys in hydraulic systems for naval applications.[7] The rationale behind these studies was that the inhibitor(s) might block the initiation of pitting, preventing the introduction of stress raisers, or might interfere with the electrochemical reactions taking place at the tip of an advancing crack, if one originated. In another study,[8] the use of inhibitors to control SCC of brass, titanium, aluminum alloys, austenitic stainless steels, and carbon and low alloy steels in a variety of environments was discussed, and a number of instances were described where positive effects were obtained.

The concept of using inhibitors to control SCC is relatively simple. The presence of inhibitors on the surface of a metal can modify the potential, modify the film forming processes resulting from corrosion, serve as a barrier to isolate the metal from the environment, and in some cases, modify the effect of cathodic protection. The approach used in Reference 7 was to address the first two effects. Others[9,10] have pointed out that, if the effect of the inhibitors is to shift the potential of the metal out of the cracking range, then this might be considered dangerous, because stray currents or changes in the environment could cause the potential to move back into the cracking range. On the other hand, a "safe" inhibitor is one that is effective at all potentials. No systematic investigation of the use of inhibitors to control SCC in specific environments has been reported. The purpose of this paper is to describe some of the results that have been obtained over a number of years on a research program conducted by Battelle Columbus for the Pipeline Research Committee of the American Gas Association (Project NG-18). The objective of this particular program was to investigate the use of inhibitors to control SCC of line pipe steel under conditions likely to be encountered in the field.

Technical Background

The first occurrence of SCC in gas transmission pipelines was discovered in 1965.[13] Shortly thereafter, the NG-18 Committee of the Pipeline Research Committee of the American Gas Association initiated a research program at Battelle's Columbus Laboratories to investigate the causes of, and the means of mitigating, SCC.

Stress corrosion cracks initiate on the outside surface of the pipe. Cracks are intergranular, and usually contain magnetite (Fe_3O_4) and $FeCO_3$, on occasion. Most cracks have occurred on coated pipelines, and most are surrounded by some degree of disbonding, although a few cracks have occurred on bare pipe. Cathodic protection has been applied to all lines that have exhibited SCC failures, although not in all cases have the lines been polarized to, or more negative than, -0.85 V (Cu/$CuSO_4$) as usually measured over the buried pipe. Only 2 to 4 service failures per year have been reported, largely because of a concerted effort by the transmission companies to locate and replace areas of pipeline exhibiting SCC.

In attacking the problem, methods for preventing SCC of line pipe steels may be grouped into three general categories: 1) modifying the steel from which the pipes are made; 2) modifying the operating practices involving stress, temperature, and cathodic protection; 3) modifying the environment. In the latter category, various approaches have been considered. One approach is the use of inhibitors to alter the environment known to promote SCC. In order to be able to select inhibitors that might be effective, it is impor- tant to know what types of environments are responsi-

---

*Battelle, Columbus Laboratories, Columbus, Ohio

ble for SCC in buried pipelines. Chemical analyses of liquids under disbonded coatings near locations of SCC in the field have shown that the principal constituents are sodium carbonate and sodium bicarbonate, and that the pH of such solutions is typically in the range of 9 to 11. Other constituents include small amounts of nitrate in some locations. Also, iron carbonate and magnetite have been found in SCC.[11] The carbonate/bicarbonate solutions are thought to be formed as a result of electrochemical and chemical reactions initiated by the impressed cathodic protection current in the presence of soil moisture and CO_2 from the air or organic matter in the soil. First, electrolysis of the moisture yields hydroxide, and raises the pH at the metal surface. Subsequent reactions with dissolved atmospheric carbon dioxide, or that resulting from decaying organic matter, yields the carbonate/bicarbonate solutions observed.

There are several ways in which an inhibitor may be introduced into an environment, namely: 1) dissolved directly in the liquid phase; 2) added to the soil above a pipe; 3) incorporated into the pipe coating system.

Direct Addition to the Environment

The first way is experimentally easy to accomplish and control, but in most cases does not represent situations likely to be encountered in practice. Nevertheless, the approach is useful for screening candidate inhibitors for their effectiveness.

Although a water solution of sodium carbonate and sodium bicarbonate is thought to be the most likely chemical environment that has been involved in SCC failures in pipelines, the possibility of other environments, such as hydroxides and nitrates, has not been completely eliminated. Thus, attempts were made to find inhibitors that would be effective in all three environments, but the bulk of the experimental work has been with the carbonate/bicarbonate solution.

In the hydroxide, carbonate/bicarbonate, and nitrate aqueous solutions, the results of previous research indicated that seven chemicals were safe inhibitors of SCC, namely:

calcium monobasic phosphate (monocalcium orthophosphate)
sodium monobasic phosphate (sodium dihydrogen orthophosphate)
sodium tripolyphosphate (sodium triphosphate)
potassium silicate
sodium chromate
sodium dichromate
potassium chromate

Concentrations of 0.01, 0.1, and 1.0 Wt.% of the chemicals were used, and in general, the 1.0 percent level was the most satisfactory. However, as a result of the evaluation, it was shown that those lower levels of these inhibitors did not increase the susceptibility to SCC. Thus, while some of the chemicals might not be effective as inhibitors at low concentrations, in general they would not be harmful to the line pipe steel. Recent work, to be described later, has shown that there are narrow ranges of chromate and phosphate contents that can promote SCC, but these ranges are so narrow that the beneficial effects would be expected to far outweigh the harmful effects during inhibitor depletion.

The earlier evaluation also showed that, in the nitrate and carbonate/bicarbonate environments, phosphate inhibitors also were effective in controlling crack growth when added after SCC had been initiated.

Subsequent experimental work was concerned with the further evaluation of chromate inhibitors. These chemicals were not investigated in detail in the earlier work directed at additions to the soil above the pipelines because of their known toxicity.[12] However, in several industrial applications, these chemicals were added to coating systems to prevent general corrosion and pitting attack. Incorporation of chromates into pipeline coating systems, therefore, was considered to be an attractive approach for controlling SCC without undue pollution of the environment.

Chromates were found to be effective inhibitors in carbonate/bicarbonate, hydroxide, and nitrate aqueous environments. In the former solution, concentrations of 1.0, 0.10 and 0.01 Wt.% were found to be effective, while lower levels (0.001 and 0.0001%) produced little or no effect on the susceptibility to SCC. The behaviors of potassium chromate, sodium chromate, and sodium dichromate were similar, and all three chromates were more effective inhibitors than silicates and the phosphates at equal concentrations.

In summary, the Battelle experimental data indicated that certain inhibitors can prevent the initiation of SCC in line-pipe steel in aqueous environments, and can stop the growth of cracks that were initiated prior to the addition of the inhibitors.

Addition to the Soil

Additions of inhibitors to the soil above the pipe, whether bare or coated, were investigated in the laboratory and in the field. It was found that there are some practical constraints on accomplishing protection by this method.

In the studies[12] to determine if benefit could be derived from adding inhibitors to the soil above a pipeline, addition of a soluble phosphate inhibitor was shown to be ineffective because of several factors, including the slow diffusion of the phosphate ions through the soil to the pipe surface, and precipitation of calcium phosphate through reaction with calcium present in the soil. The agricultural literature indicates that inorganic phosphates migrate through soil at the rate of about 1 inch per year. Similarly, the movement of silicates through the soil is slow, and these also have the tendency to precipitate as insoluble compounds, a form in which they are no longer useful. As mentioned above, little consideration was given to the use of chromates because of the toxicity of certain chromium compounds. In summary, the feasibility of using inhibitors for presently buried pipelines, by adding them to the soil above, is questionable, at best.

Addition to Coating Systems

The third way, adding inhibitors to coating systems, also was investigated in the laboratory and in the field, and is attractive in that additives may be made in a controlled manner to new pipe before burial. The rationale behind this approach is that pipeline with intact coatings do not require protection, but if the coatings were to break down in service, it would be beneficial to protect the pipe at those defects. If the inhibitor were incorporated into the coating system, and could be leached out at holidays, for example, then the bare steel could be protected.

In general, one or more inhibitors may be used in a given environment to afford protection. However, it is convenient to think of a single compound. The selection of a suitable inhibitive system will depend on several factors. The inhibitors must not in themselves cause any undesirable effects on the steel, the environment, or the coating system if one is used. The inhibitor, however, must react with the environment next to the pipe to provide a chemistry that does not favor SCC. Thus, the inhibitor must be soluble in the environment. If the inhibitor is lost slowly by diffusion, migration or chemical action with the environment, it must be replenished to maintain at least the minimum level of concentration considered necessary for protection. By inference, it is anticipated that different inhibitors would be compatible only with certain environments or operating conditions. Thus, each inhibitor addition would have to be tailored for the specific set of circumstances for optimum results. For example, for pipelines that are cathodically protected, the inhibitor selected should be effective in the environment produced by the cathodic current at holidays.

Techniques for Evaluating the Influence of Inhibitors

It is necessary to evaluate quantitatively the effects of inhibitors on SCC susceptibility. Test methods have been developed[11] to show the short term effectiveness of inhibitors, particularly those incorporated into the pipeline coatings. These methods are the following:

1. The slow strain rate technique with a small volume cell involves slowly straining a coated steel tensile specimen under conditions that would be expected to promote SCC in the absence of the inhibitor.

2. The slow strain rate technique with a leach extract involves slowly straining a bare steel specimen in a cell containing a leach of the inhibitor containing coating.

3. The constant load technique involves subjecting a bare steel specimen to a constant, high tensile load in a cell containing a leach extract of the inhibitor containing coating.

These test methods are useful for screening purposes. The specimens may be any typical line pipe steel that is susceptible to cracking in the absence of an inhibitor; for example, X52 grade, machined into tensile type specimens and pretreated to provide material with reproducible properties.[11] If the small volume technique is used, the specimen is coated. When the coating is dry, narrow circumferential strips are removed to expose 2 to 5% of the area of the gauge section. The solution used for the electrolyte for the small volume technique, and for obtaining the leach extracts, is an aqueous solution containing 4.8% sodium carbonate and 7.6% sodium bicarbonate (one normal solution of each). It should be made up fresh each week (detailed experimental procedures are given in Reference 11). Duplicate or triplicate tests for each condition are recommended, because of the variability often experienced, and control tests should be conducted in the solution without the inhibitors.

The resulting test data may be reported in one or more of several ways. One way is to report values of t_i/t_o, the ratio of the time to failure with an inhibitor to the time without the inhibitor. Values greater than unity indicate an inhibition effect, and the ratio may be used for all three types of tests listed. Another way is to report values of RA_i/RA_o, the ratio of the reduction in area obtained with an inhibitor to that without one. This ratio also is appropriate for the constant strain rate, leach extract, or small volume experiments. Again, values greater than 1.0 indicate inhibition. The third way is to report values of CV_o/CV_i, the ratio of calculated crack velocity without an inhibitor to the crack velocity with an inhibitor. Again, inhibition is represented by ratio values greater than unity, and this ratio may be applied to both of the slow strain rate types of experiments.

Historically, RA values have been used for the slow strain rate tests, while t values have been used for both the constant load and slow strain rate tests. With the use of CV values, it is possible that errors might be introduced, because it must be assumed that, during the subsequent metallographic examination, the deepest crack has been located, and that cracking initiated at the beginning of the test and proceeded at a uniform rate. However, the CV values have been included in this report for comparative purposes and serve to indicate severe SCC has occurred if the ratio is large. The advantage with the use of CV values is that a measurement of crack length is used with the time to failure values in order to compute average velocities; hence, the results more directly reflect the extent of SCC and not some other embrittling mechanism.

Preliminary Experiments with Inhibitor Containing Coating Systems

Experiments performed at Battelle with the cooperation of pipeline coating system manufacturers were directed at determining in a preliminary manner the effectiveness of inhibitors added to pipeline coatings, in particular to the primers.[13] The manufacturers had determined that phosphates and chromates could be satisfactorily mixed with coal tar primers. However, silicates could not be added because of a gelling effect. No effort was made at Battelle to measure the effect of the additives on other coatings properties, such as adhesion, resistance to cathodic disbonding, or resistance to water penetration, although such experiments were conducted by the manufacturers. For example, sodium phosphates in some coating systems promoted disbonding and water penetration, possibly because of their high solubility. A limited number of leaching experiments did prove that phosphate inhibitors could be leached from primers into solutions that could cause SCC, thus setting the stage for the study in which the manufacturers submitted samples of their inhibited primers or coatings for evaluation. The objective of this study was to look at a representative selection of systems to determine if they were effective as SCC inhibitors. A more detailed characterization could be achieved for the most promising candidates by the manufacturers themselves.

Many types of coating systems exist for pipeline service in soil. Choate, for example, lists bituminous coatings (various coal tar and asphalt enamels) thin films (coal tar epoxy and thermosetting powdered resin) tape systems, and mastics.[14] Berger adds to this list wax coatings (hot or cold applied), and extruded plastic coating systems.[15] In obtaining the cooperation of coating system manufacturers, Battelle was fortunate in receiving specimens representing several of the generic types of coatings listed above. Details of coating systems evaluated are given in Table 1. The type of inhibitor present, and its concentration in the coating on a weight percentage basis, is given when available.

Bituminous coatings. Coating System A, listed in Table 1, consisted of a coal tar base primer containing 1% of sodium tripolyphosphate based on the solids content. System B was an asphalt base primer containing 0.5 or 5% sodium tripolyphosphate, or zinc chromate. Systems A and B were supplied by different manufacturers.

Thin Film coatings. Two manufacturers supplied specimens coated with thin film epoxy systems. One manufacturer had added calcium monobasic phosphate to its barrier type coating, while the other had added an unspecified amount of zinc chromate to its polyamide epoxy primer for general corrosion protection.

Tape systems. Coating System A consisted of a natural rubber based primer containing a tackifying resin, carbon, and inorganic fillers for a total solids content of 19%. To this basic primer 5 or 20% of calcium monobasic phosphate, sodium monobasic phosphate or sodium tripolyphosphate was added. System B was also a primer, but was supplied by a different manufacturer, and contained 2% zinc chromate. The third system listed is an adhesive undercoating supplied by another source, containing 2% of either sodium tripolyphosphate or sodium monobasic phosphate.

Extruded Plastic coatings. The last system listed in Table 1 was a chlorinated rubber base primer for an extruded plastic coating containing 1% of sodium tripolyphosphate based on the solids content of the coating.

Results from Preliminary Experiments

Detailed results of both the slow strain rate and constant load techniques are not given here because of space limitations; however, Table 2 presents some of the data grouped according to the type of inhibitor used. Presentation of the results in this fashion is useful for comparing the relative effectiveness of both types of inhibitor.

In summary, with only a few exceptions, inhibitor containing coating systems reduced the severity of

TABLE 1 — Inhibitor Containing Coating Systems Evaluated

Type of Coating System[1]		Inhibitor Present[2]	Inhibitor Concentration, Wt.%
Bituminous	A(1)	None	None
	(2)	$Na_5P_3O_{10}$	1.0
	B(1)	None	None
	(2)	$Na_5P_3O_{10}$	0.5
	(3)	$Na_5P_3O_{10}$	5.0
	(4)	$ZnCrO_4$	0.5
	(5)	$ZnCrO_4$	5.0
Thin Film	A(1)	None	None
	(2)	$Ca(H_2PO_4)_2 \cdot H_2O$	5.0
	B(1)	$ZnCrO_4$	Unknown
Tape Systems	A(1)	None	None
	(2)	$Ca(H_2PO_4)_2 \cdot H_2O$	5.0
	(3)	$Ca(H_2PO_4)_2 \cdot H_2O$	20.0
	(4)	$NaH_2PO_4 \cdot H_2O$	5.0
	(5)	$NaH_2PO_4 \cdot H_2O$	20.0
	(6)	$Na_5P_3O_{10}$	5.0
	(7)	$Na_5P_3O_{10}$	20.0
	B(1)	None	None
	(2)	$ZnCrO_4$	2.0
	C(1)	None	None
	(2)	$Na_5P_3O_{10}$	2.0
	(3)	$NaH_2PO_4 \cdot H_2O$	2.0
Extruded Plastic	A(1)	None	None
	(2)	$Na_5P_3O_{10}$	1.0

[1] Details of the various types listed under each heading are given in the text.
[2] "None" represents a control system containing no inhibitor.
$Na_5P_3O_{10}$ = sodium tripolyphosphate (sodium triphosphate).
$ZnCrO_4$ = zinc chromate.
$Ca(H_2PO_4)_2 \cdot H_2O$ = calcium monobasic phosphate (monocalcium orthophosphate).
$NaH_2PO_4 \cdot H_2O$ = sodium monobasic phosphate (sodium dihydrogen orthophosphate).

TABLE 2 — Comparison of Test Data According to Type of Inhibitor Incorporated into the Coating System

Coating System Code[1]		t_i/t_o	Test Data[2] RA_i/RA_o	CV_o/CV_i
I: Phosphate Type				
Bituminous	A(2)	1.47	1.48	—[3]
	B(2)	1.12	1.24	2.12
	(3)	1.38	1.68	∞
Thin Film	A(2)	1.03	0.96	0
Tape Systems	A(2)	1.53	1.47	1.69
	(3)	—	—	—
	(4)	1.02	1.08	1.27
	(5)	1.13	1.60	
	(6)	1.19	1.47	2.13
	(7)	—	—	—
	C(2)	0.95	0.74	—
	(3)	0.97	0.63	—
Extruded Plastic	A(2)	1.09	1.09	0.80
II: Chromate Type				
Bituminous	B(4)	1.20	1.72	—
	(5)	1.78	1.84	∞
Thin Film	A(3)	1.50	1.06	—
Tape Systems	B(2)	1.15	1.21	3.25

[1] Table 1 and text for description of coating systems.
[2] Constant strain rate test data were used to compile tables. See text for explanation of ratios.
[3] These data not available in quantifiable form.

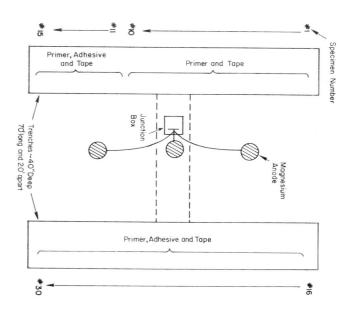

FIGURE 1 — West Jefferson Field site plan for long term exposures of pipe specimens.

SCC (t, RA, and CV ratios greater than unity, compared with control coating systems without added inhibitors). At least for the bituminous and thin film coating systems, chromate inhibitors appeared to be more effective than phosphate inhibitors. Concentrations between 2 and 5% zinc chromate gave the best results in this study. For a given concentration of inhibitor in the coating system, chromate type inhibitors reduced the severity of SCC by greater amounts than did the phosphate type inhibitors. This conclusion may be demonstrated by comparing the experimental data for bituminous coating systems B(4) and B(2), both of which containing 0.5% inhibitor; tape coating systems B(2) and C(2) or C(3), which contained 2.0% inhibitor; and bituminous coating systems B(5) and B(3), which both contained 5.0% of the inhibitors.

Long Term Exposure Studies

This brief review of research on inhibitors demonstrates that adding a suitable inhibitor to the coating system will decrease the severity of SCC for short periods of exposure. Behavior during long term exposure to underground environments is not known at the present time; however, some preliminary results of some field experiments are described next.

Experiements with Pipe Specimens

The field experiments were begun in late 1977 to evaluate inhibitor containing coating systems exposed to natural and simulated soils for extended periods of time. Two coating system manufacturers have cooperated by supplying coated 2 inch diameter, 24 inch long pipe specimens for burial in the field at West Jefferson, Ohio (Figure 1), and in simulated soil in the laboratory at Battelle. One coating system contained a primer incorporating 10 Wt.% of a tripolyphosphate inhibitor, while the other incorporated 6 Wt.% of a dichromate inhibitor. In addition, coated tensile specimens were buried in simulated soil. The purposes of using the pipe specimens, which contained artificial holidays, were: to determine if the inhibitor could be leached from the coating to protect those holidays; to measure how the extent of leaching varied with exposure for periods of 2 years. The tensile specimens also had bare regions, and were used to determine if the inhibitor concentration would remain sufficiently high to reduce the severity of SCC in the slow strain rate experiments. All specimens, both pipe and tensile, were fabricated from X52 steel and were cathodically protected with magnesium anodes.

Specimens were removed at approximately 3 or 6 month intervals from the natural and simulated soils, and were tested to determine SCC severity. Analyses to determine any loss of inhibitor from the coatings through the holidays were conducted.

Figure 2 shows the results of 24 months exposure on the rate of dichromate inhibitor loss from the coated pipe specimens containing holidays, 0.125 inch in diameter, in natural and simulated soils. Two curves are shown because half of the number of exposed pipes were coated with a primer, adhesive, and tape (PAT) system, while the other half were coated with the primer and tape (PT) to simulate a disbond

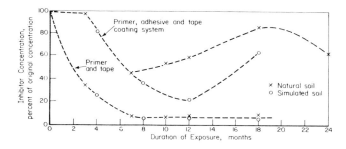

FIGURE 2 — Rate of dichromate inhibitor loss from artificial holidays in coated pipe specimens in natural and artificial soils as a function of time of burial.

situation. As can be seen from Figure 2, there was a marked difference in behavior between the PAT and the PT coated specimens. Inhibitor was lost quickly by leaching from the holidays in the PT systems, attaining a concentration of about only 7% of the original level after 6 to 8 months, then remaining at that level for more than a year. In contrast, with the PAT coating system after 6 to 8 months, approximately 40% of the inhibitor remained in the primer. After 8 months exposure, a difference was observed between pipe specimens buried in natural and simulated soils, although the apparent inhibitor concentration increased in both instances. The reason for the difference is attributed to the type of calcareous deposit that formed after about 8 months exposure. A more compact deposit was obtained in the natural soil, where the concentration of metal cations (for example, Ca^{++}, Mg^{++}) was higher. The plugging of the holiday by the deposit acted as a mechanical barrier to reduce inhibitor migration and diffusion. However, because a concentration gradient had been established around each holiday, even though the holiday was plugged, the inhibitor in the primer is thought to have redistributed itself. Thus, when an analysis was made, a higher concentration was measured.

A scanning electron microscope/energy dispersive X-ray (SEM/EDAX) analysis of samples of the deposits from the holidays showed no evidence of chromium in the PT natural or simulated soil deposits, nor in the PAT natural soil deposit. However, chromium was detected in the holiday deposit from the PAT simulated soil specimen. These results confirm the hypothesis that, in the PT specimens, there was a disbond around each holiday where it was possible for a liquid phase to be present. Under those circumstances, the inhibitor concentration was quickly leached out, and, if and when a deposit formed, there were few dichromate ions left to be incorporated in it. With the PAT specimens in simulated soil, if the deposit was not very compact, then some inhibitor could have been lost, and this accounts for the detection of chromium. If no migration or diffusion were possible through the compact film from natural soil exposures, then chromium would not be detected. Further studies are required to verify these conclusions. However, a preliminary mathematical model[16] has predicted a 5 orders of magnitude difference between the rate of inhibitor loss from intact coatings, and that from primed but uncoated steel. The PAT system resembles the intact coating situation because the adhesive slows the lateral diffusion of inhibitor to the holidays. The PT system, if a liquid phase forms because of a disbond at the holiday, will resemble more closely the uncoated pipe situation. Thus it is to be expected that differences in rates of leaching should be observed.

Experiments with Tensile Specimens

Concurrent experiments were conducted with coated tensile specimens in the simulated soil, and the slow strain rate (small volume) technique was used to determine the effect of inhibitor leaching on SCC susceptibility. Both RA and CV data were obtained. Typical results for the tripolyphosphate inhibited primer are depicted in Figure 3, where measured crack velocities are shown as a function of duration of exposure. As the time of exposure increased, the severity of SCC increased, rapidly at first, then leveling off as it approached values for cracking when no inhibitor was present and no film was on the steel surface. As with the pipe specimens, the presence of deposits on the bare steel caused a marked deviation in behavior in the slow strain rate tests. Again, the deposit appeared to exert a blocking action, thus protecting the bare steel surface. The curves for dichromate inhibitor are similar to those shown for the phosphate type inhibitor.

Discussion

Few detailed studies of the effects of inhibitors on SCC susceptibility have been found in the literature, although several references[12,17,19] acknowledge that the use of inhibitors may have a beneficial effect on line pipe. One commercial producer in England has reported that leachable primers may offer a solution to SCC of line pipe steel[19], while it is believed that one United States producer of coating systems also has

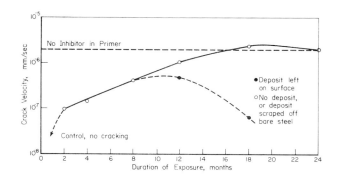

FIGURE 3 — Effect of exposure and inhibitor loss on stress corrosion cracking susceptibility for tensile specimens coated with a tripolyphosphate containing primer.

available on a trial basis an inhibited primer for pipe coatings. Both producers are using a chromate type inhibitor addition, which has been found to be effective over a wide range of potentials likely to be encountered on a cathodically protected system. In Reference 19, phosphate and silicate additions are mentioned, and, although not found to be as effective as chromates, they could be added to conventional primer formulations. Disbonding tests with the preferred formulations (which were not identified) showed that the adhesion of the coating was improved, and the resistance to disbonding increased, compared with the controls. Reference 19, however, does state that with some formulations problems were encountered. Thus, each coating system should be evaluated on its own merit, to ensure its applicability to a given situation/environment.

The experimental data on inhibitors for "leach primers" summarized in Reference 18 also confirm the general findings of Battelle and suggest that, for short term exposures at least, chromates are more effective than phosphates which, in turn, are more effective than silicates. Frequently, the amounts of these inhibitors required to control SCC are greater than those adequate for preventing general corrosion. Screening experiments used in Reference 18 were based upon potentiokinetic polarization experiments, current decay measurements, and selective etching tests. While these proved useful for screening potentially useful inhibitors, for the evaluation of performance, conventional SCC tests were required. The concentrations of inhibitors found to be useful in aqueous solutions ranged from about 0.2% for sodium chromate, up to about 3.0% for sodium monobasic phosphate. Smaller concentrations would reduce the amount of cracking observed in SCC tests, but would not completely inhibit SCC.[18] When incorporated into leachable primers and coated on steel specimens, much higher concentrations of inhibitor were found to be necessary to control SCC.

During laboratory experimentation[18,20] with inhibitors in aqueous solutions (for example, chromates in caustic or carbonate/bicarbonate solutions) it has been observed that, at low concentrations, there are narrow ranges where increases in cracking severity are found. These increases are typically of the order of X2 or X3, and are considered to be relatively insignificant when changes of an order of magnitude or more are usually of concern in environment sensitive fracture phenomena. Because of the narrow ranges of compositions, and the relatively small increases in severity of SCC seen within these ranges, compared with the much larger range of concentrations where no effects or beneficial effects are seen, this phenomenon is considered of little practical concern. During exposures, or while buried, the inhibitors will be leached out at a finite rate (depending on size and type of holiday/disbond, soil chemistry, applied cathodic protection current, soil moisture content, type of coating, etc.). At some point in time, the inhibitor concentration at the exposure pipe surface will fall into these critical ranges and may remain there for a short period of time. It is anticipated that the proportion of time spent in the critical range will be small compared with the time the inhibitor falls within acceptable ranges; therefore, the benefits are expected to outweigh any disadvantages.

Summary and Conclusions

Experiments have shown that the use of appropriate inhibitors to control SCC of line pipe steel is a practical approach. Early experiments in aqueous solutions of aggressive chemicals (hydroxides, nitrates, carbonates/bicarbonates) were useful for determining which inhibitors would be useful candidates. Preliminary experiments with coated specimens showed that certain inhibitors could be incorporated into coating systems (especially primers), and that these inhibitors could be leached out in simulated service.

However, at about the same time it was demonstrated, that the addition of inhibitors to the soil above buried pipelines was impractical. One reason was that the rate of migration or diffusion to the pipe surface was slow, and the inhibitor could be immobilized by reaction with soil constituents and subsequent precipitation. Another reason was that some of the effective inhibitor systems, such as chromates, could introduce toxic materials into the soil.

Long term exposure tests, with specimens buried in natural and simulated soils, have been performed in cooperation with two coating system manufacturers. Results to date have shown that inhibitors may be incorporated into practical coating systems, and that beneficial effects can be obtained. Inhibitors are leached from the coatings at a rate that depends, for example, on defect size and nature and soil composition. A preliminary mathematical model has been established to predict the rate of inhibitor loss for various limiting conditions. Migration or diffusion, and the formation of insoluble compounds, will play an important role in practical applications. The formulation of compact, calcareous deposits at holidays, or other defects, can act physically to prevent further inhibitor loss through leaching, and can separate the pipe from the environment. These actions favor reduction in the severity of SCC.

There appear to be narrow ranges of inhibitor concentrations that promote increased severity of SCC in aggressive environments of interest (for example, crack velocity increased by a factor of 2). In practical situations, this phenomenon is considered to be cause for little concern, because the time periods during which beneficial effects are observed are much longer than the time periods during which the inhibitors may be in their critical ranges.

The main practical limitation to the use of inhibitors is that they may only be incorporated in coatings for future pipelines or for replacement pipes, unless existing pipelines can be dug up, cleaned and recoated.

Acknowledgements

The work on inhibitors has been sponsored by the NG-18 Committee of the Pipeline Research Committee of the American Gas Association. Many people have contributed to the various investigations, but special thanks are due to Tom Barlo, Joe Payer, John Hassell, and John Parent (Battelle), and to Professor Redvers Parkins (University of Newcastle, England) for their helpful discussions, and assistance with the experimental aspects of the work.

References

1. Ranney, M. W., Corrosion Inhibitors: Manufacture and Technology, Noyes Data Corp., NJ (1976).
2. McCafferty, E., Mechanisms of Corrosion Control by Inhibitors, Proc. Mtg., Corrosion Control by Coatings, H. Leidheiser, Jr. (Ed.) Scientific Press, NJ, p. 279 (1979).
3. Miller, R. N., Corrosion Inhibitive Sealants and Primers, ibid, p. 325.
4. Clark, K. G., A New Method to Solubilize Inhibitors in Coatings, ibid, p. 349.
5. Scantlebury, J. D., The Mechanisms of Corrosion Prevention by Inhibitors in Paints, ibid, p. 319.
6. Hamner, N. E., Inhibitors in Organic Coatings, in Corrosion Inhibitors, C. C. Nathan (Ed.), Natl. Assoc. of Corr. Engineers, TX (1973), p. 190.
7. Brown, C. L. and Rehn, A. F., Jr., Inhibition of Stress Corrosion Cracking, Paper No. 96 presented at Corrosion/73, (March, 1973).
8. O'Dell, C. S. and Brown, B. F., Control of Stress Corrosion Cracking by Inhibitors, Proc. Mtg., Corrosion Control by Coatings, H. Leidheiser, Jr. (Ed.), Scientific Press, NJ, p. 339, (1979).
9. Scully, J. C. Stress Corrosion Cracking, Paper No. 10 in Corrosion Chemistry, Brubaker and Phipps (Eds.), Proc. of the 1976 ACS Symp. (1979).
10. Fessler, R. R., Berry, W. E., Wenk, R. L. and Parkins, R. N., Inhibition of Stress Corrosion Cracking of Steel Pipeline, U.S. Patent No. 3,973,056 (August 3, 1976).
11. Berry, W. E., Barlo, T. J., Payer, J. H., Fessler, R. R., and McKinney, B. L., Evaluation of Inhibitor Containing Primers for the Control of SCC in Buried Pipelines, Paper No. 64, presented at Corrosion/78, (March, 1978).
12. Davis, J. A. and Fessler, R. R., Addition of Inhibitors to the Soil Above Pipelines to Minimize Stress Corrosion Cracking Topical Report No. 110 (June, 1977), AGA Catalog No. L11777.
13. Proc. 5th Symp. on Line Pipe Research, (November, 1974), The American Gas Association, Inc., VA: Catalog No. L-30174.
14. Choate, L. C., New Coating Developments, Problems and Trends in the Pipeline Industry, *Materials Perf.*, (April, 1975), p. 15.
15. Berger, D. M., Selection and Application of Coatings for Underground Steel Pipe is a Real Challenge in Metal Finishing, *Metal Finishing* (October, 1975) p. 34.
16. Hassell, J. A. and Brooman, E. W., Calculation of Movement of Ions Influenced by Diffusion and Cathodic Protection at and Near the Pipe Surface, unpublished data (1979).
17. Parkins, R. N. and Fessler, R. R., Stress Corrosion Cracking of High Pressure Gas Transmission Pipelines, *Materials in Eng. Applic.*, **1** (December, 1978) p. 80.
18. Parkins, R. N. and Tems, R. D., Inhibitors for Leach Primers in the Control of Stress Corrosion Cracking of Underground Pipelines, Paper No. 212, presented at Corrosion/79, (March, 1979).
19. Mange, E. A. O., Leach Primer Offers Solution to Stress Corrosion Cracking, *Pipe Line Ind.*, **45,** No. 3 (September, 1976) p. 45.
20. Brooman, E. W., Stress Corrosion Cracking Inhibition, Paper presented at 6th Symp. on Line Pipe Research, Houston, TX (October/November, 1979), AGA Proc., Paper No. O, The American Gas Association, Inc., VA.

The Paintability of High Strength Cold Rolled Steels

J.A. Kargol, D.L. Jordan*

Introduction

Emphasis on weight reduction has generated increased interest in high strength cold rolled (HSCR) steels for exposed automobile body panel applications.[1,2] However, little is known concerning the underfilm corrosion resistance of these steels when painted. A number of studies have been made in an attempt to isolate the surface and microstructural features of low carbon cold rolled (CR) steel that influence paint adhesion and underfilm corrosion resistance.[3-9] As these studies have progressed, it has become more and more evident that surface carbon is the most important factor that controls underfilm corrosion resistance. The surface carbon is apparently formed from decomposed tandem mill oils during batch annealing, and possibly by dissociation of carbon monoxide in annealing atmospheres that contain this constituent. Surface carbon apparently degrades underfilm corrosion resistance by preventing the formation of quality zinc phosphate pretreatment coatings.

No clear cut evidence exists concerning the underfilm corrosion resistance of HSCR steels. Unlike low carbon CR steels, HSCR steels may differ from each other and from low carbon CR steel in both chemical composition and batch annealing cycles. Therefore, in addition to surface carbon, other factors must be considered that may influence their underfilm corrosion resistance.

There are a number of methods for producing HSCR steel sheet with the two important properties needed for auto body panels; that is, high yield strength (38 to 45 ksi or 260 to 310 MPa) and stamping press formability. Each approach incorporates microalloy additions and/or specialized annealing cycles.[1,2] Commercially available HSCR steels, which employed each of the strengthening mechanisms, were selected for this study.

It is generally accepted that there is considerable difficulty in assessing the underfilm corrosion resistance of a painted steel surface by methods other than the time consuming exposure of a large number of samples to actual field conditions.[9] The most widely used accelerated test method has been salt spray (ASTM B117); other accelerated tests have been considered, including exposure to severe atmospheric environments and simulated severe service environments.[9]

Low carbon CR steel studies have indicated that factors which influence zinc phosphate pretreatment coatings also, in turn, influence underfilm corrosion resistance of painted surfaces.[3-8] This paper describes the results of a study that employed a similar approach to gain a better understanding of the underfilm corrosion resistance that can be anticipated for several different types of HSCR steels. The observed correlation in the low carbon CR steel studies between zinc phosphate coating quality and underfilm corrosion resistance was the basis for the experimental program to study the paintability of HSCR steels. Stated simply, the experimental program was designed to compare surface chemistry and microstructural features of a number of HSCR steels and, for comparison, several low carbon CR steels to their zinc phosphate coating quality. Zinc phosphate coating quality has been measured in terms of a number of phosphate characteristics.[3,5-7,10,12-16] One of the characteristics noted, phosphate porosity, was given the most emphasis because it has been shown to be the most representative measure of zinc phosphate coating quality.[3,6,7,10,11]

Before reviewing the results, it must be emphasized that the study was made to determine what factors influenced the paintability of HSCR steels and, once identified, if the factors are the same as have been observed for low carbon CR steels. The study was not intended to determine if HSCR steels provide better or worse underfilm corrosion resistance than low carbon CR steels.

Steel Characterization

Three HSCR steels, each produced by a different manufacturer and representing a different alloy addition approach for strengthening by solution hardening and/or precipitation hardening, were used in the study. The chemical compositions (heat or ladle analyses) of the three steels, designated 1H, 2H and 3H, are given in Table 1. Table 1 also gives the chemical compositions of three low carbon drawing

---

*Department of Metallurgical Engineering and Materials Science, University of Notre Dame, Notre Dame, Indiana

TABLE 1 — Chemical Analyses of HSCR and Low Carbon CR Steels

Steel[1]	C	Mn	P	S	Si	Al	Other
1	.06	.32	.011	.012	.021	.064	—
1H	.05	.45	.075	.018	.026	.066	—
2	.05	.29	.011	.021	.02	.046	—
2H	.04	.37	.012	.020	.02	.065	.10 Ti
3	.05	.33	.005	.021	.014	.063	—
3H	.03	.34	.072	.008	.010	.030	.01 Cb

[1]H designates HSCR grade.

quality aluminum killed (DQAK) CR steels, designated 1, 2, and 3. Low carbon CR and HSCR steels with the same number designation were produced by the same manufacturer. It can be seen that the HSCR alloy addition approach included: a) rephosphorization (1H); b) titanium addition (2H); c) rephosphorization plus columbium addition (3H). The steels differed not only in chemical composition, but also somewhat in their respective annealing atmospheres and thermal cycles, as shown in Table 2.

All the HSCR and low carbon CR steels had been processed for exposed automotive body panel application. Tests were carried out on the steels received in the as annealed (AN), and annealed plus temper rolled (TR) conditions, both of which were obtained from the same coil of each steel. The TR samples were received in mill oiled condition. The AN samples were also coated with mill oil to prevent surface oxidation during shipment. To assure that the results of individual tests could be validly compared, all tests were performed on samples sheared from the same location (with respect to both width and length) of a given AN or TR coil.

TABLE 3 — Grain Sizes and Grain Boundary Densities for HSCR and Low Carbon CR Steels

Steel	$d_L^{(1)}$ (μm)	$d_T^{(1)}$ (μm)	$(d_L + d_T)/d_L d_T$ ($\mu m/\mu m^2$)
Annealed			
1	14.5	11	.160
1H	17	11	.150
2	24	17	.100
2H	9	9	.222
3	36	20	.078
3H	18	12	.139
Temper Rolled			
1	11	10	.191
1H	14	12	.155
2	33	24	.072
2H	13	9	.188
3	27	22	.082
3H	14	14.5	.136

(1) Grain dimensions parallel to surface for longitudinal, d_L, and transverse, d_T, sections.

Microstructural Analysis

Grain size measurements were made for the AN and TR samples of each steel. Two of the steels, 1 and 1H, were observed to have a large gradient in grain size near their surfaces. Therefore, for consistency, the grain size at the surface was calculated for each steel.

Calculated grain dimensions for longitudinal sections, d_L, and transverse sections, d_T, of each AN and TR sample are listed in Table 3. Small differences in grain dimensions between AN and TR samples of the same steel were attributed to the samples having been sheared from different locations along the coil.

TABLE 2 — Box Annealing Cycles of HSCR and Low Carbon CR Steels

Steel	Atmosphere Type	Atmosphere H_2	N_2	Composition (%) CO	CO_2	Dew Point (°F)	Annealing Cycle
1	HNX	10[1]	90	—	—	−100	1370-1390 F soak, 45 hour total cycle
1H	HNX	10[1]	90	—	—	−100	1330-1350 F soak, 45 hour total cycle
2	DX	7	76	11	6	+48	1330 F, 28 hour soak, 68 hour total cycle
2H	HNX	7	93	—	—	−35	1330 F, 27 hour soak, 51 hour total cycle
3	HNX	15	85	—	—	+65	1280 F minimum, 4 hour soak, 50 F/hour heat up and cool down
3H	HNX	15	85	—	—	+65	1280 F minimum, 4 hour soak, 50 F/hour heat up and cool down

(1) Changed to 4% HN on cooling.

Grain boundaries have been predicted to influence phosphate coating formation.[7] Therefore, grain boundary (GB) density, that is, the GB length per unit area, was calculated from the grain dimension measurements according to: GB density = (0.5 average grain perimeter) / (average grain area) = $(d_L + d_T) / d_L d_T$. The calculated GB densities are listed in Table 3.

Surface Chemical Analysis

The surface chemical analysis was obtained for each steel in both the AN and TR conditions. The analyses were made at Inland Steel Research Laboratories, East Chicago, Indiana, with a Physical Electronics Auger electron spectrometer (AES). Prior to analysis, all AN and TR samples were alkaline cleaned and rinsed according to the procedure used for the zinc phosphate coating tests that will be described later.

Also analyzed were several as annealed samples that were alkaline cleaned, rinsed and chemically etched for 1 minute in a Zn free H_3PO_4 plus HNO_3 solution (ET). The ET samples were intended to represent the steel surface condition at the onset of zinc phosphate coating formation in the acid phosphate bath that will be described later.

Previous studies have shown that elements are not uniformly distributed, on a microscopic scale, on steel surfaces.[3,4] Therefore, differentiated Auger electron energy spectra were obtained at two different locations for each AN and TR steel sample. The average measured elemental peak heights, converted to surface Wt.% values according to empirical atomic sensitivity factors,[17] are listed in Table 4 for elements that were obtained in measurable quantities on most of the samples. Those elements included Fe, C, O, P, N and S. It is interesting to note that Cb (Nb) was detected on several of the samples that did not contain the element as an intentional alloy addition (1, 1H and 2H), while it was not detected on the steel to which it was added for strength (3H). Similarly, Ti was not found in measurable amounts on the steel to which it was added for strength (2H). In addition to the individual elemental surface concentrations, the C/O ratio for each steel is also listed in Table 4. The concentrations of C and O have been reported to be inversely related,[3,4,9] and previously the ratio has been suggested to be related to phosphate coating quality on low carbon CR steels.[7]

Several of the steel samples were also ion sputtered for various times in order to obtain depth composition profiles for elements detected on their surfaces. Depth composition profiles for AN and ET samples of HSCR steels 2H and 3H are shown in Figure 1. Calibration of the argon ion sputtering indicated that surface material was removed at about 50Å/minute. Inspection of the depth composition profiles led to several observations.

1. C and, to a lesser degree, P were concentrated at the outermost layer (~50Å) of the AN samples. Etching for 1 minute (ET samples) did not remove the surface C layer.

2. O, as oxides, was also concentrated near the surface, but to a greater depth than the C and P on the AN samples. Etching tended to increase the oxide layer thickness.

3. Neither N nor S appeared to be surface active.

TABLE 4 — AES Surface Analyses of HSCR and Low Carbon CR Steels Values in Wt.%

Steel	Fe	C	O	C/O	P	N	S	Other
As Annealed (AN)								
1	46.1	9.4	18.0	0.52	1.1	5.2	0.2	16.8 Cb
1H	54.6	21.8	20.4	1.07	1.4	0.5	0.8	
2	58.2	8.1	21.6	0.38	1.5	0.6	0.4	
2H	63.0	8.8	24.6	0.36	2.2	0.6	0.6	
3	61.4	12.5	22.8	0.54	1.8	0.5	0.3	
3H	61.4	10.8	23.4	0.46	2.1	0.8	0.3	
Temper Rolled (TR)								
1	35.3	15.0	14.8	1.01	0.8	5.4	nd[(1)]	20.5 Cb
1H	43.8	18.2	19.0	0.96	1.4	1.1	0.4	2.8 Cb
2	54.2	15.0	22.2	0.68	1.2	0.4	0.3	
2H	45.0	18.7	16.8	1.11	0.8	2.8	0.3	9.0 Cb
3	53.4	7.6	23.6	0.32	2.4	0.4	0.2	
3H	58.0	10.4	25.8	0.40	2.0	0.3	0.2	

[(1)]Not detected

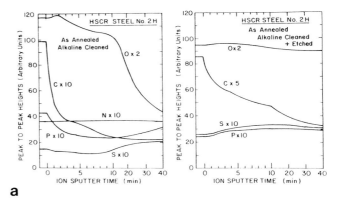

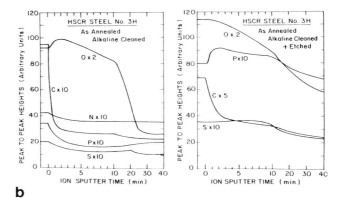

FIGURE 1 — AES depth composition profiles for alkaline cleaned and alkaline cleaned plus etched a) steel no. 2H AN; b) steel no. 3H AN.

Steel Characterization Summary

Analyses of the steels indicated several variables that should be considered in the phosphate coating quality tests.

Bulk Chemistry. Previous studies have reported that bulk chemistry does not affect underfilm corrosion resistance of painted steel sheet.[3,18,19] However, the studies were made for low carbon CR steels that did not contain the alloy additions of P, Cb and Ti.

Grain Boundary Density. Greater differences were observed among the HSCR steels than among the low carbon CR steels (Table 3). That may be attributed to differences in bulk chemistry (Table 1) and/or annealing cycles (Table 2).

Surface Chemistry. No clear cut difference was evident in surface chemistry between the HSCR and low carbon CR steels (Table 4).

Processing. The surface chemical analyses did not indicate clearly whether or not the difference in processing (AN vs TR) affected the steel surfaces.

Zinc Phosphate Coating Tests

Panels of each AN and TR steel, 3 × 4 inch (75 × 100 mm) in size, were coated by immersion in a zinc phosphate coating solution developed by Ghali and Potvin[15] and modified by Lakeman et al.[20] The phosphate coating solution contained 6.4 g/l ZnO, 10 ml/l H_3PO_4, 4 ml/l HNO_3 and 1.0 g/l $Ni(NO_3)_2$ in distilled water. This zinc phosphate coating solution has been shown previously to provide reproducible and uniform coatings when applied under laboratory conditions.[15,20] The solution contained nitrate ion to accelerate phosphate crystal nucleation and thus refine crystal size;[20] it also contained Ni^{++} ion which tends to decrease coating weight and alter crystal morphology.[20] However, this relatively simple coating solution does not contain the buffers and accelerators that are added to reduce coating time and temperature in commercial solutions.[21] The complex commercial solutions may also mask subtle differences in phosphatability among the HSCR and low carbon CR steels.

All panels were zinc phosphate coated according to the following procedure:

1. Clean for 4 minutes in room temperature alkaline cleaning solution.

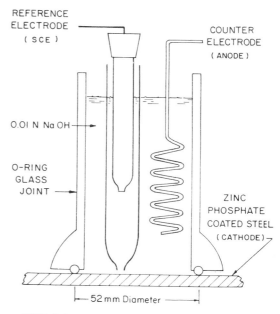

FIGURE 2 — Electrochemical cell for measuring phosphate coating porosity.[11]

TABLE 5 — Oxygen Reduction Current Densities for Zinc Phosphate Coated HSCR and Low Carbon CR Steels

Steel	Oxygen Reduction Current Density, i_c (µAmp/cm²)[1]									Average
	Individual Test Panels									
As Annealed (AN)										
1	2.7	2.7	2.7	2.7	3.5	3.5	4.4	4.4		3.32
1H	4.5	8.1	8.1	8.4	8.9	8.9	10.6	12.1		8.82
2	2.5	3.0	3.7	3.8	3.9	4.2	4.5	4.8	5.7	4.01
2H	2.7	2.9	3.2	3.3	3.4	3.8	3.9	4.1	4.3	5.9	3.75
3	2.1	2.1	3.0	3.1	3.6	4.3	5.7	8.4		4.04
3H	4.8	6.2	7.2	7.2	7.9	8.0	8.4			7.10
Temper Rolled (TR)										
1	3.0	3.1	4.6	4.9	6.4	6.9	7.2	8.1		5.52
1H	7.8	8.1	8.3	8.9	8.9	9.5	11.8	12.3		9.45
2	3.0	3.1	3.7	3.8	4.6	5.4	5.9	5.9		4.42
2H	2.6	3.5	3.5	3.5	3.5	4.2	4.2	5.5	5.9	6.1	4.25
3	2.1	2.3	3.0	3.1	3.6	3.6	3.7	3.7		3.14
3H	4.8	5.5	5.9	6.1	8.0	8.1	8.4	8.9		6.96

[1]Bare steel equals 30 to 35 µA/cm²

2. Clean for 10 minutes by immersion in 70 C (160 F) alkaline cleaning solution.

3. Rinse for 2 minutes in 70 C (160 F) distilled water. A water break test was performed to assure proper cleaning of each panel; panels that did not pass the test were rejected.

4. Coat by 4 minutes immersion in zinc phosphate solution at 90 ± 1 C (195 F) in a thermostatically controlled water bath.

5. Dry with compressed air and oven dry for 10 minutes at 135 C (275 F).

Following oven drying, all panels were placed in a desiccator until they were evaluated. All phosphate coating evaluation tests were carried out within 8 hours after oven drying.

Zinc Phosphate Coating Porosity

Zinc phosphate coating porosity for each coated panel was electrochemically measured by a method developed by Zurilla and Hospadaruk.[11] The electrochemical cathodic polarization method for measuring porosity is based on the principle that oxygen will be reduced from a dilute NaOH solution on uncoated steel surface sites, but not on inactive (insulated) phosphate coated sites.[11] Therefore, at a given cathodic potential, the oxygen reduction current density is a measure of porosity.

The electrochemical cell shown in Figure 2 was used to obtain the oxygen reduction current densities for the phosphate coated panels.[11] A 52 mm ID O-ring glass joint was clamped to the phosphate coated surface and filled with 100 ml 0.01N NaOH in distilled water solution. This cell design enables current density values to be obtained for a large surface area (~20 cm^2) which minimizes microscopic variability that may be present on phosphate coated panels. The panels were potentiodynamically polarized from −200 to −700 mV SCE at a scan rate of −2 mV/second with a voltage driven Wenking Model LT73 potentiostat. The oxygen reduction current values for all the panels were compared at the same cathodic potential of −550 mV SCE. At this potential, oxygen reduction is the dominant reaction at the steel surface.

The −550 mV SCE oxygen reduction current density values, i_c, are given in Table 5. As noted in Table 5, bare (not phosphated) samples of each steel yielded i_c values of 30 to 35 μAmp/cm^2. The i_c values exhibited only small variation among the individual phosphaste coated test panels for each steel, as can be seen by comparing the individual test panel i_c values with their respective average values in Table 5. The individual panel i_c values for each steel (listed in increasing order, not in the order by which they were run) illustrate the high phosphate coating reproducibility. They also indicate the consistency in the reaction of each steel with the phosphate coating solution.

Scanning Electron Microscopy

The zinc phosphate coated panels were analyzed with an ISI Super II scanning electron microscope (SEM). The SEM analyses were made on 1 cm diameter samples punched from each phosphate coated panel.

A qualitative assessment of the phosphate coatings indicated that the phosphate crystal size was relatively uniform on each coated panel, and that the average crystal size varied considerably from panel to panel. There did not appear to be any relationship between phosphate crystal size and phosphate coating porosity for either the HSCR or the low carbon CR steels. That observation is consistent with the results reported by Iezzi and Leidheiser for various low carbon CR steels.[3]

X Ray Diffraction

Each of the phosphate coated panels was analyzed with a Diano Model XRD 8535D semiautomatic X ray diffractometer. X ray diffraction spectra were obtained with a monochromatic 45 kV, 30 mAmp Cu K_α X ray beam that was scanned from 7° to 67° (2θ) at 2°/minute. All X ray diffraction analyses were performed within 6 hours after the phosphate coatings were applied.

No major differences were observed in the X ray diffraction spectra of the phosphate coated panels. The coatings on all panels consisted predominantly of phosphophyllite [$Zn_2(Fe,Mn)(PO_4)_2 \cdot 4H_2O$], zinc orthophosphate dihydrate [$Zn_3(PO_4)_2 \cdot 2H_2O$], and a small amount of hopeite [$Zn_3(PO_4)_2 \cdot 4H_2O$]. The predominance of the zinc orthophosphate dihydrate over the hopeite (tetrahydrate) is expected for freshly baked phosphate coatings; the dihydrate tends to gain water to form hopeite on aging.[12]

Zinc Phosphate Coating Summary

Zinc phosphate coatings were applied by an immersion method to alkaline cleaned panels of each AN and TR HSCR and low carbon CR steel. Several means were used to evaluate the zinc phosphate coating on each test steel.

Porosity. An electrochemical technique was used to measure porosity in terms of the oxygen reduction current on the coated test panels. The test method yielded reproducible results for each test steel, and indicated that variations in phosphate coating porosity were obtained among the test steels (Table 5).

SEM. A qualitative SEM examination of the coated test panels indicated that although the coatings differed considerably in appearance, no relationship was apparent between phosphate crystal size and shape and phosphate coating porosity.

X Ray Diffraction. No appreciable differences in crystal structure or composition in the phosphate coatings were indicated by X ray diffraction spectra.

As was expected, the primary difference among the phosphate coatings was porosity. That variable, measured in terms of oxygen reduction current density, was correlated with each steel variable (Steel Characterization) in an attempt to isolate the factors that influence the phosphatability of the steels.

TABLE 6 — Analysis of Variance of Phosphate Coating Porosity Data

Source of Variation	Degrees of Freedom, d	Sum of Squares, s	Mean Square, $\sigma^2 = s/d$	F[1]
A: HSCR vs CR	1	18.620	18.620	6.102
B: AN vs TR	1	1.124	1.124	0.368
E: Error	6	18.309	3.051	—

[1] $F = \sigma_A^2/\sigma_E^2$ and σ_B^2/σ_E^2, at $F \geq 5.99$, less than 5% chance that difference between data sets is due to random error.

Test Result Correlations

Analysis of Variance

Each of the steels had been classified earlier in two ways: 1) HSCR or low carbon CR; 2) AN or TR. Before establishing correlations among the test results, an analysis of variance was made of the porosity data (Table 5) to determine if these classifications were significant. If the difference in the variances between the two sets of porosity data within a classification were found to be large, then the data would be analyzed separately; if the difference were small, then the classification would not be significant, and the two sets of data could be grouped together for the correlation analyses.

The analysis of variance[23] shown in Table 6, indicated:

1. The variation in the porosity data between the HSCR and low carbon CR steels was large (based on the F value defined in Table 6) when compared to the random error (error in measurement). The analysis indicated less than a 5% chance that the HSCR vs low carbon CR difference was due to random error.

2. There was little difference (small F value) between the AN and TR porosity data, which is ascribed to random error. The significance of the observed similarity between the AN and TR condition for each steel will be discussed later in the paper. Based on the results of the analysis of variance, the HSCR and the low carbon CR steel test results were considered separately (unless noted otherwise) when determining possible correlation in the results. The AN and TR data were, however, grouped together for each steel.

Sample Correlation Coefficients

The sample correlation coefficient, R, was used to determine the strength of linear relationships[24] between the oxygen reduction current density data, i_c (that is, phosphate coating porosity), and each of the various steel characterization data. Although it is possible (and, as will be seen, likely) that two or more variables in combination may influence porosity, multiple comparisons were not performed because the data sets were not large enough. For the same reason, only linear relationships between variables, obtained by linear regression analysis, were considered.

R values were calculated for i_c (identified as response Y) vs each HSCR and low carbon CR steel characteristic (identified as variable X). The calculations that yielded noteworthy results are listed in Table 7. R values always fall in the range of zero (little or no correlation) to 1 (strong correlation) or −1 (strong negative correlation). Also listed is a measure, according to the T test[7,23] of the significance of the calculated linear relationship, T.

Several of the steel surface variables, surface C, O, C/O and Fe, are noteworthy in that they have been related previously to low carbon CR steel paintability.[3-9] The low carbon CR steel results in Table 7 give correlations similar (although the R values are low) to the previous studies. That is, i_c or porosity increases as: 1) surface C increases; 2) surface O decreases and; therefore, 3) as surface C/O ratio increase and; finally, 4) as surface Fe decreases. This supports the previous suggestion that the surface of alkaline cleaned low carbon CR steel consists of Fe in the form of a thin oxide layer. A thin (50Å thick) layer of C, which interferes with phosphate coating formation, and also masks the Fe and O when the surface is analyzed, may be deposited on the surface.

Similar, though not quite as strong, correlations were obtained for the HSCR steels, which suggests that surface C may also be a source of high porosity in phosphate coatings on HSCR steels. However, the linear relationships shown in Figure 3 suggest that other factors must also be considered for the HSCR steels. For any surface C value (or O, C/O or Fe), the linear relationships indicate a higher porosity for the HSCR steels. Several factors were considered to account for the higher HSCR porosity; three of these factors are listed in Table 7: grain boundary (GB) density, surface P, and bulk P.

GB density has not been noted to influence paintability in previous low carbon CR studies, but was considered here because the HSCR steels included greater GB densities (smaller grain dimensions) than

TABLE 7 — Sample Correlation Coefficients for Oxygen Reduction Current vs HSCR and Low Carbon CR Steel Characteristics

Variables		HSCR Steel		Low Carbon CR Steel	
Y	X	R	T[1]	R	T[1]
i_c - Surface C		.386	2.933	.415	3.126
i_c - Surface O		−.089	−.627	−.312	−2.250
i_c - Surface C/O		.250	1.807	.463	3.584
i_c - Surface Fe		−.284	−2.011	−.522	−4.156
i_c - GB Density		−.775	−8.313	.445	3.371
i_c - Surface P		−.069	−.487	−.344	−2.513
i_c - Bulk P		.924	16.360	.442	3.341

[1] At $T \geq [1.99]$, less than 5% chance that linear relationship is due to random error.

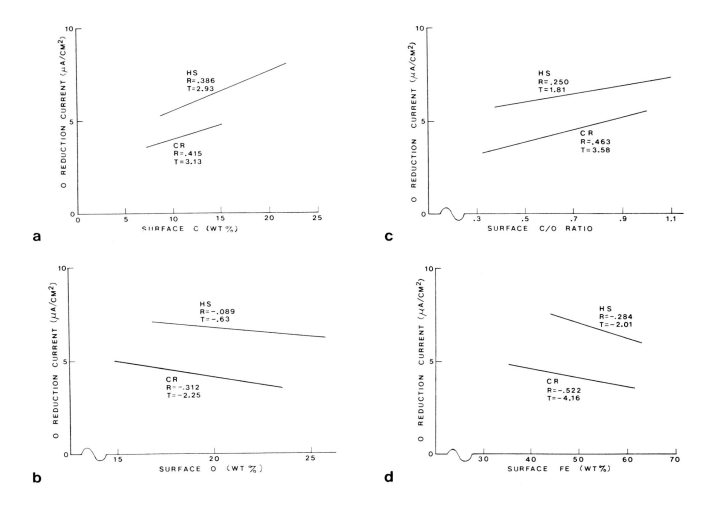

FIGURE 3 — Linear relationships and correlation coefficients between oxygen reduction current and a) surface C; b) surface O; c) surface C/O ratio; d) surface Fe.

the low carbon CR steels. However, the HSCR and low carbon CR steels were found to have opposite correlation coefficients (Table 7), and at overlapping GB densities (.135 to .190 $\mu m/\mu m^2$) they differed considerably in their predicted porosities (Figure 4).

Surface P has also not been noted in previous low carbon CR studies. However, it was considered here because the AES analysis indicated P to be somewhat surface active (alkaline cleaned samples in Figure 1) and also because two of the HSCR steels (1H and 3H) were rephosphorized (Table 1). As shown in Table 7, no significant correlation was obtained between surface P and i_c for the HSCR steels.

The final factor that was considered was bulk P, again because two of the HSCR steels were rephosphorized. As shown in Table 7, a strong correlation of 0.924 was obtained between bulk P and i_c for the HSCR steels; the predicted linear relationship is shown in Figure 5. The low carbon CR data combined with the HSCR data also yields a relationship (Figure 5) with a high correlation of 0.927. The high correlation between bulk P and i_c indicated that differences in bulk P should be considered to explain the observed

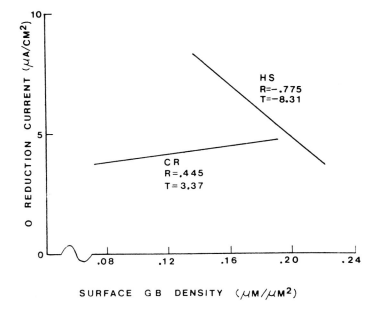

FIGURE 4 — Linear relationships and correlation coefficients between oxygen reduction current and grain boundary density.

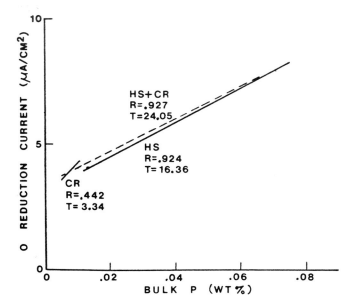

FIGURE 5 — Linear relationships and correlation coefficients between oxygen reduction current and bulk P.

HSCR and low carbon CR steel porosity differences. Because of the high correlation between bulk P and i_c, the SEM results were reexamined to determine if there was a relationship between bulk P and phosphate crystal size that had been overlooked in the original SEM analysis. However, the reexamination, in which samples were grouped according to bulk P content, did not indicate clearly any relationship. It can be seen in Figures 6 and 7 that both large and small phosphate crystal sizes were obtained, at both the high and low i_c values, for steels with low bulk P (Figure 6) and high bulk P (Figure 7). If anything, it appeared that the phosphate crystal sizes were, in general, smaller on the high bulk P steels. However, a quantitative analysis to determine the crystal size difference was not performed because of the wide range in crystal size on both high and low bulk P steels.

Discussion

The test result correlations indicated that surface C, the primary factor that influences phosphatability

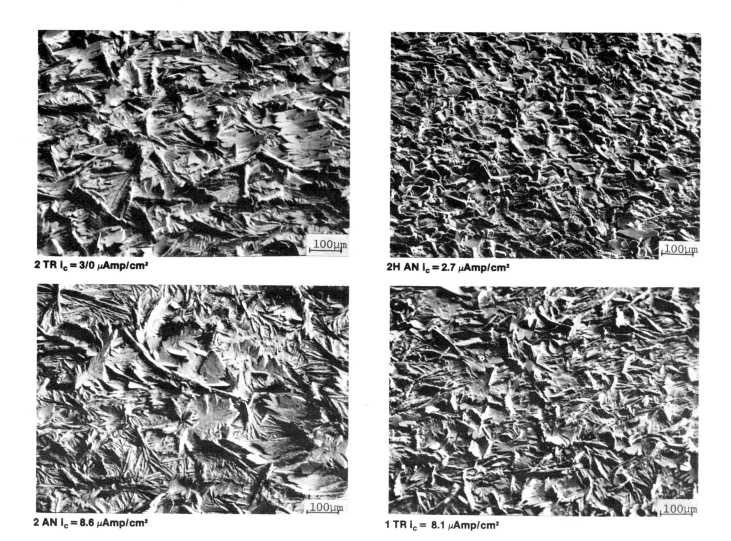

FIGURE 6 — SEM photomicrographs of zinc phosphate coated steels with low (0.005 to 0.012 Wt. %) bulk phosphorous contents.

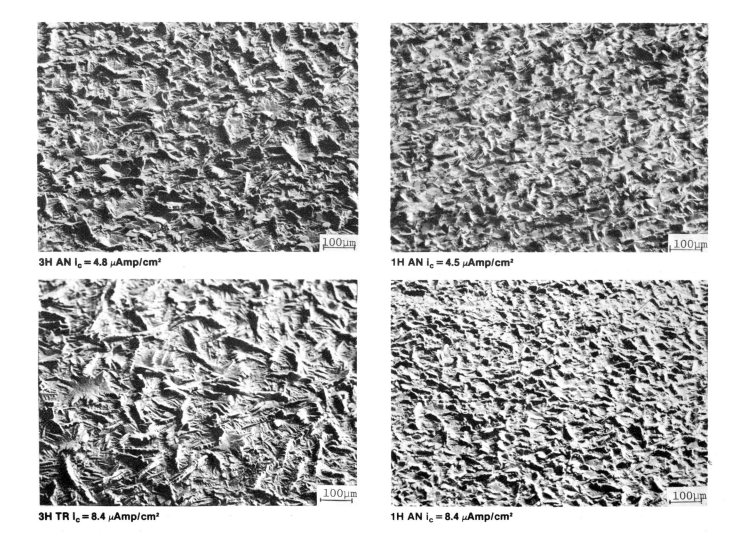

FIGURE 7 — SEM photomicrographs of zinc phosphate coated steels with high (0.072 to 0.075 Wt.%) bulk phosphorous contents.

of low carbon CR steels, also affects the phosphatability of the HSCR test steels. Surface C most likely increases porosity by acting as a barrier coating[3-8] that interferes with the sequence of electrochemical reactions, shown in Figure 8, that are necessary for the development of a dense protective zinc phosphate coating. Surface analyses of acid etched samples (alkaline cleaned plus etched samples in Figure 1) clearly indicated that the surface C is not readily removed by zinc phosphate solutions, and it may thereby prevent the phosphate reactions from proceeding. Analysis of variance indicated no difference in phosphatability between the as annealed (AN) and annealed plus temper rolled (TR) samples of each steel. This result supports previous studies that concluded that the source of surface C is decomposed tandem mill oils.[3-8] Temper rolling apparently has no further effect on the surface C con-

$$Fe^0 + 2 H_3PO_4 \longrightarrow Fe(H_2PO_4)_2 + H_2\uparrow$$

$$2\ Zn(H_2PO_4)_2 + Fe(H_2PO_4)_2 + 4 H_2O \longrightarrow \underbrace{Zn_2Fe(PO_4)_2 \cdot 4H_2O}_{COATING} + 4 H_3PO_4$$

$$3\ Zn(H_2PO_4)_2 + 4 H_2O \longrightarrow \underbrace{Zn_3(PO_4)_2 \cdot 4H_2O}_{COATING} + 4 H_3PO_4$$

FIGURE 8 — Fundamental zinc phosphate coating reactions on steel.

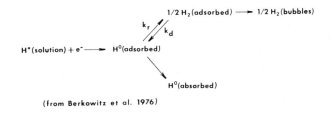

(from Berkowitz et al. 1976)

FIGURE 9 — Hydrogen reduction reactions at a metal surface.[28]

tamination (that is, it neither removes it nor contributes to it).

Linear regression analysis indicated that, for a given amount of surface C, a greater phosphate porosity developed on the HSCR steels. This result suggested that another factor or factors, in addition to surface C, influence the HSCR steel phosphatability. The factor that exhibited the greatest correlation with HSCR steel phosphate porosity was bulk P, with a correlation coefficient, R, of 0.924 (Figure 5). This high correlation does not, of course, imply a causal relationship between porosity and bulk P; an explanation must be given to support any such claim.

One explanation for the bulk P effect may be that which has been offered to explain electrochemically induced hydrogen embrittlement (HE) of several alloys, including steels.[25-27] Phosphorous that has been segregated to grain boundaries and free surfaces (as a result of thermal treatments) has been found to promote HE.[25-27] The segregated P has been suggested to promote HE by inhibiting recombination of H atoms following H+ discharge at the alloy surface and, thus, increasing H absorption. The P acts to decrease the rate constant for H recombination, k_r, as illustrated in Figure 9.[28] The P apparently need not be concentrated at the alloy's free surface; P concentrated at grain boundaries that intersect the free surface has the same effect.[26] For the sheet steels used in the current study, some degree of P segregation is likely during slow cooling following box annealing, particularly for the two rephosphorized HSCR steels (1H with 0.075 P and 3H with 0.072 P).

Inhibition of H recombination by the segregated P may also interfere with phosphate coating formation by supressing H_2 formation, thus interfering with the first reaction shown in Figure 8, (Fe dissolution). Reaction of the steel with phosphoric acid, which activates the steel surface and increases pH, is a critical step in the formation of the zinc phosphate coating. If this reaction proceeds slowly and nonuniformly over the steel surface due to the effect of bulk P, the resultant zinc phosphate coating may also be nonuniform and more porous.

The effect of bulk P has not been noted in previous studies of low carbon CR steels; that may be attributed to the low, and relatively constant, P levels in low carbon steels, as illustrated in Figure 5. However, when the low carbon CR and HSCR steel test results are combined, the same high correlation, R = 0.927, is obtained between porosity and bulk P (dashed line in Figure 5).

The increase in porosity due to dissolved P will not necessarily degrade underfilm corrosion resistance, as has been shown to be the case for surface C. The suggested effect of the dissolved P on phosphate formation is more subtle than that of surface C; that is, P affects the H+ reduction reaction, while surface C acts as a barrier coating. As a result, the increase in porosity due to P may be dispersed throughout the phosphate coating and not, as has been suggested for surface C, concentrated at the phosphate/steel interface. Therefore, the P induced porosity will not necessarily degrade underfilm corrosion resistance. Corrosion tests of painted HSCR steels containing various P contents will have to be performed to determine the effect.

Finally, it should be noted that oxidizing agents are added to phosphate coating solutions primarily to oxidize H to form H_2O and, thus, suppress H_2 bubble formation.[12] If, indeed, the porosity due to inhibition of H recombination by dissolved P is found to degrade underfilm corrosion resistance, greater oxidizing agent additions to the phosphate coating solution may counteract the P effect.

Conclusions

The porosity that develops in zinc phosphate conversion coatings formed on high strength cold rolled (HSCR) steels was found to increase as surface contamination, in the form of surface carbon, increased. That is the same result obtained in a number of previous studies of low carbon cold rolled (CR) steels. The carbon most likely serves as a barrier to phosphate coating reactions; the resultant porosity is concentrated at the phosphate/steel interface, and is likely to degrade underfilm corrosion resistance.

The zinc phosphate porosity on HSCR steels also was found to be higher on the rephosphorized grades. The increase in porosity due to dissolved phosphorous was attributed to the interference of phosphorous with the hydrogen reduction reaction that occurs during the first stage of phosphate coating formation. Unlike the effect of surface carbon, the porosity caused by dissolved phosphorous will not necessarily lead to a degradation in underfilm corrosion resistance. Corrosion tests of painted samples must be carried out to make that determination.

Acknowledgments

The authors express thanks to the American Iron and Steel Institute for support of this work. The authors also express thanks to the AISI member companies that provided the steel test materials, and to Dr. K. L. Slepicka, University of Notre Dame, for his assistance in making the statistical analyses.

Finally, the authors express thanks to Dr. Dennis Quinto, Mrs. Mary Anton and Mr. Fred Nietzel, Inland Steel Research Lab, for performing the AES surface analyses.

References

1. P. R. Mould, *Mat. Eng. Quart.*, pp. 22-31, August 1975.
2. P. B. Lake and J. J. Grenawalt, *Mat. Eng. Quart.*, pp. 9-21, August 1975.
3. R. A. Iezzi and H. Leidheiser, Jr., Corrosion, in press.
4. S. P. Clough and W. C. Johnson, Perkin-Elmer Corp., Physical Electronics Div., 6509 Flying Cloud Dr., Eden Prairie, MN 55344, unpublished research, 1979.
5. J. Rones, BISI 12774, Nov. 1974, from German, *Galvanotech*, **65,** No. 1, pp. 19-27, 1974.

6. V. Hospadaruk, J. Huff, R. W. Zurilla, and H. T. Greenwood, SAE Tech. Paper 780186, (1978).
7. J. J. Wojtkowiak and H. S. Bender, Research Publ. GMR-2446, General Motors Research Laboratories, June 1977.
8. J. A. Slane, S. P. Clough and J. R. Nappier, *Met. Trans.*, **9A**, 1839 (1978).
9. J. Westberg and L-G. Börjesson, Corrosion -80 Paper No. 279, NACE, 1980.
10. G. D. Cheever, *J. Paint Tech.*, **39**, No. 1, p. 13 (1967).
11. R. W. Zurilla and V. Hospadaruk, SAE Tech. Paper 780187, (1978).
12. R. L. Chance and W. D. France, Jr., *Corrosion*, **25**, p. 329 (1969).
13. S. L. Eisler, J. Doss and W. D. McHenry, *Organic Finishing*, pp. 5-9, May 1956.
14. W. L. Henesley, *Industrial Finishing*, pp. 24-25, February 1975.
15. E. L. Ghali and R. J. A. Potvin, *Corrosion Science*, **12**, p. 583 (1972).
16. J. B. Mohler, *Metal Finishing*, **73**, No. 6, p. 30 (1975).
17. L. E. Davis, N. C. MacDonald, P. W. Palmberg, G. E. Riach and R. E. Weber, Handbook of Auger Electron Spectroscopy, Physical Electronics Corp., Eden Prairie, MN 55344, (1976).
18. E. Bronder and D. Funke, *Trans. Inst. of Metal Finishing*, **51**, 201, (1973).
19. W. Weimann and W. Rausch, Stahl U. Eisen, **95**, p. 750 (1975).
20. J. B. Lakeman, D. R. Gabe and M. D. W. Richardson, *Trans. Inst. Metal Finishing*, **55**, p. 47 (1977).
21. D. R. Gabe, Principles of Metal Surface Treatment and Protection, 2nd Ed., Pergamon Press, New York, pp. 150-154, 1978.
22. G. D. Cheever, *J. Paint Tech.* **41**, p. 259 (1969).
23. I. Miller and J. E. Freud, Probability and Statistics for Engineers, Prentice-Hall, Englewood Cliffs, NJ, pp. 261-293, 1965.
24. ibid., pp. 254-258.
25. B. J. Berkowitz and R. D. Kane, *Corrosion* **36**, p. 24 (1980).
26. R. M. Latanision and H. Opperhauser, Jr., *Met. Trans.* **5A**, p. 483 (1974).
27. C. L. Briant and S. K. Banerji, *International Metals Review* **23**, No. 4, p. 164 (1978).
28. B. J. Berkowitz, J. J. Burton, C. R. Helms, and R. S. Polizzotti, *Scripta Met.*, **10**, p. 871 (1976).

Relation of Steel Surface Profile to Coating Performance

*Lee K. Schwab, Richard W. Drisko**

Introduction

Inadequate surface preparation is probably the most frequently reported cause of premature coating failure on steel surfaces. This inadequacy usually manifests itself by adhesion loss of the cured coating or by pinpoint rusting where the surface profile heights extend through the coating. A study was initiated by the Civil Engineering Laboratory (CEL) to determine the effects of surface profile, as compared to cleanliness, in affecting coating performance.

The work of preparing and coating the steel surfaces with different profiles was contracted by CEL to the Steel Structures Painting Council (SSPC), a recognized authority[1] in this field. SSPC became so enthusiastic about the project that it expanded the proposed program to obtain additional important information. Thus, in addition to preparing specimens for CEL adhesion and exposure studies at Kwajalein Atoll in the Marshall Islands, SSPC prepared for itself similar specimens for studies of the uncoated surfaces, laboratory salt fog exposure, field exposure in an industrial environment at Pittsburgh and in an atmospheric marine environment at the International Nickel Company exposure site at Kure Beach, N.C. The CEL exposure study at Kwajalein is being conducted in triplicate so that reliability of field data, as well as performance, in a severe marine atmospheric environment will be determined. The SSPC salt fog study was conducted to obtain: 1) early performance data; 2) a measure of correlation between accelerated laboratory and field exposure performances. The latter is important because the federal government is changing its specifications on coatings from a compositional type to a performance type, which will require the use of accelerated laboratory tests. The SSPC field exposures will provide data on coating performances under different environments. Although CEL and SSPC are independently conducting studies on similarly prepared specimens, each will exchange experimental data, so that all possible correlations of experimental findings can be made.

The first report of these studies[2] concerned the adhesion of the coating systems to the steel substrates prior to laboratory or field exposure. The present paper concerns the adhesion, blistering, and rusting of coated specimens after salt fog testing.

Experimental

Structural steel panels were cut to size, and their surfaces were prepared for coating by abrasive blasting with conventional equipment to a white metal finish (SSPC-5 or NACE 1). The use of the eight different abrasives of Table 1 (which also includes

TABLE 1 — Identification of test abrasives

Abrasive	Resultant Profile Height
Steelgrit G14	Very high
Steelgrit G40	Medium
Polygrit 40	Medium
Polygrit 80	Low
Black Beauty 400	Medium
Black Beauty 4016	High
Flintshot	Low
Steelshot S280	Medium

profile heights assigned by SSPC) resulted in eight different surface profiles. In addition to these, two of the abrasives (Black Beauty 4016 and Polygrit 40) were used to clean panel surfaces to a commercial finish (SSPC-6 or NACE 3) to give a total of 10 different surface variations.

Each of the six coating systems of Table 2 was spray applied to panels with each of the 10 surface variations. Thus, each complete set of test panels totaled 60. The average dry film thicknesses of the test panels are listed in Table 3.

One set of panels was exposed in a standard salt fog cabinet by SSPC according to method 6061 of the Federal Test Method Standard No. 141B.[3] Periodically, standard ASTM ratings for blistering[4] and rusting[5] were made by SSPC personnel. After 8,383 hours of salt fog testing, the remaining panels[(1)] were sent to CEL to determine the bonding strength of the coating systems after this artificial weathering.

[(1)]The acrylic latex and alkyd systems had previously failed and had been removed from the test.

*Naval Civil Engineering Laboratory, Port Hueneme, California

TABLE 2 — Identification of test coating systems

System Type	Identification of Different Coats in System		
	Primer	Intermediate	Finish Coat
Alkyd	TT-P-86, Type III	TT-P-86, Type III	SSPC-Paint 104
Acrylic Latex	SSPC-Paint XWB1X	SSPC-Paint XWB1X	SSPC-Paint XWB2X
Vinyl	SSPC-PT 3 Wash Primer Plus Mil-P-15929	MIL-P-15929	SSPC-Paint 9
Epoxy	SSPC-Paint XEP1X	SSPC-Paint XEP2X	SSPC-Paint XEP3X
Coal Tar Epoxy	SSPC-Paint 16	—	SSPC-Paint 16
Inorganic Zinc/Vinyl	SSPC-Paint XZ1X	SSPC-PT 3 Tie Coat	SSPC-Paint 9

TABLE 3 — Thickness of test coatings

System Type	Average Dry Film Thickness (micrometers/mils)		
	Primer	Intermediate	Finish Coat
Alkyd	41/1.6	102/4.0	137/5.4
Acrylic Latex	56/2.2	114/4.5	132/5.2
Vinyl	53/2.1	104/4.1	135/5.3
Epoxy	56/2.2	119/4.7	147/5.8
Coal Tar Epoxy	145/5.7	—	287/11.3
Inorganic Zinc/Vinyl	71/2.8	—	114/4.5

TABLE 5 — Levels of significance of variables after salt fog testing

Experimental Variable	Level of Signficance
Coating system on bonding strength	0.999
Coating system on blistering	varied with time
Coating system on rusting	varied with time
Abrasive on bonding strength	0.999
Abrasive on blistering	0.99
Abrasive on rusting	not significant
Profile height on bonding strength	0.999
Profile height on blistering	0.95
Profile height on rusting	not significant
Cleaning level on bonding strength	0.95
Cleaning level on blistering	not significant
Cleaning level on rusting	not significant

In determining the bonding strengths of the coating systems to the 10 different surfaces, steel probes were bonded to the finish coats with an epoxy adhesive (Hysol EA 9309). The circular probe ends, 1 cm² in area, were abrasive blasted to a white metal finish before bonding. After three days of curing, the probes were pulled in tension at a rate of 0.5 cm/minute in a table model Instron testing machine until failure occurred. The coating surrounding the bonded probes was routinely cut to the bare metal before testing even though preliminary experimentation indicated that this had little effect. Both the magnitude and the type of failure (adhesive bond to topcoat, system failure, or primer bond to steel) were recorded. Breaking strengths were recorded to the nearest 0.5 kg/cm².

Discussion of results

Table 4 lists the blistering and rusting ratings after selected periods of salt fog exposure. As shown in this table, the acrylic latex and alkyd specimens failed at relatively early dates and were removed from test. It can also be seen in Table 4 that the blistering ratings had a much greater variation (level of significance) than did the rusting ratings, and that these variations and the rankings of the coating systems changed

TABLE 4 — Blistering and rusting ratings after selected periods of salt fog explosure

Coating System	Average Blistering Rates					Average Rusting Rating				
	1,625 hr	4,111 hr	5,175 hr	6,319 hr	8,383 hr	1,625 hr	4,111 hr	5,175 hr	6,319 hr	8,383 hr
Coal Tar Epoxy	10.0	9.4	8.7	8.9	7.0	10.0	10.0	10.0	10.0	10.0
Vinyl	10.0	9.0	6.8	7.0	6.8	9.6	9.1	9.0	9.5	9.1
Epoxy	10.0	7.5	5.1	5.0	4.9	9.6	9.2	9.4	9.2	8.7
Inorganic Zinc/Vinyl	10.0	6.8	6.8	7.1	6.3	9.2	8.5	8.3	8.0	7.0
Alkyd	9.7	2.7	4.9	—	—	9.4	8.1	5.6	—	—
Acrylic Latex	2.9	—	—	—	—	3.6	—	—	—	—

Note: — indicates removed from test after complete failure.

TABLE 6 — Ranking of bonding strengths of different coating systems to white metal surfaces after 8,383 hours of salt fog exposure

Rank	Coating System	Average Bonding Strength (kg/cm^2)
1	Coal Tar Epoxy	45.3
2	Epoxy	36.9
3	Inorganic Zinc/Vinyl	26.6
4	Vinyl	14.7
5	Alkyd	(a)
6	Acrylic Latex	(b)

(a) failed and removed from test after 5,175 hours.
(b) failed and removed from test after 1,625 hours.

TABLE 7 — Ranking of initial bonding strengths of different coating systems to white metal surfaces

Rank	Coating System	Average Bonding Strength (kg/cm^2)
1	Epoxy	180
2	Vinyl	109
3	Coal Tar Epoxy	98
4	Alkyd	92
5	Acrylic Latex	57
6	Inorganic Zinc/Vinyl	22

TABLE 8 — Ranking of bonding strengths after 8,383 hours of salt fog exposure associated with different abrasives on white metal surfaces

Rank	Coating System	Average Bonding Strength (kg/cm^2)
1	Polygrit 40	60.2
2	Polygrit 80	60.2
3	Flintshot	37.7
4	Black Beauty 4016	27.1
5	Black Beauty 400	21.6
6	Steelshot S280	15.9
7	Steelgrit G40	15.5
8	Steelgrit G14	10.9

TABLE 9 — Ranking of initial bonding strengths associated with different abrasives on white metal surfaces

Rank	Coating System	Average Bonding Strength (kg/cm^2)
1	Black Beauty 4016	108
2	Flintshot	99
3	Steelgrit G40	99
4	Steelshot S280	92
5	Black Beauty 400	91
6	Polygrit 80	87
7	Polygrit 40	86
8	Steelgrit G14	82

greatly with time. Thus, any conclusions of a statistical analysis of the ratings of the different coating systems would vary with the time period chosen. The data in Table 4 permit the reader to determine the relative rankings and ratings after various time intervals. To determine the effects of abrasive, profile height, and level of cleanliness, analyses were made of the data for each coating system at the time of its maximum variation. The levels of statistical significance of the test variables are listed in Table 5.

It can be seen from Tables 4 and 5 and subsequent tables that the effects of the four variables (coating system, abrasive, profile height, and cleaning level) were, in all cases, greatest on bonding strength, followed by blistering and rusting in that order.

Table 6 lists the bonding strength to white metal surfaces of the four coating systems that completed 8,383 hours of salt fog testing. These ratings were much lower than the initial bonding strengths (Table 7), except for the inorganic zinc/vinyl system, and of a different ranking. Nevertheless, the levels of significance were the same (0.999). None of the bonding failures after salt fog exposure were related to adhesive failure. Before exposure, all of the bonding failures of the coal tar epoxy system were, in part, due to adhesive failure. Except for the coal tar epoxy and epoxy specimens with surfaces cleaned with Polygrit 40 and Polygrit 80, and all of the inorganic zinc/vinyl specimens which failed by loss of topcoat, all bonding failures were of primer to steel.

Table 8 lists the bonding strengths to white metal surfaces associated with the eight different abrasives used. The bonding strengths are much lower, and the variation is much greater, than were measured before salt fog exposure (Table 9). The rankings were also different, but the levels of significance were the same (0.999). The great variation in rankings before and after salt fog exposure was due, in large part, to the much greater ranking of Polygrit 40 and Polygrit 80 after exposure.

There was a large interaction between coating system and abrasive, in that bonding strength ratings for the coal tar epoxy, epoxy, and vinyl systems were highest with Polygrit 80 and Polygrit 40, and the inorganic zinc/vinyl system ratings were highest with Steelshot S280.

The effect of four different profile heights, as received from four different abrasives on the bonding strengths of the four coatings systems after receiving 8,383 hours of salt fog exposure is shown in Table 10. There was an interaction of coating system and profile height, in that the coal tar epoxy, epoxy, and vinyl

TABLE 10 — Ranking of bonding strengths associated with four profile heights on white metal surfaces after 8,383 hours of salt fog exposure

Rank	Profile Height	Abrasive	Average Bonding Strength (kg/cm²)
1	Low	Flintshot	37.7
2	High	Black Beauty 4016	27.1
3	Medium	Steelgrit G40	15.5
4	Very high	Steelgrit G14	10.9

TABLE 11 — Ranking of bonding strengths associated with four profile heights on white metal surfaces

Rank	Profile Height	Abrasive	Average Bonding Strength (kg/cm²)
1	High	Black Beauty 4016	108
2	Low	Flintshot	99
3	Medium	Steelgrit G40	99
4	Very high	Steelgrit G14	82

TABLE 12 — Relating level of cleanliness to bonding strength after 8,383 hours in salt fog

Rank	Level of Cleanliness	Average Bonding Strength (kg/cm²)
1	White Metal Finish	43.7
2	Commercial Finish	31.2

TABLE 13 — Relating level of cleanliness to initial bonding strength

Rank	Level of Cleanliness	Average Bonding Strength (kg/cm²)
1	White Metal Finish	97
2	Commercial Finish	90

TABLE 14 — Blistering ratings associated with different abrasives

Rank	Abrasive	Average Blistering Rating[1]
1	Steelgrit G40	9.16
2	Polygrit 40	9.16
3	Flintshot	8.83
4	Black Beauty 400	8.00
5	Polygrit 80	7.50
6	Steelgrit G14	6.33
7	Black Beauty 4016	5.50
8	Steelshot S280	4.66

[1] Ratings for each coating system at time of maximum variation.

TABLE 15 — Rusting ratings associated with different abrasives

Rank	Abrasive	Average Blistering Rating[1]
1	Steelgrit G40	9.16
2	Steelgrit G14	8.83
3	Flintshot	8.83
4	Steelshot S280	8.66
5	Black Beauty 400	8.50
6	Black Beauty 4016	8.50
7	Polygrit 40	8.16
8	Polygrit 80	8.00

[1] Ratings for each coating system at time of maximum variation.

systems rated highest on a low profile, and the inorganic zinc/vinyl system rated highest on a high profile. Although the ranking of Table 10 is somewhat different than a rank of the same type before salt fog exposure (Table 11), the levels of significance were the same (0.999).

The effect of level of cleanliness (white metal or commercial finish) on bonding strength after 8,383 hours of salt fog exposure is shown in Table 12. The level of significance of cleanliness was much greater after (0.95) than before (0.90) salt fog exposure, as seen in a comparison of Tables 5, 12, and 13.

The effect of abrasive on blistering was much greater (0.99 level of significance) than on rusting (no statistical significance) during salt fog exposure. This can be seen in Tables 14 and 15. To a lesser extent, this was also true for profile height, which had a significance level of 0.95 on blistering, but no statistical significance on rusting (Table 16). All coating systems applied to surfaces cleaned with abrasives Black Beauty 4016 and Steelshot S280 had significant blistering. As seen in Table 17, the average blistering and rusting ratings were slightly higher on white metal than on commercial blasted surfaces, but the difference was not statistically significant.

General Discussion

Concern about the effect of profile of a steel surface on coating performance has generally centered on: 1) inadequate bonding with too low a profile, or 2) incomplete covering of peaks on too high a profile. Test results suggest that a low profile will usually provide adequate texture for bonding. However, for an in-

TABLE 16 — Relating four profile heights to salt fog ratings

Rank	Profile Height	Abrasive	Average Salt Fog Rating[1] Blistering	Rusting
1	Medium	Steelgrit G40	9.16	9.16
2	Low	Flintshot	8.83	8.33
3	Very High	Steelgrit G14	6.33	8.33
4	High	Black Beauty 4016	5.50	8.50

[1] Ratings for each coating system at time of maximum variation.

TABLE 17 — Relating level of cleanliness to salt fog ratings

Rank	Level of Cleanliness	Average Blistering Rating	Average Rusting Rating
1	White Metal Finish	7.33	8.33
2	Commercial Finish	6.75	7.83

dividual coating system, there may well be a preferred profile height or a preferred abrasive. The preferred abrasive may be associated with total surface area (or another measure of profile other than height, which is only one measure of profile), and not clearly defined. Also, abrasives may have been embedded on surfaces, or may have reacted chemically with them.

The environment found in a salt fog chamber is not similar to that found anywhere in nature, so caution must be exercised in applying salt fog data to other environmental exposures. Thus, the acrylic latex and alkyd coating systems that had relatively early failure during salt fog exposure may perform quite well in many natural environments. Also, the methods for rating salt fog performance (blistering and rusting) are not completely satisfactory. Thus, the coal tar epoxy coatings had the highest average bonding strength, blistering, and rusting ratings after salt spray exposure, but were extremely brittle. Also, some of the epoxy coatings, that looked good superficially, had moisture penetration to the steel, which was found during adhesion testing. It is believed that the data from slower, natural exposure will be much more meaningful.

As previously noted, the variations in coating, abrasive, profile height, and cleanliness had the greatest effect on adhesion, followed by blistering and rusting in that order. Certainly, coating adhesion is only one requirement for coating performance, but it is an important one. The initial low bonding strength of the inorganic zinc/vinyl system and the reported difficulty[6] in topcoating inorganic zinc primers did not present a serious problem during salt fog exposure; indeed, its average bonding strength was slightly higher after exposure, while those of the other coating systems were greatly reduced. Also, the previously reported[2] solvent retention in this coating system had little effect, despite the reported[7] association of solvent entrapment and blistering. The inorganic zinc/vinyl specimens are performing well under natural exposure. In a previous CEL study,[8] it was shown that the bonding strength of alkyd coatings to steel increases with weathering, probably due to further crosslinking of polymer; this was not shown to be the case for alkyd specimens after 8,385 hours of salt fog exposure, because of their early removal from tests.

The rankings associated with profile heights in Tables 10 (bonding strength) and 16 (blistering) would have been somewhat different if other than the chosen abrasives were selected for the analysis. (For example, if Polygrit 40 had been used instead of Steelgrit G40 in Table 10, and Steelshot S280 instead of Steelgrit G40 in Table 16.)

The salt fog study, like the initial adhesion study, suggests that profile is much more important than cleanliness in adhesion of coatings. This may be due in large part, however, to the specific levels of profile and cleanliness that were used. It should be pointed out, however, that the range of surface preparation studied (white metal or commercial blast) covers the recommended preparation for the coatings of Table 2. Also, the standards of cleanliness, as determined by the naked eye, may not be a good measure of the necessary degree of cleanliness. On badly pitted steel the extensive abrasive cleaning required for a white metal finish may leave a high or nonuniform profile. It will be interesting to observe the relative importance of profile and cleanliness on actual field performance.

Acknowledgment

The statistical analysis of the test data was made by Mr. I.W. Anders of CEL. The blistering and rusting ratings were made by SSPC personnel.

References

1. Keane, J.D., J.A. Bruno, and R.E.F. Weaver. Surface Profile for Anti-Corrosive Paints, Steel Structures Painting Council, 1976.
2. Schwab, L.K. and R.W. Drisko. Relation of Steel Surface Profile and Cleanliness to Bonding of Coatings, Paper No. 116, Corrosion/80, Chicago, March 1980, National Association of Corrosion Engineers.
3. General Service Administration. Federal Test Method Standard No. 141B: Paint, Varnish, Lacquer, and Related Materials; Methods of Inspection, Sampling, and Testing. Washington, D.C., Feb 1979.
4. ASTM D714-56, Evaluating Degree of Blistering of Paints, Part 27, 1980 Annual Book of ASTM Standards, American Society for Testing Materials, Philadelphia, 1980.
5. ANSI/ASTM D610-68, Evaluation of Degree of Rusting on Painted Steel Surfaces, Part 27, American Society for Testing and Materials, Philadelphia, 1980.
6. Hendry, C.M. Inorganic Zinc Rich Primers — Fact and Fancy, Materials Performance, **17**, No. 5, p. 19, (1978).
7. Funke, W. Blistering of Paint Films, this volume.
8. Civil Engineering Laboratory. Technical Note N-1516: Repair System for Damaged Coatings on Navy Antenna Towers — Part I. Port Hueneme, Calif., March 1978.

Evaluation of Surface Pretreatment Methods for Application of Organic Coatings

F. Mansfeld, J. B. Lumsden, S. L. Jeanjaquet and S. Tsai*

Introduction

As the first part of a program of which the objective is to develop a methodology for predicting the lifetime of organic coatings/metals systems, the effects of surface pretreatment on adhesion of organic coatings and corrosion protection have been evaluated. In studying the effects of various pretreatment and cleaning procedures for steel and Al alloys, it was considered important to use methods that find wide practical application. The procedures defined in MIL-S-5002 and TT-C-490 were, therefore, chosen as a base line. To obtain more detailed information about surface characteristics resulting from various pretreatment procedures, additional methods (such as electropolishing, activation in acids, passivation by immersion or under applied potentials, and other methods that can be considered to produce a wide spectrum of surfaces with different activity) were used also.

The surfaces were characterized initially by the scanning electron microscope (SEM), scanning Auger microscopy (SAM), surface energetics measurements, and by surface mapping techniques. The next step was the application of phosphate coatings, the morphology of which was characterized by observation in the SEM.

Corrosion tests and determination of the electrochemical characteristics of coated surface by AC impedance techniques, to evaluate the effects of the various surface pretreatment procedures, will be carried out next. After coating the conversion coated samples with a suitable organic compound, corrosion tests, such as the cyclic humidity/temperature test per MIL-STD-202E, Method 106D, will be performed. AC impedance measurements will be carried out to determine the coatings properties before exposure to corrosion tests. In parallel, the adhesion of the coating system will be determined for the different surface pretreatment procedures. The coatings properties will again be analyzed by AC impedance measurements after the corrosion tests in order to determine the changes in coatings properties and the extent of corrosion under the coating system. Attempts will be made to separate the coating from the substrate and to analyze the chemical composition of both surfaces by SAM, SIMS, and/or XPS. A final analysis, which considers the results obtained in the various stages of this task, will be made, and conclusions will be drawn as to which surface properties lead to the optimum coating adhesion and corrosion protection for steel and Al alloys.

The progress made until now includes evaluation of pretreatment procedures for mild steel, determination of surface properties after coating, application of zinc phosphate coatings, and evaluation of surface morphology of these coatings.

Experimental Results and Discussion

Surface Cleaning Methods

It is the intent of this study to correlate surface properties after pretreatment with coating adhesion and corrosion protection. For this reason, this study includes a number of surface pretreatments which find practical use, as well as some treatments that are expected to result in surfaces having quite different chemical structures and activity. The cleaning methods for steel were selected from specifications such as U.S. Federal Specification TT-C-490B and the British Code of Practice CP3012:1972. In addition, a number of other surface cleaning procedures, such as pickling, passivation and electropolishing, were used.

Table 1 lists sixteen procedures initially selected after a review of government specifications and various monographs on this subject.[1-4] Federal Specification TT-C-490B, entitled "Cleaning Methods and Pretreatment of Ferrous Surfaces for Organic Coatings," covers six pretreatment processes for chemical conversion coatings without, however, specifying the exact solution composition or the steps in a pretreatment procedure. British Code CP3012 has sections on preparation of steel surfaces prior to the application of surface coatings; methods no. 6 to 9 were taken from this document; but only the anodic etch and the chemical smoothing treatments were used more extensively.

For solvent cleaning, the samples were degreased by wiping them with trichlorethylene followed by liquid/vapor degreasing for 15 minutes. Some samples

*Rockwell International Science Center
Thousand Oaks, California

TABLE 1 — Surface Preparation Methods for Steel

No.	Treatment	Spec.	Solution	Time/Temperature	Remarks
1	Sandblasting	TT-C-490 Method I			
2	Degreasing	TT-C-490 Method II	Trichloroethylene	15 minutes	
3	Hot Alkaline	TT-C-490 Method III		5 minutes/100 C	
4	Alkaline derusting	TT-C-490 Method V		10 minutes/50 C	Type III
5	Phosphoric Acid	TT-C-490 Method VI		5 minutes/70 C	
6	Anodic Etch	CP3012 Method H.1	H_2SO_4	2 minutes/RT	0.1 A/cm^2
7	Chemical Smoothing	CP3012 Method L	$H_2C_2O_4$, H_2O_2 H_2SO_4	10 minutes/RT	
8	Acid Pickling	CP3012 Method F.6	HF/HNO_3	2 minutes/65 C	
9	Acid Dipping	CP3012 Method G.1	H_2SO_4	<2 minutes/70 to 85 C	
10	Pickling		42 v/o HCl + Inhibitor	5 minutes/RT	2-butyne-1,4-diol
11	Passivation I		Borate buffer	1 h/RT	+ 0.60 V vs SCE
12	Passivation II		1N H_2SO_4	1 h/RT	+ 1.00 V vs SCE
13	Passivation III		50 w/o HNO_3	30 minutes/RT	
14	Electropolish I		HNO_3/AC_2O	30 s/RT	
15	Electropolish II		$HClO_4/BuOH/MeOH$	1 minute/ −20 C	
16	Sulfuric dichromate		$H_2SO_4/Na_2Cr_2O_7/H_2O$	10 minutes/60 C	

TABLE 2 — Film Thickness, Film Compositions, and Surface Morphologies Resulting from Various Treatments

Treatment	Polished and Degreased			Degreased Only		
	Thickness (Å)	Composition	Morphology	Thickness (Å)	Composition	Morphology
Sandblasting	>50	Oxide, Cl, S, C, Ca, Si, Al	Very Rough	>50	Oxide, Cl, S, C, Ca, Si, Al	Very Rough
Degreasing	<50	Oxide + Cl + C	Original	<50	Oxide + Cl + C, Ca	Original
Hot Alkaline	—	—	—	>50	Oxide + SiO_3	Original
Alkalin derusting	—	—	—	>50	Oxide, low S, low Cl	Original
Phosphoric Acid	—	—	—	<50	Oxide, high S, high P	Original
Anodic Etch	<50	Oxide, no S	Pits*	<50	Oxide, no S	GB etched, smoothed
Chemical Smoothing	—	—	—	<50	Oxide, no S	GB etched, smoothed
Acid Pickling	>50	Oxide + Cu + S	Smooth, inclusions not dissolved	>50	Oxide + Cu + S	Smooth, inclusions not dissolved
Acid Dipping	<50	Oxide, high S, Cu	Etched	>50	Oxide + high S	Etched
Pickling	<50	Oxide, high S, low Cl	Smooth	<50	Oxide, high S, low Cl	Smooth
Passivation I	<50	Oxide	Smooth	—	—	—
Passivation II	<50	Oxide + high S	Etched & Pits[1]	—	—	—
Passivation III	<50	Oxide, low S	Etched	>50	Oxide, low S	Etched
Electropolish I	>50	Oxide	Etched	>50	Oxide	GB etched
Electropolish II	<50	Oxide, high Cl + S	Etched	—	—	—
Sulfuric + Dichromate	<50	Oxide, high Cl + S	Smooth	>50	Oxide, high S, Cl, Cu	Etched
42% HCl	<50	Oxide, low Cl	GB Etched	—	—	—
Mechanical Polish	<50	Oxide, low S, low Cl	Smooth	—	—	—

GB = grain boundary
[1]Pits are apparently due to dissolution of sulfides.

were wet polished first with 600 SiC paper. Sandblasting was carried out with 10 μm Al_2O_3 powder.

The steel used throughout this study was 1010 steel (0.5 mm thick) in the as received condition.

Surface Analysis

The effects of the various surface pretreatments in Table 1 on surface morphology and the chemical composition of surface films were investigated with

SEM and surface analysis techniques such as SAM and XPS. The pretreatments were applied to both surfaces, which were degreased only, and to mechanically polished surfaces. The experiments with mechanically polished samples were carried out to evaluate whether the various pretreatment procedures would leave any residues on the sample surfaces, and the "degreased only" samples were tested to evaluate the effectiveness of the cleaning procedures. The cleaning procedures include both those used as pretreatment prior to the application of a coating and those that produce surfaces having known corrosion properties. The latter will provide baseline data for assessing the pretreatment procedures.

Table 2 gives film thickness, composition, and surface morphology resulting from the various treatments for both the "polished and degreased" and "degreased only" conditions.

In general, both types of samples had the same surface conditions when the same pretreatments were followed. There were three exceptions: "degreasing only" failed to remove some calcium salts, which were removed, however, by additional mechanical polishing; the acid dipping and the sulfuric and dichromate pretreatments left different amounts of copper on the surface, depending on whether the surface was polished or not.

From Table 2, on the basis of thickness, the films resulting from the various treatments can be divided into two groups: those of less than 50Å thickness, and those thicker than 50Å. In most cases, the films having thicknesses of less than 50Å resulted from passivation treatments or from treatments in which the surface was undergoing active dissolution. The films resulting from the latter treatments were formed by air oxidation after the samples were removed from solution. The films labeled as greater than 50Å ranged from 75Å to 250Å in thickness.

All films were iron oxides or hydroxides, with the exception of the film resulting from exposure to the hot alkaline cleaning solution. In this case, large amounts of silicate were incorporated in the film because of the presence of the inhibitor (sodium metasilicate). All samples exposed to sulfate solutions, with the exception of the anodic etch treatment, had films with high sulfur concentrations. This suggests that iron sulfate salts are present on the surface. Most films resulting from other treatments had low to moderate sulfur concentrations, which probably resulted from the dissolution of sulfide inclusions in the steel. In some cases, pits could be observed on the surface from which inclusions had been dissolved. Copper was found on the surfaces of samples that had been exposed to HNO_3 + HF (acid pickling) and hot H_2SO_4 (acid dipping). The copper was dissolved from solution and then redeposited onto the surface. Its presence would have a deleterious corrosion effect on an uncoated surface. The most contaminated surface was produced by sandblasting with apparently impure material. High concentrations of salts, SiO_2, and Al_2O_3 were

FIGURE 1 — SEM of steel after hot alkaline treatment.

FIGURE 2 — SEM of steel after degreasing.

detected. Subsequent cleaning by sandblasting was performed with clean Al_2O_3. It is also of interest to note that, after degreasing, a 3 to 4 monolayer coverage of organic material from the trichloroethylene degreasing solution remained on the surface.

The morphology of the surface depends on the treatment used. The hot alkaline treatment (Figure 1) does not change the surface morphology when compared to the "degreased only" (Figure 2), but does

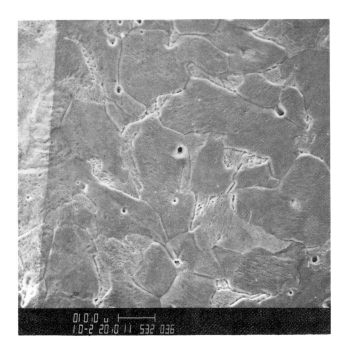

FIGURE 3 — SEM of steel after the chemical smoothing process.

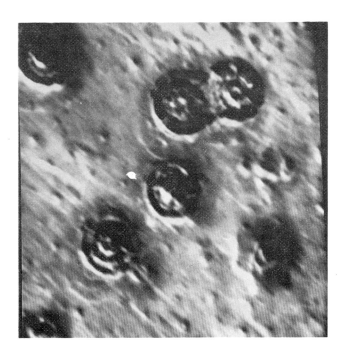

FIGURE 5 — SEM of steel after mechanical polish and exposure to RN = 80%, 1 ppm SO_2 for 19h.

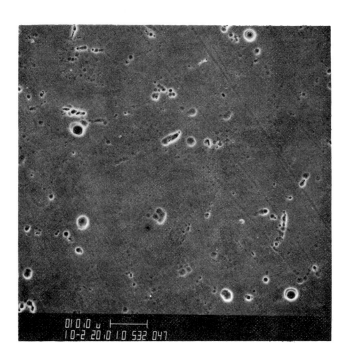

FIGURE 4 — SEM of steel after the anodic etch treatment.

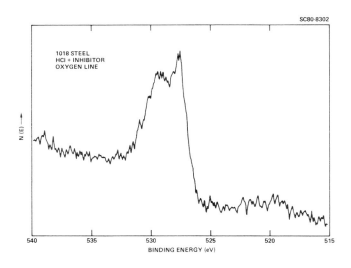

FIGURE 6 — XPS spectrum (oxygen line) for steel pickled in inhibited HCl.

leave surface irregularities, inclusions, etc. Those treatments that had high anodic potentials etched the grain boundaries (Figure 3). Passivation treatments in acidic solutions roughened the surface, whereas the passivation treatment in borate buffer left the surface morphology unchanged.

Treatments that dissolve sulfide inclusions (Figure 4), such as the anodic etch, produce corrosion resistant surfaces. The sulfide inclusions act as initiation sites for rusting. An example of this, shown in Figure 5, shows a mechanically polished steel surface after it has been exposed to an 80% relative humidity + 1 ppm SO_2 atmospheric condition for 19 hours. Disc shaped rust spots can be observed on the surface. These spots were not present after exposure in the absence of SO_2. A SAM analysis showed that a sulfide inclusions was located at the center of each disc. On the other hand, a surface that was pickled in inhibited 42 v/o HCl before exposure to the above conditions was unattacked.

Figures 6 and 7 show the X ray photoelectron spectra (XPS) of oxygen for surfaces pickled in inhibited

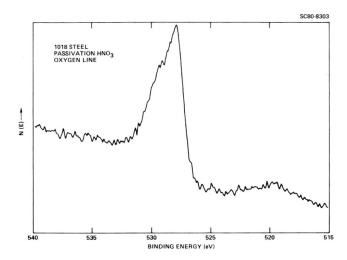

FIGURE 7 — XPS spectrum (oxygen line) for steel passivated in HNO₃.

TABLE 3 — OH Content in Surface Films After Pretreatment (XPS-Data)[1]

No OH	Low to moderate	High
Conc. HNO₃	Anodic Etch	Electropolish II
Electropolish I	Chemical Smoothing	Degreased
	H₂SO₄ + Na₂Cr₂O₇	Inhibited HCl

[1]1018 steel, all samples were polished; 1μm finish.

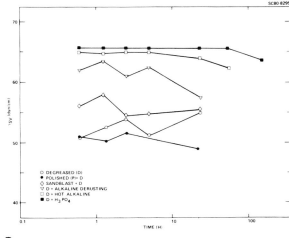

a

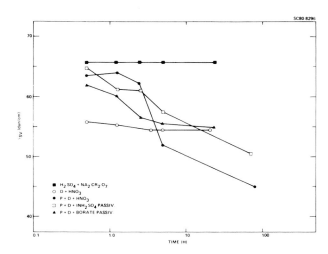

b

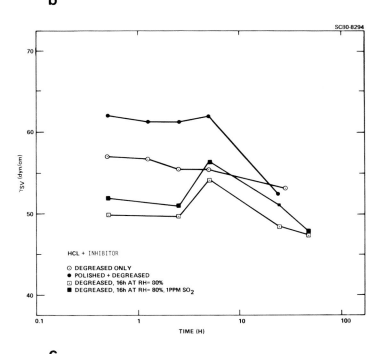

c

FIGURE 8 — Changes of surface tension γ_{SV} as a function of time for 14 different surface pretreatments.

HCl and passivated in concentrated HNO₃, respectively. A shoulder at higher binding energies is observed for the pickled specimen; this shoulder arises from the presence of OH in the film. This shoulder is not present for the sample passivated in HNO₃. As suggested by these two spectra, large variations were observed in the OH concentrations for films formed in the different pretreatment procedures. A summary for treatments which gave high, low to moderate, and no hydroxide concentrations in the surface films is given in Table 3.

Surface Energetics

As discussed by Kaelble,[5] surface energetics provides an important new tool in both materials selection and surface treatment. This technique has, therefore, been used to characterize steel surfaces after various pretreatment procedures outlined in Table 1. The wettability of such surfaces has been measured quantitatively, and the nominal values for the solid vapor surface tension γ_{SV} of such samples have been calculated. The experimental methods and analysis employed in this study have been described in detail by Kaelble and Dynes[6] who studied surface energetics of Al2024-T3. Liquid/solid contact angles are measured with calibrated liquids of known sur-

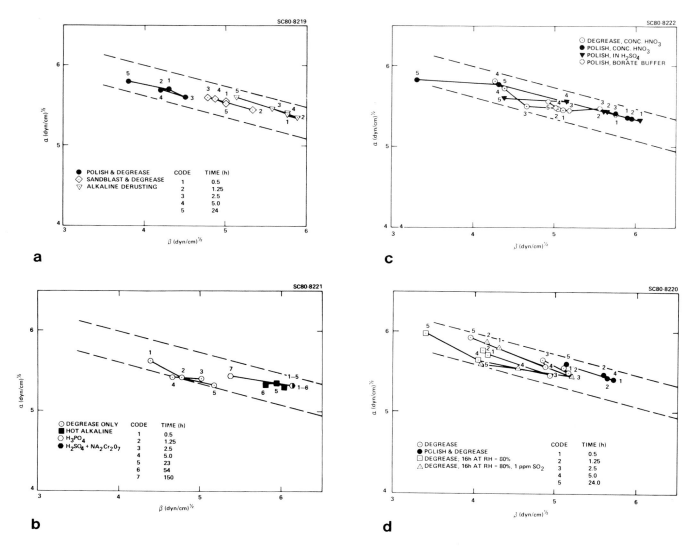

FIGURE 9 — Plot of dispersion α vs polar β components for different surface pretreatments.

face tension γ_{LV} and ratio of polar dispersion character. The surface tension properties of the test liquids in this study are those used by Kaelble.[5] Contact angles were measured with a NRL goniometer (Rame-Hart) using the sessile drop method where the drop size is small ($\leq 3\,\mu l$) to avoid gravitational flattening.

The dispersion (nonpolar) γ_{SV}^d and polar γ_{SV}^p components of $\gamma_{SV} = \gamma_{SV}^d + \gamma_{SV}^p$ have been calculated from the measured values of γ_{SV} as a function of time to detect changes in surface characteristics, while the freshly prepared surfaces age in contact with the environment (laboratory air at RH = 30 to 45%).

Figure 8a through c shows the changes of γ_{SV} with time for 14 different surface pretreatments. Most tests were terminated after 24 hours; however, some tests were conducted up to 160 h. The magnitude of γ_{SV} and its time dependence characterize the activity of different surfaces. The H_3PO_4, hot alkaline and H_2SO_4 + $Na_2Cr_2O_7$ treatments gave the highest values of γ_{SV} without any significant changes with time. The passivation treatments in HNO_3, $1N\ H_2SO_4$, and borate buffer initially produced higher γ_{SV} values; however, these surfaces were not stable (Figure 8b). A polished sample, which was pickled in inhibited HCl, initially had higher γ_{SV} values than a sample which was degreased but not polished before pickling; however, after 24 hours the surface energetics of the two samples were almost identical. Exposure to humid air and SO_2 after pickling lowered the surface tension (Figure 8c).

More detailed information concerning the relative polar character of the surface result from plots which represent the surface characteristics as $\alpha = \gamma_{SV}^d$ and $\beta = \gamma_{SV}^p$. Highly polar surfaces exhibit low α and high β values. A plot of α vs β shows how the surface properties differ for different surface treatment, and how aging effects these properties. The results for 15 different surface treatments of steel are plotted in Figure 9a through d. In general, all data fall within the band shown in Figure 9 which extends from polar characteristics at short times after treatment to more dispersive characteristics after longer times.

Figure 9a shows a pronounced difference in surface properties of steel that was polished and degreased, and steel that was sandblasted or treated by alkaline derusting. The latter two samples have a much more polar character which results from the outer surface hydroxyl ions or hydrated oxides. After 24 hours these surfaces were still much more polar than the surface which was polished and degreased only. A surface which was degreased, but not polished, changed to more polar character with time (Figure 9b). Hot alkaline cleaning, the phosphoric acid and the $H_2SO_4/Na_2Cr_2O_7$ treatments produced surfaces which initially were completely wettable with highly polar character (Figure 9b). Only small changes occurred with aging time for these treatments; after 150 hours, the α and β values had changed somewhat for the H_3PO_4 treated surface (Figure 9a).

Three different passivation treatments are compared in Figure 9c. A surface which had been polished before passivation in HNO_3 was much more polar, initially, than a corresponding sample which had been degreased only before passivation. However, after 5 hours, the surface properties were about equal. The point labelled "5" for the polished sample was taken after 78 hours. Passivation of polished samples in 1N H_2SO_4 and in borate buffer produced surfaces that were similar to those passivated in HNO_3. However, the changes to less polar characteristics were less pronounced than for this sample. The data point labelled "5" for passivation in 1N H_2SO_4 was taken after 70 hours (Figure 9b).

The data for samples treated in inhibited HCl are shown in Figure 9d. A sample which was polished and degreased before pickling was much more polar and more stable than a sample which had not been polished prior to pickling. Exposure of the unpolished, pickled sample to RH = 80% for 16 hours with or without 1 ppm SO_2 produced surfaces with a highly nonpolar character which became more polar in the first 5 hours, then changed again to the initial properties.

Surface Mapping

When surfaces are prepared for coating, they are usually assumed to be spatially uniform, and little attention is given to spatial heterogeneity. However, due to inclusions, surface history, contamination, etc., surfaces are likely to have some heterogeneity with respect to surface properties. Since the coating adhesion under stress and corrosion is directly related to flaws at the coating substrate interface, heterogeneity becomes of great significance. To discover surface heterogeneity, we have developed an automated scanning facility that will map the surface of samples that have been prepared for coating with respect to several properties. The surface properties that are mapped include optical properties (that is, oxide or contamination thickness by ellipsometry), dielectric properties (surface potential difference, SPD, that is sensitive to the outer dipole layer) and electron emission and attenuation properties (by photoelectron emission, PEE). A detailed description of the equipment used for these measurements and their interpretation is given by Smith.[7]

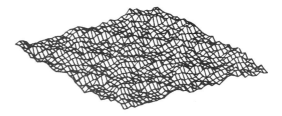

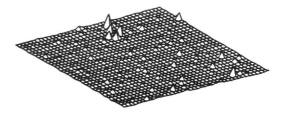

FIGURE 10 — Ellipsometry data for mapping of sandblasted sample and sample which was degreased and cleaned with hot alkaline treatment.

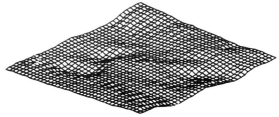

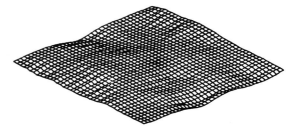

FIGURE 11 — SPD maps for samples of Figure 10.

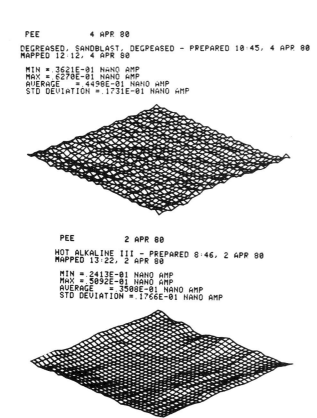

FIGURE 12 — PEE maps for samples of Figure 10.

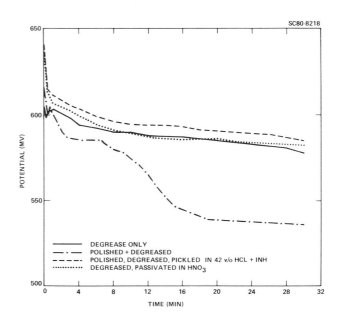

FIGURE 13 — Potential time curves for steel of four different surface pretreatments immersed in phosphating solution.

Surface mapping for different pretreatments. The mapping facility scans the samples and produces a map of the signal from each instrument (ellipsometry, SPD, PEE). A computer prints out the maximum and minimum values along with the average and standard deviation. Examples of ellipsometer maps for two surface treatments of steel are given in Figure 10. Each map covers approximately 4 cm × 4 cm with a grid of 1 mm × 1 mm. The scattered light reveals the roughness of the sandblasted sample. The hot alkaline sample is much smoother, and differs in the value of the phase shift (DELTA), which is related to the oxide film thickness. The film thickness of the hot alkaline cleaned sample is estimated to be ~10Å, and the sandblasted sample is ~80Å. Figure 11 compares SPD maps for these two surface treatments. The average SPD of the hot alkaline treated surface is −0.18 volts as compared to +0.11 volts for the sandblasted surface. Auger spectroscopy has shown that silicate contamination is at the outer surface of the hot alkaline cleaned samples (Table 2). It is postulated that the Si^{O}\Si structure exposes the electronegative oxygen which causes SPD to be negative. The

TABLE 4 — Coating Weight, Metal Loss and Potential for Zinc Phosphating of 1010 Steel

Surface Treatment	Coating Weight (mg/cm²)	Metal Loss (mg/cm²)	−E_{10} (mV)	−E_{30} (mV)
Degrease (d)	1.78	0.74	579	578
Polish (p) & d	0.83	0.36	580	536
Sandblasted & d	3.04	1.18	—	591
d & Passivate in HNO_3	2.68	1.19	576	582
p & d Passivate in HNO_3	3.40	1.27	593	—
d & 42 v/o HCl & Inhibitor	3.29	1.43	591	—
p & d & 42 v/o HCl & Inhibitor	4.30	1.39	584	585
d & hot alkaline	2.05	1.31	591	558
d & alkaline derusting	4.03	1.49	591	577
d & H_3PO_4	2.30	1.32	588	573
d & anodic etch	4.41	1.42	588	584
d & electropolishing I	2.91	1.27	583	588
d & chemical smoothing	3.43	1.26	593	578

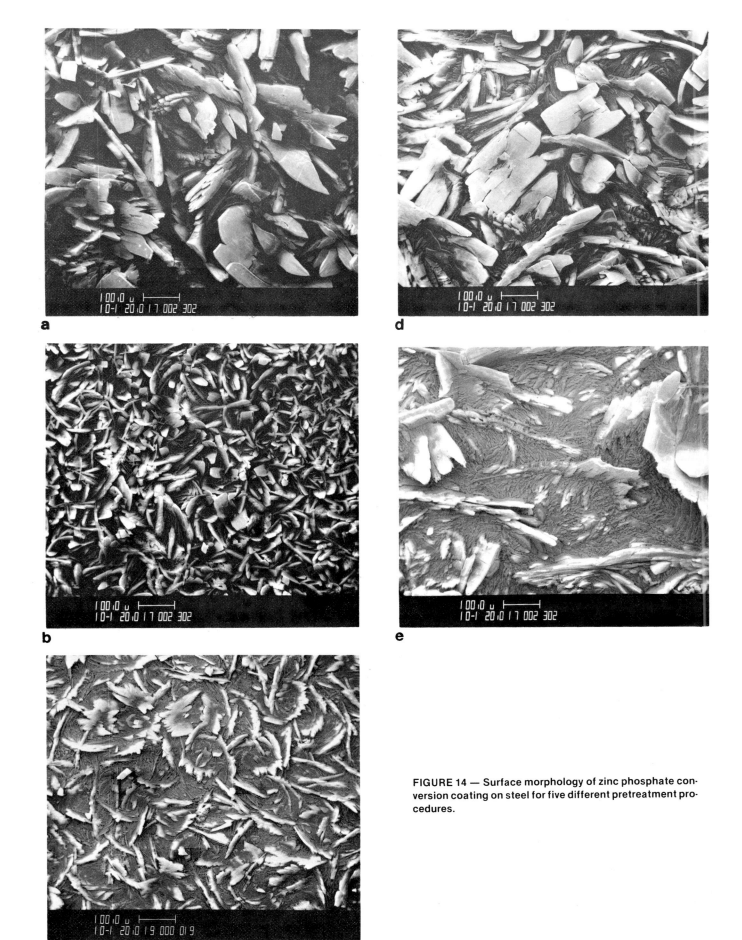

FIGURE 14 — Surface morphology of zinc phosphate conversion coating on steel for five different pretreatment procedures.

sandblasted samples, on the other hand, are heavily contaminated with organics which usually expose hydrocarbons at the outer surface and oxygen bods to the substrate, leaving the dipole orientatioon with the positive end outward. This accounts for the positive SPD for the sandblasted samples.

Figure 12 compares PEE maps for the same two surface treatments. The average PEE for the hot alkaline surface is 0.035 nA, as compared to 0.044 nA for the sandblasted surface. The larger PEE for the sandblasted surface may be associated with the thicker oxide layer. However, the attenuating properties of the outer contamination are not known at this point.

Each map demonstrates heterogeneity over the surface. More detailed studies will reveal the cause of the heterogeneity. The utility of these maps will be revealed fully when they can be compared with functional maps that indicate such things as corrosion resistance, coating adhesion strength, or durability.

Phosphate Conversion Coatings

After pretreatment with some of the procedures of Table 1, zinc phosphate coatings were applied, and the deposition kinetics, the coating weight, and the coating morphology were determined. The phosphating solution was that used by Ghali and Potvin[8] and others (6.4 g/l ZnO, 14.9 g/l H_3PO_4, 4.1 g/l HNO_3 at 95 C). Potential time curves were measured while the sample was immersed in the bath. After application of the coating, the samples were immediately neutralized in 0.3 w/o CrO_3 at 65 to 90 C for 3 minutes and aged at room temperature for at least 48 hours. For determination of the coating weight, the coating was stripped in 1980 g/l NaOH + 90 g/l NaCN by repeated immersion until the weight of the sample did not change anymore.

Deposition kinetics. As discussed by Ghali and Potvin,[8] the change of the potential of steel in the course of phosphating is related to the nature of the compactness of the coating formed during the treatment. Gabe and Richardson[9,10] have attempted to correlate potential time measurements with the physical properties of phosphate coatings.

Potential time measurements have been performed for steel with different surface treatments in the zinc phosphate bath. The steel potential was measured vs SCE which was connected to the bath by a long Luggin capillary. A compensation circuit was used to record accurately the small potential changes during phosphating accurately. Typical potential time curves for four different treatments are shown in Figure 13 for a 30 minute immersion period. The polished and degreased sample shows the largest potential change during phosphating (about 80 mV), similar to the schematic curves shown by Ghali and Potvin[7] for an electropolished mild steel. For the other treatments used here, the potential change was about 40 mV or less, which is more in agreement with the results of Gabe and Richardson.[9,10] The potentials after 10 minutes and 30 minutes taken from two different series of measurements are shown in Table 4.

According to TT-C-490, the minimum coating weight for zinc phosphate is 300 mg/ft², or 0.32 mg/cm² for dip processes. The coating weights for the 13 different pretreatment procedures are also given in Table 4 together with the weight loss of the steel surfaces during the coating formation. All coating weights are in excess of the minimum weight of 0.32 mg/cm². The anodic etch, pickling in inhibited HCl, alkaline derusting, chemical smoothing, and passivation in HNO_3 produced coatings that had ten times the minimum weight, while the thinnest coatings resulted after pretreatment by polishing and degreasing and by degreasing only. These surfaces also had, by far, the lowest metal loss. For all the other surface treatments, the metal loss was between 1.18 and 1.49 mg/cm². Gabe and Richardson[10] reported a coating weight of 3.5 mg/cm² for mild steel pickled in 20 v/o HCl.

Morphology. The morphology of the zinc phosphate coatings formed on steel after the 13 different pretreatment procedures of Table 4 were examined by SEM. Significant differences were noted for different pretreatments. The coating after degreasing only was fairly coarse, consisting of a plate like structure (Figure 14a). Polishing (Figure 14b) and sandblasting (Figure 14c) produced a finer structure. The morphology of the coating after degreasing and the hot alkaline treatment as similar. Pickling in inhibited HCl after degreasing and polishing produced the coarsest coating (Figure 14d). The coatings produced by alkaline derusting, H_3PO_4, anodic etch, electropolishing (Figure 14e) and chemical smoothing were all very similar.

More testing of the properties of the iron phosphate coating formed on steel surfaces with different pretreatment procedures is necessary to understand the effects of surface structure on adhesion and corrosion protection of these coatings. Such tests are being initiated at present. At this time, it is interesting to note that the polish degrease pretreatment which showed the largest potential change (Figure 13) produced the finest phosphate structure with the lowest coating weight, while additional pickling in inhibited HCl resulted in the coarsest surface structure with the second highest coating weight.

Summary

The surface morphology, the chemical composition of surface films and the activity of surface structures resulting from a number of different pretreatment procedures have been evaluated using the scanning electron microscope (SEM), Auger electron spectroscopy (AES, SAM), XPS, surface energetics measurements, and mapping of surface properties with ellipsometry, surface potential difference (SPD), and photoelectron emission (PEE) measurements. The deposition kinetics and surface morphology of zinc phosphate coatings have been determined. In the next phase of this program, organic coatings will be applied to the phosphated steel, and adhesion and corrosion tests will be performed for different surface pretreatment procedures. AC impedance measurements will be carried out before and after these tests to determine the changes of the coating properties and correlate these changes with surface conditions. Similar experiments will be carried out for aluminum alloys.

Acknowledgement

This project is being supported by the Office of Naval Research under Contract No. N00014-79-C-0437. Dr. T. Smith and D. H. Kaelble provided invaluable inputs for the surface energetics and mapping studies. P. J. Stocker and G. Lindberg assisted in the experimental work.

References

1. S. Spring, Preparation of Metals for Painting, Reinhold Publ. Corp., New York, 1965.
2. R.L. Snogren, Handbook of Surface Preparation, Palmerton Publ. Co., New York, 1974.
3. J.A. Murphy (editor), Surface Preparation and Finishes for Metals, McGraw-Hill Book Co., New York, 1971.
4. R.M. Burns and W.W. Bradley, Protective Coatings for Metals, Reinhold Publ. Co., New York, 3rd Edition 1967.
5. D.H. Kaelble, *Polymer Eng. and Sci.* **17** (7), (1977).
6. D.H. Kaelble and P.J. Dynes, *J. Colloid and Interface Sci.* **52**, p. 562 (1975).
7. T. Smith, *J. Appl. Phys.* **46**, p. 1553 (1975).
8. E.L. Ghali and R.J.A. Potvin, *Corros. Sci.* **12**, p. 583 (1972).
9. N. Helliwell, D.R. Gabe and M.O.W. Richardson, *Trans. Inst. Metal Finish.* **54**, p. 185 (1976).
10. J.B. Lakeman, D.R. Gabe and M.O.W. Richardson, ibid. **55**, p. 47 (1977).

The Interaction of Chlorinated — Rubber Based Lacquers With Abraded Mild Steel Substrates

D.H. Smelt*

Introduction

The initial purpose of the experiments carried out was to investigate factors affecting the adhesion of lacquers, based on chlorinated rubber, to mild steel. However, during the course of the work, some particularly interesting features appeared due to chemical interactions occurring between the coating and the substrate. This paper discusses the work in an attempt to identify the nature of these interactions, and the factors affecting their occurrence.

System Under Investigation

The coating was composed of a chlorinated rubber binder dissolved in Analar toluene in a 50:50 ratio by weight, and plasticized with a chlorinated paraffin, in the ratio 70:30 binder to plasticizer by weight. The structure of the chlorinated rubber is:

$$\underset{OH}{\overset{O}{\underset{\|}{\overset{\|}{C}}}} - (C_5H_{5.5}Cl_{3.5})_n - \underset{O}{\overset{CH_3}{\underset{\|}{\overset{/}{C}}}}$$

The substrate was prepared by abrading with 280 grade emery paper until there was no preferred direction of surface abrasion, ultrasonically degreased for 5 minutes in boiling trichlorethylene, washed with distilled water to remove any remaining surface deposits, rinsed with acetone, and allowed to dry in air. The cleaned substrate was coated as soon as it was dry with a film of lacquer approximately 150 μm thick.

The substrate was in the form of discs 2 cm in diameter, which was the appropriate size for the adhesion test in use.

Observations

It is recognized that both chlorinated rubber, and plasticizer may degrade in the presence of iron,[1] and it is generally accepted that an unpigmented lacquer composed of these two components, applied to a mild steel plate, and exposed to sunlight, will discolor badly, turning dark brown.

*Oxford, England

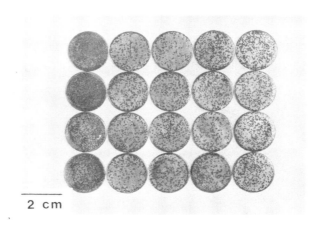

FIGURE 1 — Examples of "cell" like discoloration.

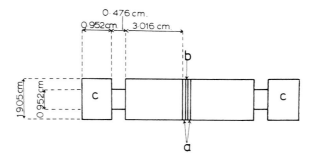

FIGURE 2 — Preparation of specimen for test. a) "Araldite"; b) steel disc with lacquer on one face; c) mild steel cylinders or "dollies".

Two examples of discoloration were observed in the specimens prepared for the test. Coated discs were stored either in a closed wooden box, or on glass plates on the laboratory bench. Some specimens developed a rust like discoloration, which was continuous across the surface of the disc; other specimens developed a cell like discoloration, as may be observed in Figure 1. The former occurred most frequently in specimens stored on the laboratory bench, while the latter occurred more often in those stored in darkness, in the box. However, there were one or two exceptions to this rule.

Another interesting observation was made regarding the discs used for the adhesion test, which involved bonding the two surfaces of each disc to the ends of steel cylinders of the same diameter, as shown in Figure 2, thus excluding the air. In some cases, specimens were maintained in this condition for several weeks. During testing, the two cylinders were pulled apart in an Instron machine, thus causing failure within the coating, or at the coating/substrate interface. The fractured specimens then were stored in open wooden boxes, in the light. Lacquers which had shown no discoloration before bonding were in the same state after failure; however, rust like discoloration soon developed on storage (on one occasion it appeared overnight).

A few specimens which had been stored in the dark developed the cell like structure. In general, discoloration occurred in areas of thick lacquer remaining on the substrate.

In the following sections, the examination of these two types of discoloration will be discussed, and as far as possible, their nature described.

Examination of the Continuous Discoloration

Visual examination

A batch of lacquer was prepared and divided into two; one half had hydrogen chloride gas passed through it for about three minutes, and films were cast from both batches in the usual way. The discs were stored on the laboratory bench, and their gradual discoloration observed. The lacquer which had been subjected to hydrogen chloride discolored rapidly compared with the untreated lacquer; however, after three months storage all specimens were the same rust like color, and there was no perceptible change during another month of storage. When a scored specimen was left in dilute hydrochloric acid overnight, the film was completely detached from the substrate, and was found to be translucent with the same "pale sand" color as a freshly cast film. The discoloration was, therefore, at the lacquer/substrate interface.

One set of tested specimens developed a continuous rust like substance in a region where the film had failed close to the coating/substrate interface. The tests which were carried out on this set revealed that it was representative of the interfacial discoloration. These tests will be described in the following sections.

Electron Spectroscopy for Chemical Analysis (ESCA or XPS)

A strip, 19 × 4 mm, was cut from the substrate using a jeweler's saw, washed in methanol, and stored in a desiccator at atmospheric pressure to await examination.

The ESCA technique is used to study only the surface of the specimen, as the penetration is of the order of 30 Å in organic materials; the relative atomic concentration at the surface can be measured to a standard error of 10%.[2] Spectra were recorded on an AEI ES 200B ESCA spectrometer; correction for sample charging effects was achieved by using the hydrocarbon signal at 285.0 eV binding energy as a reference. Line shape analysis was carried out using a Dupont 310 Analogue Curve Resolver. The spectra were recorded at electron take off angles of 30 and 70°, the latter corresponding to grazing electron take off, which enhances surface features.

It was found that the rust layer contained a higher concentration of chlorine and oxygen, in the surface regions, than did the interfacial layer of an undiscolored film. The amount of this chlorine, which was in the form of chloride ions, was the same in each instance, and the excess occurring in the rust was combined with carbon.

The contribution due to oxygen 1_s electrons in the spectrum was complex, with a large component occurring at a high binding energy. This energy suggests that it may be associated with a metal carbonate feature, by analogy with the high electron binding energy of the singly bonded oxygens in organic carbonates (in the region 533 to 535 eV).[3] Iron was also present in its highest oxidation state.

The signal intensities changed slightly with angular variation, thus demonstrating that the layer of rust is relatively thick, greater than 100 Å. It appears that the layer contains a ferric compound, a metal chloride, some form of metal carbonate, and a compound containing C-Cl groups.

Auger Analysis

A disc, 6 mm in diameter, was punched from the specimen, washed in methanol, and stored in a desiccator at atmospheric pressure. When mounted in the analyzer chamber, the specimen was bombarded first with argon ions, to remove any surface contamination: a layer 20 to 30 Å thick was removed.

Figure 3 shows the spectrum obtained from the specimen using an incident beam energy of 2.5 KeV, and sensitivity 10 mV. The relative concentrations of atomic species were calculated to be: chlorine, carbon = 1.34; oxygen, carbon = 0.59; chlorine, oxygen = 2.29. By comparison, the corresponding results obtained for the interfacial region of an undiscolored specimen were: chlorine, carbon = 0.94; oxygen, carbon = 0.18; chlorine, oxygen = 5.10.

Clearly, these results are consistent with those obtained using ESCA; the chlorine and oxygen contents both increased, relative to the carbon content, and the oxygen appears to have increased relative to the chlorine concentration. Sulphur and silicon both appear in the spectrum, although neither was present in the undiscolored film. The sulphur was a contaminant derived from a fluorescent specimen also in the chamber, and the silicon remained as silicon carbide after the abrasion and washing of the substrate. The

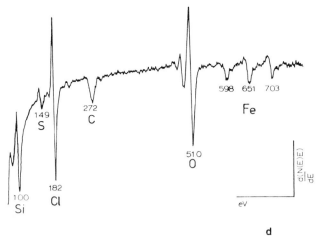

FIGURE 3 — Plot of Auger electron energy (E) vs $\frac{d}{dE}(N(E))$ for a specimen of continuous discoloration. Incident beam energy ~2.5 keV. (N(E) is the number of electrons with energy E).

FIGURE 5a — Hexagonal spot diffraction pattern obtained from specimen of continuous discoloration.

FIGURE 4a — Diffraction pattern obtained from crystals of β.FeO.OH.

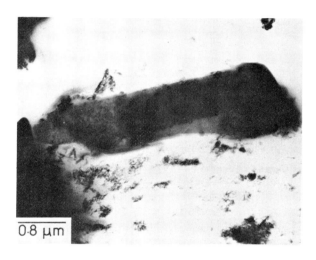

FIGURE 5b — Plate like crystal from which 5a is obtained.

relative concentration of silicon to carbon at the surface was 2.35:1.

Transmission Electron Microscopy (TEM)

Areas of discolored film were removed from a disc using a scalpel, taking care to include as much of the discoloration from the substrate as possible. Specimens were taken from lacquers which had been prepared in the same way; however, some had been stored for six months before films were cast, and some had been subjected to hydrochlorination.

A thin film of carbon was evaporated onto the lower side of the pieces of lacquer; that is, the face covered in the rust like deposit. The carbon provided a thin, continuous, conducting film to support the deposit. The lacquer was dissolved away in toluene, and fragments of the remaining deposits were mounted on copper grids for examination in the transmission electron microscope.

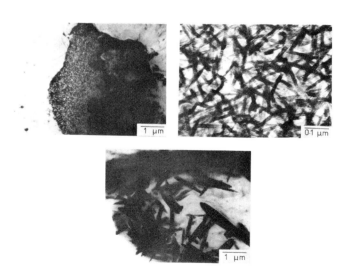

FIGURE 4b — Needle shaped crystals of β.FeO.OH.

TABLE 1 — Plane Spacings Corresponding to Hexagonal Spot Pattern

(1) 4.49 Å
2.65
2.55 (clear spot, not on the hexagonal net)
2.23
1.71
(1) 1.52
1.31
1.28
1.13
0.91
0.86
0.75
0.64
0.57

(1) Brightest spots

Two different diffraction patterns were obtained from lacquer which had been stored before the films were cast. The first (Figure 4a) derived from needle shaped crystals (Figure 4b), which lay in the plane of the specimen, which was originally the plane of the mild steel substrate. These crystals ranged in size from approximately 0.1×0.02 to 3×0.4 μm. The crystal plane spacings were calculated, and were found to coincide with those of βFeO.OH, or akaganeite.

The second diffraction pattern is shown in Figure 5a, and the type of area from which it is derived in Figure 5b. These appeared to be platelets, which were single crystals lying in the plane of the specimen. The hexagonal spot pattern corresponded with plane spacings (Table 1).

However, these spacings do not coincide with any of those given in the ASTM index,[4] so that further investigations had to be made to attempt to identify the chemical elements present in the crystal. The specimens from the hydrochlorinated lacquer contained the same two compounds, but the platelets were much more abundant, and the specimens obtained from a lacquer, which had not been stored, appeared to contain only βFeO.OH.

Energy Dispersive Analysis of X-ray (EDAX)

The same specimens were examined, using the EDAX technique, to identify the elements present by the X rays emitted from them. Figure 6 shows a superposition of the EDAX plot, obtained from both the β ferric oxide/hydroxide, and the unknown compound. Clearly, the latter contains silicon, iron, and chlorine, although oxygen, hydrogen, and carbon may be present, but cannot be detected as their X ray wavelengths are too small.

Thus, it appears that some chemical interaction is occurring between the lacquer, the iron of the substrate, and silicon carbide remaining on the substrate

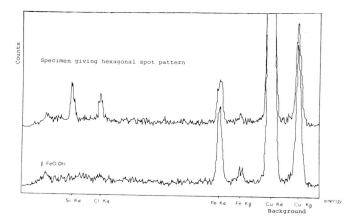

FIGURE 6 — Spectra obtained using EDAX for specimens of β.FeO.OH. and the unidentified compound.

surface after abrasion and cleaning. The small quantity of silicon present in mild steel (0.05% maximum by weight)[5] is too small to produce the comparatively large amount of this compound observed. The compound produced forms crystals in the region of 4×1 μm in dimension, which are less than about 0.1 μm in thickness. They lie in the plane of the film/substrate interface.

The quantity of this compound is increased by the presence of free hydrogen chloride in the film, as shown by its increased concentration in the specimen from the hydrochlorinated lacquer. It should be noted that the specimen from the stored lacquer also contained this compound, but in smaller quantities, and the fresh lacquer did not contain any of it. This implies that free hydrogen chloride may be produced by the decomposition of the lacquer on storage.

Remy[6] records that although silicon carbide is extremely chemically resistant, at high temperatures it is attacked slightly by chlorine. Bailar, et al,[7] however, record that chlorine attacks silicon carbide vigorously at 100 C, producing silicon tetrachloride and carbon. Thus, it is possible that at room temperature, and in the unusual and confined environment at the lacquer/substrate interface, a reaction involving chlorine ions and silicon carbide may occur, at a fairly low rate. Some compound other than silicon tetrachloride is formed, though, as the iron from the substrate is also involved in the reaction. The results of the investigation using ESCA suggest that the compound may contain chloride or carbonate features, and may not have a simple structure.

Examination of "Cell Structure"

Visual Examination

Figure 7 shows four discs which were stored in a closed wooden box for three months after coating. Cell like discoloration developed on these discs, and the surface of the film, which was originally smooth after storage showed an unevenness similar to quilting, the surface being raised at points directly

FIGURE 7 — Examples of "cell" like discoloration, showing the unevenness which develops in the surface of the film.

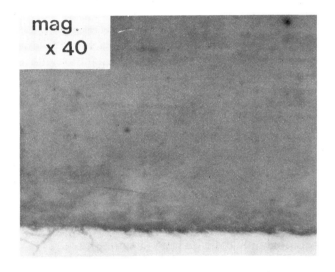

FIGURE 8a — Optical micrograph of a cross section through the specimen, showing the contours of the surface of the substrate.

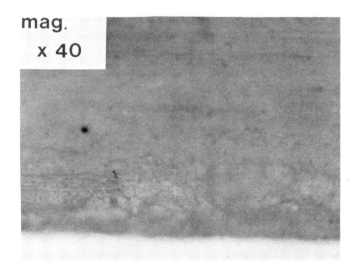

FIGURE 8b — Optical micrograph showing the "rust" colored compound occurring in the region shown in 8a.

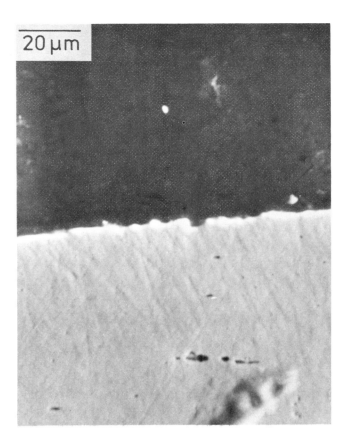

FIGURE 9 — Scanning electron micrograph (SEM) of a cross section through the specimen in an area where discoloration had not occurred.

above the circumference of each cell. The discoloration was dark brown around the edge of each cell, and a lighter and less dense color towards the center. It was also most severe around the edge of each disc, and on further storage, the number of cells increased until each disc was completely discolored. Reference should be made to Figure 1, where the discs in the left hand column are most severely discolored. The discoloration was not removed when the lacquer was dissolved in toluene, and was once again occuring at the lacquer/metal interface.

Optical and Scanning Electron Microscopy

Cross sections of the specimens were prepared to investigate the cause of the change in the contours of the film surface. Figures 8a and b are optical micrographs of the interfacial region; Figure 8a shows the contours of the edge of the mild steel substrate; Figure 8b shows the rust colored compound in this region. The discoloration in Figure 8b appears to extend into the lacquer, and there is no clear point where the compound ends, and the lacquer begins, as there is in the case of the substrate in Figure 8a. Figure 9 is a scanning electron micrograph (SEM) of an area of the interface where discoloration had not occurred.

Figure 10 shows SEM of two regions of the film, each at two magnifications. The edge of the substrate

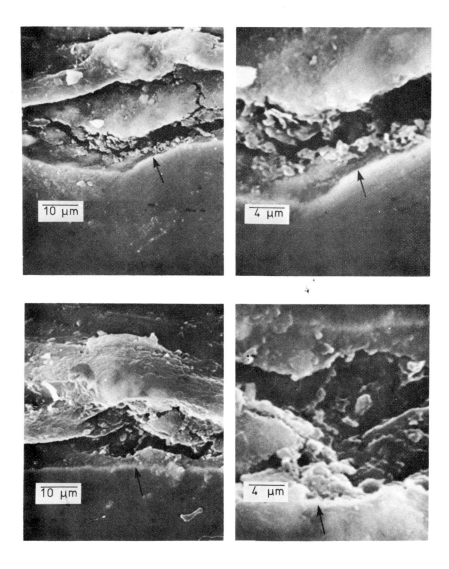

FIGURE 10 — SEM of two sections of the interfacial region. Arrows indicate the edge of the substrate.

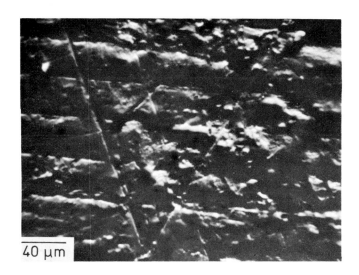

FIGURE 11 — SEM of the surface of a discolored specimen after tensile testing.

is indicated by the arrows, but a layer of encrustation may now be observed above this, which was not visible in the optical micrographs. However, there is no clear boundary to the lacquer region.

There appear to be large areas where there is no contact between the upper and lower regions of the interface. However, it is suggested that interaction products had, in fact, been carried away, either by the polishing action or the washing process involved in preparing the sample. Anything which was water soluble would be removed. If such large voids were present at the interface, the failure stress of such systems would be much lower than for specimens which had not discolored. However, from the tensile tests which were carried out on such specimens, this did not appear to be the case. After testing, the surface of this kind of specimen contains pits (Figure 11), which are not observed in tested, undiscolored films.

Transmission Electron Microscopy (TEM)

Areas of lacquer were removed from the substrate using a scalpel, as described before; a carbon film

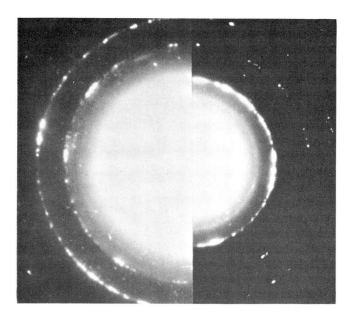

FIGURE 12 — Dominant diffraction pattern obtained from "cell" like discoloration.

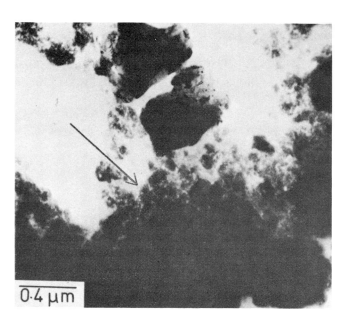

FIGURE 14 — Area of the specimen from which the spot pattern in Figure 12 is obtained.

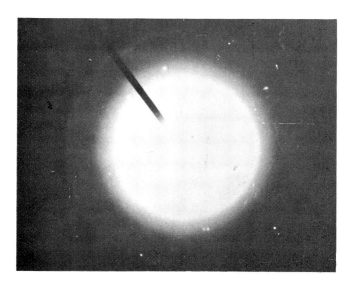

FIGURE 13 — Ring pattern component of Figure 12.

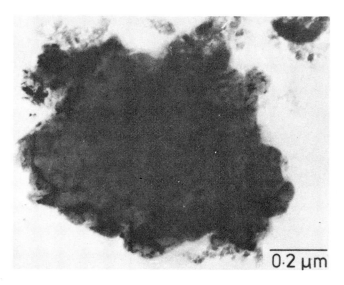

FIGURE 15 — Arrow indicates the area of the specimen from which the ring pattern in Figure 13 was obtained.

was evaporated over the discolored surface, and the lacquer dissolved away in toluene. Small pieces of the deposit were mounted on copper grids for examination in the electron microscope.

Once again, the crystals of β.FeO.OH were detected, as were the platelets, which gave a hexagonal, spot diffraction pattern. However, these were present only in small quantities. The dominant diffraction pattern is shown in Figure 12. This pattern is a combination of the ring pattern shown in Figure 13, and a spot pattern. Both patterns occurred alone occasionally, and Figures 14 and 15 show examples of the areas on the specimen from which each pattern derived. However, the patterns occurred together most frequently. Table 2 shows the plane spacings calculated from the diffraction patterns. It has not been possible to identify either of these compounds, as their plane spacings do not coincide with any recorded in the ASTM index.

However, Figure 16 shows the EDAX plots obtained from each compound, and it may be observed that the compound giving the ring pattern contains at least iron, chlorine, and a trace of silicon. The compound giving the spot pattern does not contain chlorine or

TABLE 2 — Plane Spacings Calculated from the Diffraction Patterns

Spot Pattern Å	Ring Pattern Å
2.73	2.59
2.09	1.52
1.65	
1.56	
1.44	
1.18	
1.02	
0.91	
0.84	
0.77	

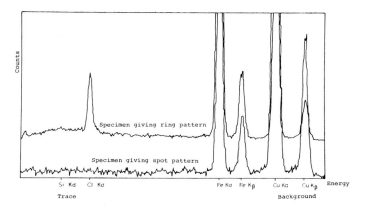

FIGURE 16 — Spectra obtained using EDAX for compounds occurring in the "cell" like discoloration.

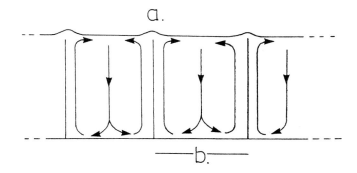

FIGURE 17 — Schematic representation of the formation of "Benard Cells". a) Unevenness of the surface occurring at the circumference of each cell, where pigment may be deposited. b) Diameter of the cell. The arrows indicate the direction of motion within the convection cell.

silicon, but does contain iron. Oxygen, hydrogen, or carbon may also be present in either of them, but cannot be detected using this technique.

Discussion and Conclusions

Some comparison may be drawn with the formation of "Benard cells" in pigmented films. The mechanism of their formation is described in Figure 17. Small convection cells develop in the paint as it cures, or dries. These convection cells carry pigment particles upwards to the film/air interface. As the film becomes more viscous, and finally sets, the particles are held at the circumference of each cell. Thus, markings similar to those observed in the cell like discoloration are formed on the surface of the film. It is possible that a similar mechanism may be involved in the case of the unpigmented film; loose particles may be lifted from the substrate surface, into the film, but as the lacquer is fairly viscous when the film is cast, they remain close to the interface. Here, the conditions are such that some chemical interaction takes place between the film and the particles (whether fragments of the substrate material, or silicon carbide), thus forming a cell like discoloration in the interfacial region. Several points have not been resolved; for example, how the presence of excess hydrogen chloride in the lacquer accelerates the formation of the continuous film of discoloration, and in particular the compound which gives a hexagonal, spot diffraction pattern. Also unresolved, is how the conditions of storage affect the interactions between the film and the substrate.

Although the structures of several of the compounds present do not match any of those given in the ASTM tables, it is possible that they may be compounds which normally have a particular crystal structure, but that the unusual environment at the interface promotes the formation of different crystal structures, or even nonstoichiometric compounds.

It is clear that silicon carbide plays an important part in these reactions, which suggests that some other abrasive medium should be used. However, as it does not appear to have a destructive effect on the bonds between the film and the substrate, this may not be necessary from a practical point of view.

Although these interactions have not, to the author's knowledge, been reported previously, it is likely that they also occur between pigmented films and a mild steel substrate, and their presence is masked by the pigmentation.

Acknowledgments

The author would like to express gratitude to Prof. Sir Peter Hirsch, F.R.S. and ICI Ltd. (Mond Division, Runcorn, England) for laboratory facilities; also to the Science Research Council and ICI for financial support. Thanks are also due to Dr. D.A. Smith (now of IBM) and Dr. W.W. Harper (now of International Paints Ltd., Felling, England) for their supervision and advice during the course of this work.

This work forms part of a D. Phil. thesis prepared in the Department of Metallurgy, University of Oxford, England. Present address: A.E.R.E., Harwell, Didcot, Oxfordshire, England.

References

1. ICI Internal Report No. MD 14,656/3/B.
2. R.S. Swingle + W.M. Riggs. — *Critical Reviews in Anal. Chem.* **5** (3), p. 267 (1975).

3. A. Dilks — University of Durham, Chemistry Department, England. Private Communication.
4. ASTM index to Powder Diffraction File 1968 + File.
5. Metals Handbook, American Soc., for metals, 8th ed. **1**, 1961.
6. H. Remy — Treatise on Inorganic Chemistry, **1**. Elsevier, 1956.
7. J.C. Bailar, *et al*, Comprehensive Inorganic Chemistry, **1**, Pergamon 1973.

Is the Salt Fog Test an Effective Method to Evaluate Corrosion Resistant Coatings?

Tony Liu*

Introduction

Although the subject may be familiar to most, I believe the findings that have resulted from our work with the salt fog test will provide useful information and thought provoking questions about other methods of testing as well.

Is the salt fog test an effective method to evaluate corrosion resistant coatings? I have been using the test for over ten years. Every time I think about this question, I manage to avoid a definitive answer. Now, I would like to present some facts that I have observed, give my answer to this question, and leave the final answer to the reader.

The salt fog (spray) test was invented in 1939 and was issued under the designation B117 in the ASTM specification book. It is one of the most widely used accelerated tests for coatings. When we want to sell a coating or a system, we never ask our customers to wait for five or ten years. We show them the results of the salt fog test along with any other test results. We mention magic figures, such as 500 and 1000 hours. What do these figures actually mean?

Before an attempt is made to answer these questions, a quick review of the details of the salt fog test is in order. According to ASTM specifications, the size and detailed construction of the salt fog cabinet are optional. Generally, there is one atomizing nozzle; however, if preferred, there may be more than one, as long as the conditions comply with the specifications. The nozzles can be above the trough, or a central or side spray tower. To run the test properly, attention must be given to a number of factors.

Salt Solution

The salt used to make the salt solution is sodium chloride. It should be nickel and copper free, and contain less than 0.1% sodium iodide, and less than 0.3% total impurities. There are a few salt suppliers; before you buy, you should check to determine if they meet the specifications. The water should be distilled, or contain less than 200 ppm of total solids. Deionized water is used by many persons for economical reasons. The purity of a deionized water depends on the process used. The salt solution is made by dissolving 5 parts by weight of the salt in 95 parts of water. When the solution is atomized at 95 F, the pH should be between 6.5 and 7.2. When the pH is adjusted at room temperature, it should be adjusted below 6.5, so that when it is atomized at 95 F, the loss of dissolved carbon dioxide will bring it back within the specified range.

The concentration of the salt solution is important to the test. The higher the concentration, the stronger the effect of the salt on the coating. However, the solution becomes less penetrating because of its higher solids. It is a mistake to think that lowering the concentration of the salt solution will make the test less severe.

The compressed air supply should be free of oil and dirt. This can be accomplished by passing it through a water scrubber or some other cleaning material, such as wool or activated alumina. Atomizing pressure should be kept between 10 and 25 psi. It has been observed that fluctuations in air pressure may result in an increase in corrosivity of the fog from a nozzle. Atomizing may have a critical pressure at which an abnormal increase in the corrosiveness of the salt fog occurs. The critical pressure varies with the kind of nozzles used.

Cabinet Conditions

Temperature in the cabinet should be maintained between 92 and 97 F. The quantity of fog is measured by the quantity of salt solution collected as a result of condensation. A glass funnel 4 inches in diameter may be used to collect the solution in a graduated cylinder. It is suggested that at least two such collectors be used; one nearest to the nozzle and the other farthest from the nozzle. The collection within a 16 hour period should be between 1 and 2 ml/hour. To be precise, the pH and the concentration of the collected solution can be double checked.

Position of Specimens

All specimens should be supported or suspended at angles of between 15 and 30° from the vertical and there should be no direct contact between any two (Figure 1). Specimens should be spaced in such a way

---

*NL Chemicals, NL Industries, Inc., Hightstown, New Jersey
Printed by permission of NL Industries.

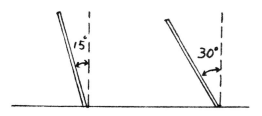

FIGURE 1 — Position of specimens: no direct contact; no shadowing.

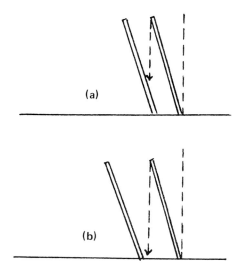

FIGURE 2 — (a) Shadowing; (b) no shadowing.

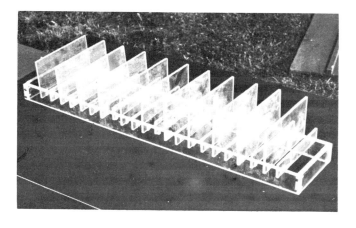

FIGURE 3 — Special rack provides uniformity, good spacing, and cleanliness.

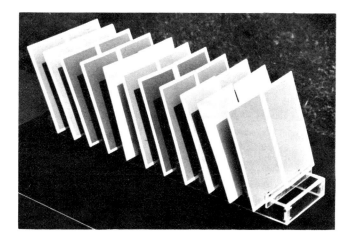

FIGURE 4 — Samples in special rack.

FIGURE 5 — Salt fog test. (a) Cold rolled steel: Q-panel, 960 hours; (b) Bonderite, 2250 hours.

that there will be no shadowing. In other words, solution must not drip from one specimen to another (Figure 2). If shadowing occurs, the specimens subjected to the excessive drip of salt solution will tend to show premature failures. We have designed a special rack which provides uniformity, good spacing, and cleanliness (Figure 3). Many people use wooden racks, in which the specimens stand at an angle in the grooves (Figure 4). Rust and rust solution accumulate in the grooves and may cause premature edge failure if the bottom edge is not well protected.

Preparation of specimens for the salt fog test will now be considered. Depending on the application or the end uses of the coatings or systems to be tested, the substrate should be chosen carefully so that the data will be more meaningful to the tester and to the consumer of the coatings. The selection of substrates is also dependent on the generic type of coatings. Different results are obtained with different substrates.

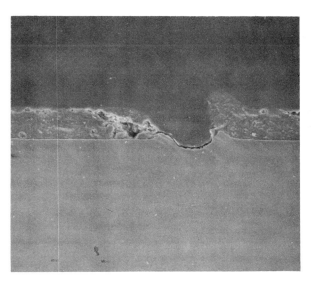

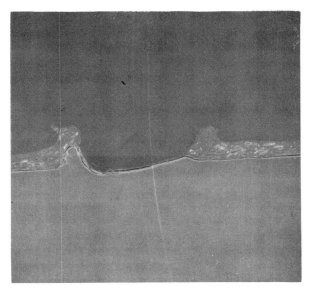

FIGURE 6 — Cross sections of three different scribes. (a) Tungsten carbide; (b) utility knife; (c) ice pick.

Many companies choose to use cold rolled steel Q-panels for their tests because they provide greater uniformity and convenience. Some companies use phosphatized steel for everything because it produces better results. Other companies use substrates, similar to those used in the field, for their tests (Figure 5).

Surface preparation plays a significant part, as does the drying time of specimens before they are subjected to the test. Scribing the panels to bare metal is a general practice to evaluate corrosion protective coatings. A scribe is meant to simulate a mechanical damage to the coating, where failure can begin easily. Undercutting is the loss of adhesion of the film at the scribe, and is the major reason for the scribe. There are various knds of scribing tools such as an ice pick, a utility knife, or a tungsten carbide tip. Different tools make different cuts on different coatings.

Figure 6 (a), (b), and (c) shows highly magnified cross sections of three different scribes. The first cut is made with a tungsten carbide tip, the second with a utility knife, and the third with an ice pick. Each tool exposes a different area of bare metal at the scribe. A blunt tool may destroy more of the coating on both sides of the scribe than a sharp tool. This preexposure destruction of the film may cause premature failures or more undercutting. In addition, a blunt tool may also push some destroyed coating into the metal, causing uneven exposure of the bare metal.

All the previously mentioned factors involved in the preparation of specimens can influence greatly the total performance of the coatings and systems. Despite all the precise measures taken in setting up the salt fog test cabinet, any one of these factors can nullify totally the effects of the equipment. Nevertheless, every precaution should be taken to maintain the specified conditions of the testing cabinet.

The difficult part of the test, that is, the evaluation of the results, will be discussed now. There are three major failures generally looked for on a test specimen after a certain number of hours in the salt fog test. They are: 1) blistering; 2) rusting; 3) undercutting. There are also many other lesser failures to look for, such as checking and cracking, which should also be noted.

For blistering resistance, ASTM D 714 offers an evaluating method. Briefly, the measurements are based upon the size and frequency of the blisters. On a numerical scale from 10 to 0, a 10 represents no blistering and a 0 represents total failure of the film. The frequency of the blisters is expressed in terms of: dense (D); medium dense (MD); medium (M); few (F). The rating system seems to be simple and explicit, but experience shows it is not. The rating numbers have no absolute values, and the frequency terms are relative. If there is only one size of blister on a panel, rating would be rather simple. However, most of the panels inspected have blisters of different sizes, and there are various frequencies of each size. Even with

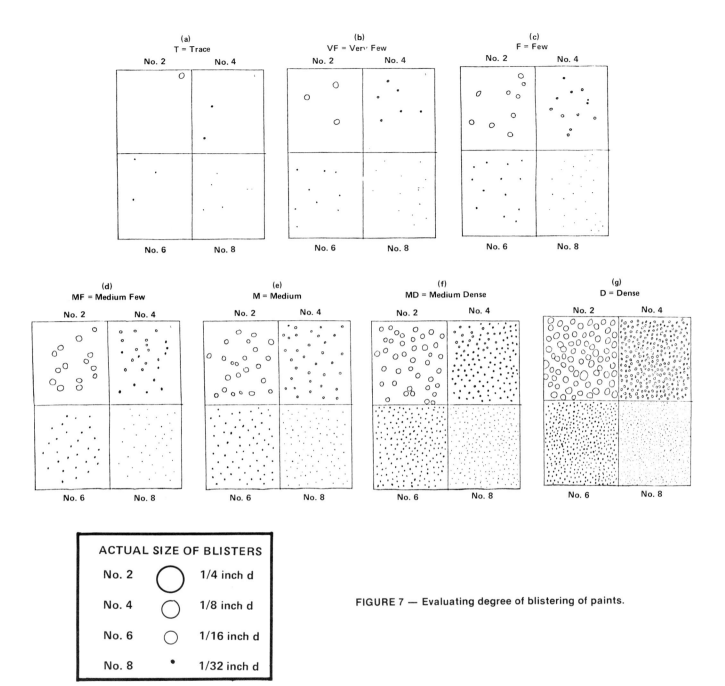

FIGURE 7 — Evaluating degree of blistering of paints.

pictorial standards, the rating is still subjective. A well trained eye can be somewhat more precise. Some companies go into great detail to separate the sizes of blisters on the same panel, and then rate each size with its frequency. This procedure ends in a list of ratings for only one panel. Granted, the accuracy is definitely improved, but it would be a tremendous job to compare 20 or 30 panels and rank them in order. For this reason, many people simply look at whatever is most obvious to their eyes, and rate a panel as 8F or 6MD as if the panel consisted of only one kind of blister.

In our laboratory, the ASTM rating guideline is modified for more accurate interpretation. Three extra steps are taken to make a more complete spectrum of the modes and degrees of blistering detected on our panels. In Figure 7, (a), (b), and (d) are the inserts. Many panels we observe belong to the classification of Figure 7 (a) and (b) and yet, according to the ASTM method, we have to call them (F) as in Figure 7 (c). There is a large gap between (F) and (M); Figure (d), as (MF), bridges them. This modified method has been used in our laboratory with excellent results. It increases the accuracy in communication between the evaluator and the report reader. Another advantage of this method lies in the fact that evaluators and readers, familiar with the ASTM method, will only have to make a small adjustment in adopting the modified method, and all evaluations in previous reports retain their same values.

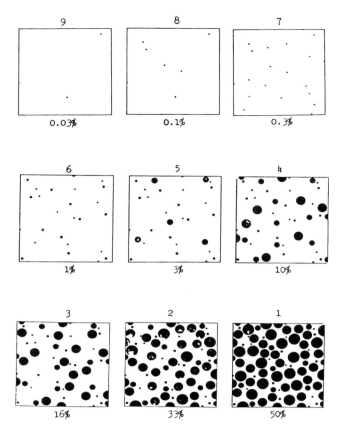

FIGURE 8 — Rating of painted steel surfaces as a function of area percent rusted.

FIGURE 9 — 340 hours in salt fog test of a 1:1 oil/alkyd primer.

For rust resistance, ASTM D 610 offers a guideline. To rate rusting, real rust must be distinguished from rust stain, which is merely the staining of the panel by real rust occurring on other portions of the panel.

According to this method, the degree of rusting is rated on a scale of 10 to 0: 10 meaning no rust at all, and 0 being total failure. In Figure 8, it can be seen that the difference between a rating of 7 and 9 is 0.27% of the total area of the panel. It is difficult for even a well trained eye to see this small difference. In an experiment, 18 laboratory technicians and scientists were asked to estimate the percentages of the nine panels shown in Figure 8; Table 1 shows the low, high, and average. Note how far off the figures are from the actual values.

Undercutting measurement is definitely more objective. However, the measurement can only be made at the end of the test because the coating at the scribe has to be destroyed. The measurement is taken by removing the affected coating, using a gentle force along the scribe until the coating is no longer removed easily with the same force, and then measuring the peeled distance from the scribe with a ruler. This method does not involve any guess work. The objectivity is obvious.

Many companies have chosen not to follow the ASTM methods of evaluation and rating. They base their systems on arbitrary point scales or summation of numerical values designated to specific failure conditions. They all work equally well; however, the subjectivity of the rating is still there. Without detailed verbal or photographic references, or interpretation by the evaluator, all the data in a report will be of little value to outside readers.

Many coatings manufacturers or consumers specify certain ratings to be acceptable and certain ratings to be unacceptable. Some say 8F (Figure 7) blistering and Number 8 (Table 1) rusting are acceptable. Should 8F blistering and Number 7 rusting be unacceptable even though the difference between Number 8 and Number 7 rusting is small (0.2% of the total area)?

Many panels are scribed diagonally on the total area of the panel. This procedure makes the ratings more difficult because blistering generally precedes undercutting. Blisters and rust may be found all over the panel because of the presence of the scribe. Is it fair to rate the panel for blistering and rusting while the panel is actually meant for undercutting?

The most controversial part of the salt fog test is discussed now. As mentioned before, 500 hours seems to be a magic figure in the coatings industries. There are many different interpretations of this magic figure. Some paint manufacturers consider one hour in the salt fog cabinet to be equal to one day outside; some say that one hour in the cabinet is equal to one week outside; and some say 500 hours is definitely good enough. Many coatings specifications call for a certain number of hours exposure in the salt fog test. Out of context, figures such as 500, 1000, 3000, etc., have no practical meaning. Many people use these figures to predict the actual performance of a coating or a system. Many people believe them. There are also many people who do not pay attention to these figures. In question is the validity of a correlation between the salt fog test and the real exposure of coatings in the field. Is it possible to have useful correlations?

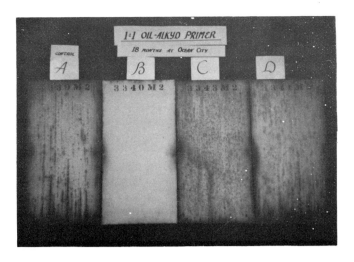

FIGURE 10 — Eighteen months exposure at Ocean City, New Jersey. 1:1 oil/alkyd primer.

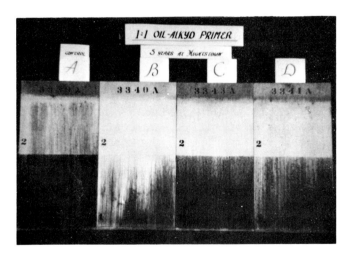

FIGURE 11 — 5 year exposure at Hightstown. 1:1 oil/alkyd primer.

FIGURE 12 — Setup for early rusting exposure.

Figure 9 shows a 340 hour exposure in the salt fog test of a 1:1 oil/alkyd primer. This test is designed to determine the performance of a nonlead and nonchromate anticorrosive pigment. Panel A is the control panel, containing no anticorrosive pigment; panel B contains the pigment to be evaluated; panels C and D contain other anticorrosive pigments. These four panels show vivid differences in their performance in this test when exposed on Bonderite panels.

Figure 10 shows the same four coating systems on hot rolled steel blasted to white metal. These panels were exposed at Ocean City, New Jersey, for 18 months. At Ocean City, the atmospheric condition is high humidity with salt spray, and the fences are about one-half mile from the water on one side and a few hundred yards from the ocean on the other side. It is a marine atmosphere. The panels were sprayed with sea water three times each day to increase severity of the exposure. After 18 months, the results are similar to those of the salt fog test, showing basically the same differences.

Figure 11 shows the same systems on sandblasted white metal in both one and two coats. The panels have been exposed in Hightstown for over 5 years. Hightstown has a rural, mild atmosphere. Again, the panels show basically the same difference.

These three sets of panels seem to show the positive correlation between the salt fog test and the exterior exposures. They indicate that System B is better than System D, System C, and the control. However, there is no relationship among the figures of 340 hours, 18 months, and 5 years. The third set of panels in Hightstown could remain for another 5 years, still showing the same differences among the panels.

Much of our current work is with latex anticorrosive coatings. There is a condition, peculiar to a latex system, called early rusting (described by Rohm & Haas a few years back). It is interesting to note that early rusting has never been observed in any of our salt fog testing because of the conditions under which the phenomenon occurs. These conditions are: 1) high humidity, above 75% RH; 2) low temperature, below 50 F; 3) cold substrate, below 50 F (ambient); 4) cold paint, below 50 F (ambient); and 5) film thickness, below 1.5 mils and over 3.5 mils in one coat.

Unlike flash rusting, which forms rapidly, early rusting can form any time after application under these conditions. Also unlike flash rusting, which is a cosmetic effect, early rusting is true rust which starts from the substrate beneath the film and continues to grow once it has formed. The damage to the coating system then becomes permanent. Recoating will only delay the failure from surfacing. From our study, it was found that for early rust to form, all the conditions mentioned need not be present at the same time. Humidity is the most important factor because it controls, to a great extent, the speed of the formation of the film. Details of this study would be time consuming, and are the subject of another paper.

FIGURE 13 — 750 hours in salt fog test. Latex maintenance primer.

FIGURE 14 — Early rusting test. Latex maintenance primer. (a) accelerated, no drying; (b) Hightstown, 4 days drying; (c) Hightstown, no drying.

There is an accelerated test for early rusting. It simulates closely the conditions just mentioned, and is found to be effective in producing early rust. A special exterior setup has been designed for early rusting exposure at Ocean City (Figure 12). The hot

FIGURE 15 — Panels exposed at Ocean City, New Jersey.

rolled steel angles are suspended on racks so that they simulate the girders under a bridge, only a few feet above the sea water. This site provides high humidity, condensation, low temperature, and salt contamination. Coatings are applied at the site on two consecutive days. The panels are not given the two or three weeks drying time under laboratory conditions before exposures. The panels are never exposed to direct sunlight. These conditions are severe yet realistic, and we believe any latex coating system that can pass this test will provide excellent performance under milder conditions.

The Hightstown exposure site provides very high humidity at certain times during the year (condensation from dew is excessive). Facilities have been set up for early rusting tests. The coatings are applied at the site in the late afternoon and are exposed immediately.

The following are actual cases. Figure 13 shows a 750 hour exposure of two panels of a latex maintenance primer on cold rolled steel Q-panels at 4 mils dry film thickness. Panel A has no anticorrosive pigment, while panel B contains the nonlead and nonchromate anticorrosive pigment being evaluated. The difference is obvious. In Figure 14, panel A is the result of the laboratory accelerated early rusting test; panel B is the result of Hightstown exposure overnight after 4 days drying time under laboratory ambient conditions before exposure; and panel C is the overnight exposure at Hightstown. This panel was not given laboratory drying time, but was exposed directly after coating application in the late afternoon. Early rust formation was evident in less than 24 hours.

Figure 15 shows panels of the same coating after being exposed for one month at Ocean City for early rusting. Both angle and flat panels in panel A were prepared in the laboratory and allowed 14 days for drying. Panels B were painted at the site to the same film thickness in two consecutive days. The angles were immediately exposed under the dock. The flat panels were exposed on a 45° fence and sprayed 3 times a day with sea water. The formation of early rusting occurred as early as 7 days on the angles.

Comparing the results of these different tests, no valid correlation can be found between the salt fog test and the actual exposures, particularly on those painted at the site, which is the condition found in the real world. And yet, one of the current national specifications for latex maintenance coatings calls for 300 hours in the salt fog test for a coating to be acceptable. This is not the only such case; in the specifications for coatings, there are many similar to it. Again, it is evident that the number of hours has no bearing on the actual length of performance of a coating or system.

The salt fog test has been used in industry for over 40 years, and it definitely has value. It is not intended to discredit it; rather, I intend to point out the proper use of the test. It is a useful accelerated test for comparing one coating with another. It offers limited value for the prediction of the actual performance of a coating. Predictions are possible only when there is a valid correlation established between the test and the real exposure. Even then, the predictions should be considered somewhat more reliable than an educated guess, especially if the varied geographic locations and climates the coatings will be used in are considered. It could be misleading to attach any absolute values to the results of this test. This is to say that when guidelines are established for a coating or a system, either for the coatings manufacturers or coatings consumers, an arbitrary figure should not be set without evidence to substantiate it.

Coatings formulation is a matter of great complexity. A skillful formulator can design a coating to pass the salt fog test or any other tests, but he should not overemphasize the significance of his success. The most reliable information still comes from actual exposure under realistic conditions for the application.

Conclusion

The following points are made regarding proper usage of the salt fog test.

Standardize testing equipment. It is doubtful if anyone in industry follows all the ASTM guidelines to the letter. However, I strongly suggest standardizing your equipment to meet the guidelines. The fewer the variables, the more accurate and more reproducible the results will be.

Select the most suitable substrate and surface preparation. Many different substrates can be used for the test. It all depends on what is looked for. I suggest you choose the substrate and surface preparation as close to reality as possible, without sacrificing the uniformity and reproducibility of the results.

Uniform application. Control the film thickness of the coatings to be tested.

Set up controls. Since the test is most useful for relative data, it is absolutely necessary to set up a standard control for comparison in every run of the test.

Minimize variables. When a comparative project is set up, limit it to one variable, if possible. The more variables, the more difficult it is to evaluate performance; therefore, the results are less conclusive.

Use duplicates. Always make duplicate panels for the same coating or system in the same test to ensure reproducibility of the results.

Determine individual drying time according to generic type. Select the most realistic drying time for the panels before exposure. Each generic type of coating requires a specific drying period for obtaining the integrity of the film.

Maintain continuity of test. The cabinet should be kept closed as long as possible. The more frequently the cabinet is opened, the more variables and interruptions are introduced to the testing conditions.

Set up exterior exposure. Try to establish a possible correlation to the field. Experienced consumers of coatings prefer to see exterior results.

Is the salt fog test an effective method to evaluate corrosion resistant coatings? I have given you my answer. What is yours?

Comparative Investigations of Corrosion Performance of Coating Systems for Automobiles by Different Methods of Accelerated Weathering

*Wolfgang Goering, Emma Koesters, Rolf Muenster**

Introduction

For a systematic development of new and more effective coating materials for cars, it is most important to have results of corrosion performance corresponding to experience in a short time. Rapid laboratory tests are indispensable, because the formulator in a paint factory cannot wait for several years to get the results of natural weathering, and to recognize the influence of variations of paint formulations upon corrosion performance. One purpose of these investigations is to get more knowledge about accelerated weathering tests and electrochemical methods. How well do results correlate between the investigated accelerated test methods and outdoor exposures? How well can results of water absorption of painted steel panels explain the results of corrosion resistance?

Experimental

This program included seven different primers (both conventional primers and electrocoats), and nine paint systems consisting of two, three, or four coatings. All steel panels for these investigations were taken from the same coil. The front face of the phosphatized steel panels was always used. We proceeded in the same way with the degreased steel panels. One important point is the pretreatment of the cold rolled steel panels. Three surface treatments were used: 1) iron phosphate coating, with chromic acid rinse treatment; 2) zinc phosphate coating, with water rinse treatment. These phosphate coatings were applied by spraying and baking; 3) degreased steel panel.

The recommended film thickness, baking time, and baking temperature characteristic for the paint material were used in each case. Altogether, 1400 painted steel panels were studied, using eight methods.

Methods A, B, C. Salt spray (fog) testing, German Standard SS DIN 50021 and DIN 53167, periods of test 96, 168, and 240 hours. This test method is similar to the American Standard Method of Salt Spray Testing (ASTM-Designation B 117-73).

Method D. This method provides for the determination of the effect of cyclic environments on coated specimens. This cyclic test consists of three parts.
 Part 1) 24 hours salt spray testing (35 C).
 Part 2) 4 × 24 hours climate KFW (40 C), DIN 50017.
  In this climate (KFW), the specimens are tested in each case for 8 hours at 100% relative humidity (40 C), and for 16 hours at a relative humidity less than 75% (23 C).
 Part 3) 2 × 24 hours standard climate (50% relative humidity, 23 C).

Method E. This test method is similar to Method D, but the employed temperatures are lower. This cyclic method consists of three parts:
 Part 1) 24 hours salt spray testing (23 C).
 Part 2) 4 × 24 hours climate KFW (23 C), DIN 50017.
  In this climate (KFW), the specimens are tested in each case for 8 hours at 100% relative humidity (23 C), and for 16 hours at a relative humidity less than 75% (23 C).
 Part 3) 2 × 24 hours standard climate (50% relative humidity, 23 C).

Method F. Electrochemical test methods.

Method G. Natural weathering with salt spray without the influence of light (for primed steel panels).

To simulate the conditions in hollow spaces or cavities of car bodies, the primed steel panels were put in special boxes, so that rain and sun could not directly influence the specimens, but air and humidity had access to the panels. There were several holes in the walls of the boxes, and the bottom was open.

Method H. Natural weathering with salt spray (for paint systems).

Results and Discussions

Natural Weathering with Salt Spray (Paint Systems)

Before testing, the painted steel panels were scribed vertically near the center, holding the scribing tool perpendicular to the surface. The scribe penetrated all coatings on the metal. The average distance between the scribe mark and the edge of the

---

*Federal Republic of Germany

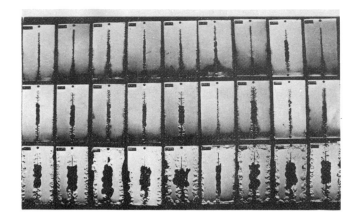

FIGURE 1 — Paint systems after 6 months of natural weathering with salt spray. Upper row: iron phosphate with chromic acid rinse treatment. Middle row: zinc phosphate with water rinse treatment. Lower row: degreased steel panel.

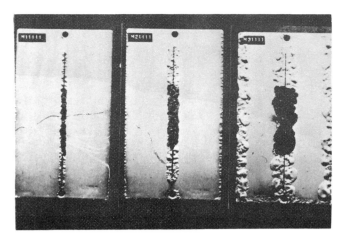

FIGURE 2 — Four coat system after 6 months of natural weathering with salt spray. Left: iron phosphate with chromic acid rinse treatment. Middle: zinc phosphate with water rinse treatment. Right: degreased steel panel.

TABLE 1 — Extent of Corrosion from the Scribe Line (W_d/mm) and the Rusting at the Edges (K_R) after 4 (6) months of Natural Weathering with Salt Spray Dependent on the Pretreatment of the Steel Surfaces (1 = Iron Phosphate Coating with Chromic Acid Rinse Treatment, 2 = Zinc Phosphate Coating with Water Rinse Treatment, 3 = Degreased Steel Panel)

	1	2	3
W_d/mm	1,8 (2,9)	4,2 (5,7)	12,5 (14,8)
K_R	1,5 (2,6)	2,6 (2,7)	4,9 (5,0)

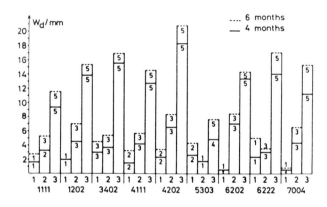

FIGURE 3 — Nine paint systems after 4 and 6 months of natural weathering with salt spray. Vertical axis: extent of corrosion from the scribe line (W_d/mm). The numbers at the upper end of the columns are the ratings for the rusting at the edges.

unaffected area of the finish was measured to determine the creepage of the under film corrosion.

The scribed painted steel panels were weathered for 6 months. The position of the painted specimens was 5° from the horizontal, facing south. The place of the weathering was near the city of Muenster in Western Germany. Once a week, the painted panels were sprayed with a 3% sodium chloride solution.

Figure 1 shows the nine paint systems in three horizontal rows. These rows differ in the pretreatment of the steel panels: upper row, iron phosphate with chromic acid rinse treatment; middle row, zinc phosphate with water rinse treatment; lower row, degreased steel panel. One can see that the extent of corrosion from the scribe line, and the rusting at the edges, are strongly dependent upon the pretreatment of the steel panels before painting. The mean values of the rust creepage from the scratch (W_d in mm), and the mean ratings for the rusting at the edges (K_R), are listed in Table 1. Here, zero indicates perfection, or absence of failure, while five represents complete failure. Figure 2 shows the dependence of the rust creepage from the scratch, and of rusting at the edges upon the pretreatment of the steel surface for a four coat system. The ratings for rusting at the edges in Figure 2, in this case, are the numbers 1, 3, and 5 (from left to right).

Figure 3 shows, on the vertical axis, the extent of corrosion from the scribe line (W_d/mm) for nine organic coating systems after 4 and 6 months of natural weathering with salt spray. The numbers at the upper end of the columns are the ratings for the rusting at the edges. The numbers 1111, 1202, 3402, etc., at the horizontal axis characterize the paint system: the 1st digit designates the primer; the 2nd digit designates the filler; the 3rd digit designates the sealer; the 4th digit designates the top coat. 4111 means that the paint system consists of primer 4, filler 1, sealer 1 and top coat 1. 1202 means, for instance, that we have a three coat system; 7004 means that we have a two coat system.

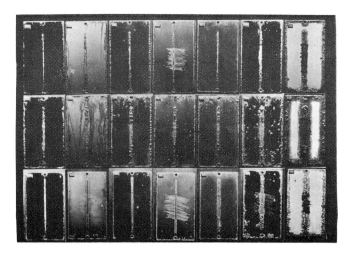

FIGURE 4 — Seven primer coats on zinc phosphatized steel panels. Upper row: cyclic test method D (6 weeks). Middle row: cyclic test method D (10 weeks). Lower row: test method G (6 months of natural weathering with salt spray without influence of light).

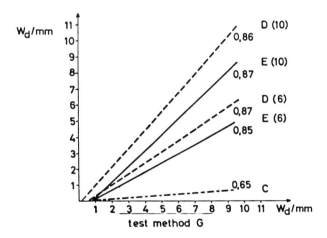

FIGURE 5 — The vertical axis shows the under film corrosion resulting from the accelerated test methods; on the horizontal axis the same values resulting from the outdoor exposure with salt spray. Next to the straight lines resulting from the analysis of regression are noted the coefficients of correlation. In brackets is the period of test in weeks.

As Figure 3 shows, the extent of corrosion at the scribe line, and the rating for the rusting at the edges, are strongly dependent upon the pretreatment of the steel panels (1 = iron phosphate with chromic acid rinse treatment; 2 = zinc phosphate with water rinse treatment; 3 = degreased steel panel). The rust creepage is small on the iron phosphatized steel panels, it is larger on the zinc phosphatized steel panels, and is most disadvantageous on the steel panels without conversion coating. The good results with the iron phosphate coating are most probably due to the chromic acid rinse treatment.

Organic Coatings Exposed Cyclically to a Series of Accelerated Tests (Methods D and E)

Some years ago, German factories producing automobiles and paints began experiments testing dynamically coated steel specimens for resistance to corrosion.[1-6] One result of this work was the development of method D, consisting of three parts, as described. The period of a cycle is 7 days. The organic systems should be exposed for 4 or 6 cycles; that means 4 or 6 weeks (or a longer time, as agreed upon between the purchaser and the seller).

Figure 4 shows, in the lower row, the results of 6 months of natural weathering in Western Germany with salt spray without the influence of light. In the upper row, the results of the cyclic accelerated test method D (exposure 6 weeks); and in the middle row, the results of the cyclic accelerated test method D, (exposure 10 weeks). From left to right, the arrangement of the seven primer coats was made so that underfilm corrosion is increasing (from 1.7 mm up to 9.2 mm, after 6 months of natural weathering with salt spray without influence of light). The pictures of corrosion in these three rows are similar, and are comparable to those seen on car bodies in the field.

Figure 5 shows the results of an analysis of regression. On the vertical axis, the under film corrosion (in mm), resulting from the accelerated test methods is noted. On the horizontal axis, the same values, resulting from the outdoor exposure with salt spray, are noted. All values were calculated on the basis of the results of the already mentioned seven primer coats on zinc phosphatized steel panels. In doing so, it was assumed that there is a linear dependence between these values.

One can see from Figure 5 that the rust creepage increases with temperature (method D gives a higher degree of corrosion than method E), and with period of test. The coefficient of the correlation has nearly the same value for methods D and E. An accelerated corrosion test method should give results in a short time, correlating well with outdoor exposure. Our experiences on the basis of the investigated primer coats show that 6 weeks by the cyclic accelerated test method D are as severe as 18 weeks by natural weathering in Western Germany with salt spray, and are also as severe as 36 weeks by natural weathering in Western Germany without salt spray.[4,6] This cyclic accelerated corrosion test method D has been standardized in Germany by the association of German manufacturers of automobiles (VDA standard test method 621-412).

Electrochemical Values

In the literature, there are numerous publications containing and discussing results of electrochemical and electrical measurements on painted panels and paint films without substrate. A good survey was given by H. Leidheiser[7] in 1979.

R.R. Wiggle, A.G. Smith, and J.V. Petrocelli[8] used electrochemical techniques to separate the corrosion reactions, and to study their separate and combined effects at the scribe line in paint films. They concluded that, for optimum corrosion protection, the conversion coating should retard anodic undercutting, and both the conversion coating and paint systems should be resistant to hydroxide produced by the cathodic reaction.

It was the aim of our experiments to see whether the measured electrochemical values can give us information about the corrosion resistance of painted steel panels. In our first experiments, we used the potentiostatic method, polarizing anodically the painted steel panels. In later experiments, we also polarized cathodically, applying the galvanostatic method.

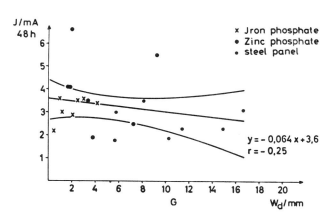

FIGURE 7 — Seven primer coats (different pretreatment of steel surface). Vertical axis: corrosion current in m amp. Horizontal axis: rust creepage (6 months natural weathering with salt spray), r = coefficient of correlation.

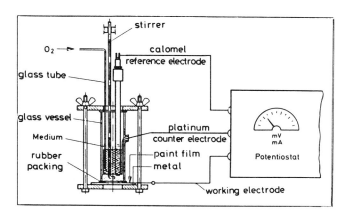

FIGURE 6 — Apparatus for electrochemical measurements.

Experimental Results of Anodic Polarization of Painted Steel Panels

Figure 6 shows the apparatus for electrochemical measurements. In these experiments, all painted steel panels had been scribed before testing. The length of the scribe line amounted to 30 mm. During the experiments at 23 C, the painted steel panels were surrounded by a 5% sodium chloride solution (pH = 7), contained in the glass vessel. The painted panel served as anode, the counter electrode consisted of platinum. For measurement of potential, a saturated calomel reference electrode was used. Starting from the open circuit potential we polarized anodically to the amount of 100 mv. During the period of testing (48 hours), the solution was stirred and aerated. The corrosion current was measured.

To gain electrochemical values for automotive organic coating systems, it is necessary to scribe the painted steel panels. When these are not scribed, and do not have any failure (for example, macroscopic pores or mechanically damaged sites), it is not possible to do such measurements because the current is too small. Painted unscribed panels give current values which indicate the number and size of the macroscopic pores and mechanically damaged sites,

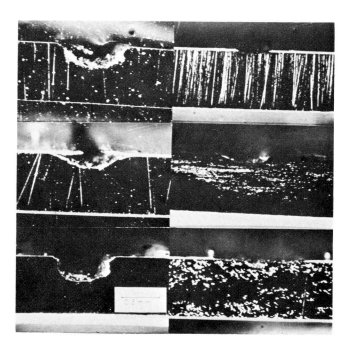

FIGURE 8 — Scribe line (cross section) Left: after the electrochemical test (48 hours). Right: after salt spray fog testing.

but do not say anything about the corrosion performance of the paint material itself.

Figure 7 shows, on the vertical axis, the corrosion current in m Amp measured after a test period of 48 hours using seven primer coats with varied pretreatment of the steel surface. On the horizontal axis, the rust creepage is recorded, after 6 months of natural weathering with salt spray (method G). It shows no correlation between the measured values of corrosion current and creepage. In Figure 7, the straight line of regression, as well as the limits of confidence are recorded. The picture is not better when recorded on the horizontal axis, the creepage resulting from salt spray fog testing (168 hours). On the vertical axis, the curves for corrosion current and the quantity of elec-

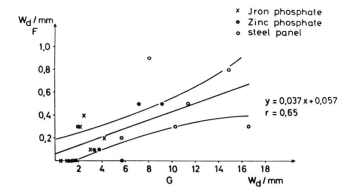

FIGURE 9 — Seven primer coats (different pretreatment of steel surface). Vertical axis: creepage after electrochemical test F. Horizontal axis: creepage after 6 months of natural weathering with salt spray.

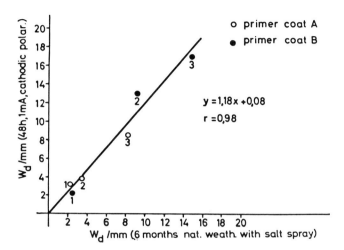

FIGURE 10 — The vertical axis plots the creepage resulting from the electrochemical test; the horizontal axis the creepage resulting from natural weathering with salt spray. 1 = iron phosphate with chromic acid rinse treatment; 2 = zinc phosphate with water rinse treatment; 3 = degreased steel panel.

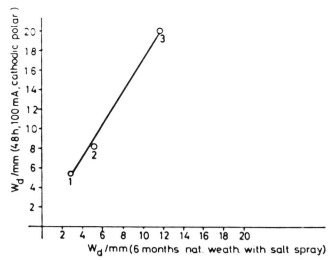

FIGURE 11 — Four Coat System (1111). Vertical axis: creepage resulting from the electrochemical test. Horizontal axis: creepage resulting from natural weathering with salt spray. Pretreatment 1,2,3, Figure 10.

creepage after 48 hours of electrochemical testing. This is an estimated value for the true, yet unknown, value of electrochemical creepage, which we can expect in the area of confidence (95% probability) between 0.29 and 0.55 mm. The resulting values from the electrochemical test (between 0 and 0.9 mm) are small; because of this, it is not possible to use this test method in the outlined way.

Experimental Results of Cathodic Polarization of Painted Steel Panels

Cathodic polarization data are given in Figure 10. On the vertical axis, the creepage resulting from the electrochemical test after 48 hours is plotted. On the horizontal axis, the rust creepage after 6 months of natural weathering with salt spray, without the influence of light, is plotted. The measured values of the primer coats A and B are seen; the pretreatment of the steel surfaces was varied in three ways, as described earlier. The strong dependence of the creepage, after 6 months of natural weathering with salt spray, both on the pretreatment of the steel panels and on the nature of the primer coats, is apparent. As Figure 11 shows, good correlation is also obtained with a 4 coat system (1111). Figure 12 gives the results for all seven primer coats with different pretreatments of the steel surface. The correlation between the values is satisfactory, in general, but there are also two high values of creepage (resulting from the electrochemical test) which do not lie on the curve.

The Water Absorption of the Primed Steel Panels

There are many publications studying the influence of water on the properties of free paint films and painted panels. Mentioned here are only the German

tricity (in coulombs) are similar, because the current was nearly constant during 48 hours. It is not possible to evaluate the corrosion resistance of painted steel panels by measuring values of corrosion current during anodic polarization.

Why is there no correlation? Figure 8 shows three examples: in each case, on the left side the cross section of the scribe line, after electrochemical testing; on the right side, after salt spray fog testing. One can see, distinctly, the anodic dissolution of the steel during electrochemical testing.

Figure 9 shows a plot of the creepage resulting from the electrochemical test (48 hours) as a function of the creepage after 6 months of natural weathering with salt spray. Beside the straight line of regression, the limits of confidence are recorded. For instance, the creepage of 10 mm after 6 months of natural weathering with salt spray corresponds to 0.42 mm

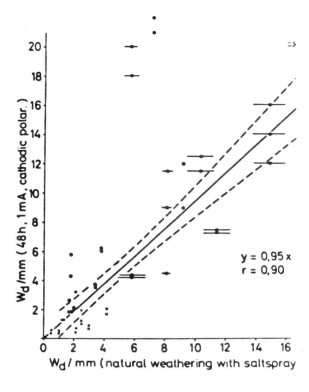

FIGURE 12 — Seven primer coats. Vertical axis: creepage resulting from the electrochemical test. Horizontal axis: creepage resulting from six months of natural weathering with salt spray without influence of light.

X   Iron phosphate with acid rinse treatment.
●   Zinc phosphate with water rinse treatment.
○   Degreased steel panel.

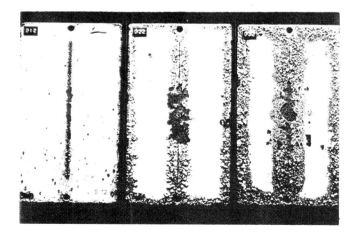

FIGURE 13 — Primer coat 2 after 6 months of natural weathering with salt spray (without influence of light).

publications of W. Funke,[9,10] who has studied, among other things, the absorption of water in the coating/metal interface, and the publications of P. Kresse,[11] which deal with the connection between corrosion and diffusion of water.

How far can values of water absorption of primed steel panels, pretreated in three different ways, help to

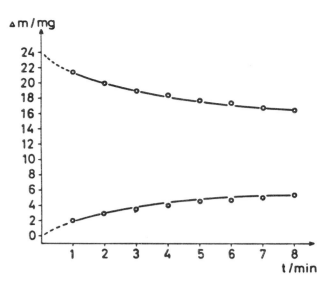

FIGURE 14 — Water absorption Δm of the primed steel panel (degreased steel panel, primer coat 1).

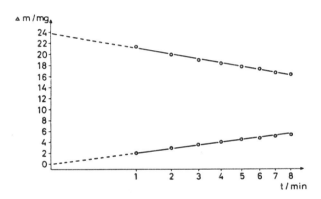

FIGURE 15 — Figure 14 ($\sqrt{t}$-scale).

interpret the results of corrosion performance? The strong influence of the pretreatment of the steel panel on the degree of corrosion is shown in Figure 13 for primer coat 2. From left to right: iron phosphate with chromic acid rinse treatment; zinc phosphate with water rinse treatment; degreased steel panel.

The water absorption of the primed steel panels was determined by weighing after the panels had been stored for 72 hours in distilled water at 23 C. Figure 14 shows, in the upper part, the weight in mg, measured after taking the painted panels out of the water and carefully drying them with fleece. Not until one minute after taking the panels out of the water could the first value be measured, due to the drying procedure. The weight of the painted panel is decreasing, because the water migrates out of the paint film and evaporates in the compartment of the analytical balance. The lower weight time curve was measured after storage of the primed steel panels for 72 hours in a closed vessel at 20% relative humidity. The curve is rising because the paint film takes up water within the

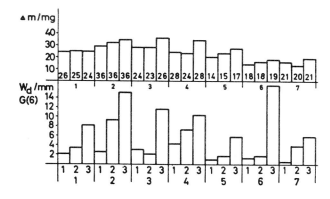

FIGURE 16 — Upper part: water uptake (△m in mg) of the primed panel. Lower part: rust creepage (W_d/mm) after 6 months of natural weathering with salt spray (without influence of light).

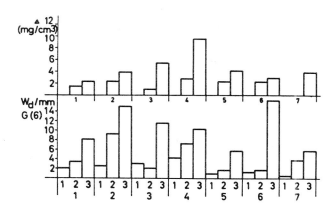

FIGURE 18 — Upper part: difference of water uptake in mg/cm³ for the primed zinc phosphatized and degreased steel panels with respect to the water uptake of the iron-phosphatized steel panels in each case. Lower part: Figure 16.

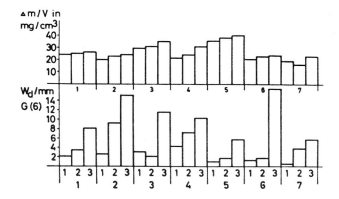

FIGURE 17 — Upper part: water uptake in mg/cm³ with respect to the volume of 1 cm³ of the coatings. Lower part: Figure 16.

compartment of the balance during weighing. To have a better survey, the absolute values of weight were not plotted on the vertical axis, but the difference values were plotted in mg. Figure 15 shows these two curves, whereby we chose on the horizontal axis the $\sqrt{t}$ scale. In this way, we got straight lines. Extrapolating them, we determined the amount of water absorption of the primed steel panels at our conditions of storage. For instance, we obtained the value 23.8 mg for panel 31. Figure 16 shows, in the upper part, the amount of water uptake △m in mg of the primed steel panels, the numbers 1 to 7 characterizing the primers. For each primed panel, the values of water uptake are plotted for the different pretreated steel panels. In the columns the mean film thickness of the coating is noted. Figure 17 shows, in the upper part, the amount of water uptake in mg/cm³, with respect to the volume 1 cm³ of the coatings. In other words, the measured values were related to the same thickness. Depending on the type of primer, the thickness was from 15 to 36 μm. In the lower part of this figure, the values of rust creepage, in mm, after 6 months of natural weathering with salt spray for the primers 1 to 7 and three kinds of pretreatment of the steel panels (1 = iron phosphate with chromic acid rinse treatment; 2 = zinc phosphate with water rinse treatment; 3 = degreased steel panel) are recorded.

If we look at these two diagrams, we notice a similar structure. The degree of corrosion, as well as the water uptake, increase if we proceed from pretreatment 1 to pretreatment 3. Figure 18 shows, in the upper part, the difference of water uptake in mg/cm³ for the primed zinc phosphatized and degreased steel panels with respect to the water uptake of the iron phosphatized steel panels in each case.

In which way can we explain the results noted in the last diagrams? For that purpose we consider the system "steel surface/pretreatment/primer coating," proceeding with the following conception: one part of the water is taken up by the primer coating itself; the other part is absorbed in the primer/steel interface or primer/phosphate coating interface. We assume that the primer coatings have nearly the same structure, especially with regard to the water absorption, independent of the kind of pretreatment of the steel surface.

We interpret in the following way. The different water uptake of the primed steel panels means that different quantities of water are absorbed in the primer/steel interface, or primer/phosphate coating interface, respectively. The quantity of absorbed water in the primer/phosphate coating interface is small in the case of iron phosphate, with chromic acid rinse treatment. The quantity is higher in the case of zinc phosphate with water rinse treatment, and is highest in the case of degreased steel panels. The values of water uptake in Figure 18 are minimum quantities in mg, which are absorbed in the interfaces. The quantity of water taken up by the primed iron phosphatized steel panels is also subdivided in two parts. One part is absorbed by the organic coating, the other in the interface primer/phosphate coating.

Therefore, it seems that there are relations between the pretreatment of the steel surface, the quantity of water absorption in the primer/phosphate coating interface, or primer/steel interface, respectively, and the degree of corrosion.

References

1. E. Kösters, H. Noack, Lecture on Corrosion Problems in the Field of Organic Coatings for Car Bodies in BASF AG, Ludwigshafen, FRG (26.11.1970).
2. D. Saatweber, DECHEMA-Symposium Korrosionsforschung für die Praxis, Braunlage, 28.03. - 01.04.1977, Tagungshandbuch, S. 289-290.
3. D. Saatweber, 2. DECHEMA-Symposium Korrosionsforschung fur die Praxis, Willingen, 21.04.-25.04.1980, Tagungshandbuch, S. 371-372.
4. W. Göring, E. Kösters, H. Noack, XIV. FATIPEC-Kongress, Budapest (1978), Kongressbuch, S. 235-240.
5. W. Göring, E. Kösters, R. Muenster, *Farbe + Lack* **85,** p. 1014 (1979).
6. VDA-Prüfblatt 621-412.
7. H. Leidheiser, *Progress in Organic Coatings,* **7,** p. 79 (1979).
8. R.R. Wiggle, A.G. Smith, J.V. Petrocelli, *J. Paint Technol.* **40,** p. 174 (1968).
9. W. Funke, H. Zatloukal, *Farbe + Lack,* **84,** p. 584 (1978).
10. W. Funke, *Farbe + Lack,* **84,** p. 865 (1978).
11. P. Kresse, *Defazet,* **26,** p. 471 (1972).

Corrosion Inhibitor Test Method

*G.A. Salensky**

Introduction

The protection of metals with coatings is a complex phenomena involving the restricted diffusion of water and oxygen to the substrate, and the influence of the environment on surface passivity.

Thick coatings can reduce diffusion significantly, but Mayne[1] has pointed out that even the best systems do not completely prevent the permeation of both water and oxygen to the metal surface. Kumins[2] has shown that both sodium and chloride ions diffuse through coatings, providing that the concentration of salt is at least 0.4 molar. Therefore, the cathodic reaction cannot be suppressed ($1/2\ O_2 + H_2O + 2e^- \rightarrow 2OH^-$). In the absence of corrosion inhibitive pigment, the corresponding anodic reaction will also take place ($Fe \rightarrow Fe^{++} + 2e$).

Chromium and lead based pigments have been used effectively in corrosion resistant coatings. Their toxicity, however, has resulted in their elimination from many paint applications. Although some relatively safe corrosion inhibitive pigments are available, there is emphasis on the use of organic corrosion inhibitors for improving the performance of coatings. The evaluation of these organic compounds, however, can be frustrating. Several established methods are listed in Table 1.

TABLE 1 — Corrosion Test Methods

Salt fog
Total immersion
  Static
  Aerated
Electrochemical
  Potential vs time
  Var. potential vs current
  Capacitance
  Resistivity
Surface effects
  Adsorption isotherms
  Gravimetric adsorption
  SEM Energy dispersion X-ray analysis
  IR — ATR
  Radio-tracer measurements
  ESCA
  Ellipsometry

---

*Union Carbide Corporation, Bound Brook, New Jersey

Salt fog has been around for many years. In spite of the fact that it does not always correlate with actual field performance, it is still widely used as a standard of comparison, particularly in the automotive industry where exposure of cars to deicing salts contributes significantly to their corrosion. We have attempted to correlate corrosion inhibitor efficiency with chemical structure, and found that salt fog could not discriminate between many compounds and that reproducibility with identical systems was poor. Total immersion tests were found to be less aggressive than salt fog, but the tests still suffered from poor selectivity and reproducibility.

Electrochemical tests were recently reviewed by Leidheiser,[3] who concluded that the corrosion rate increases as the coating resistance decreases. This method is not applicable to conductive coatings containing metallic pigments. The utility of the other procedures, such as polarization measurements, were less rewarding. Similar work at Union Carbide confirms these observations.

The surface effect measurements shown in Table 1 have been widely used by many investigators. They have proven to be valuable in specific areas, but are costly from the standpoint of instrumentation and time.

Recognizing the fact that the corrosion of coated steel, particularly in the presence of an aerated salt atmosphere, is complex, it was decided to separate the problem into parts and to study each one separately.

During active corrosion, an important process is the rate of adsorption or chemisorption of the corrosion inhibitor on the steel, and its ability to withstand desorption in the presence of the corrosive salt solution. The kinetics of the chemisorption process in a nonaqueous environment during the coating application is another important consideration, as is the transport of the corrosion inhibitor through the coating to the substrate and/or defect area.

This study was limited to a determination of the effect of various chemicals during the active corrosion phase. The corrosion process was simulated by using an organic liquid as a model compound for the coating. It is recognized that an inhibitor can more readily orient itself on a substrate from a liquid than from a solid state coating. The liquid, however, being an ideal diffusional media, should permit reproducible measurements of chemical structure/inhibitor performance relationships.

Although early work was based on diisodecylphthalate as a model compound to simulate an alkyd resin, later effort concentrated on phenyl ether (diphenyl oxide), which is not subject to hydrolysis under alkaline conditions. Phenyl ether also approximates the structure of Phenoxy® resin, a high molecular weight thermoplastic derived from bisphenol A and epichlorohydrin. This resin is used extensively in the formulation of corrosion resistant zinc rich primers. Phenyl ether, unlike Phenoxy®, does not contain a hydroxyl group, allowing an assessment of the effect of hydroxyl containing materials on corrosion rate.

Experimental

A schematic of the testing procedure is given in Figure 1. Five milliliters of the model compound, with or without the inhibitors, is placed in a test tube. A weighed steel specimen is inserted, and the tube is tipped so that the entire steel surface is wetted. Then one milliliter of a 3% sodium chloride solution is added. The test tube is stoppered with a condenser tube, then placed into a shaker water bath maintained at 40 C. A shaking rate of 250 cycles/minute with a 1 inch amplitude provides violent fluidization and aeration of the liquid mixture so that the steel panel is completely wetted at all times. The movement of the condenser tube provides a syphon pumping effect so that fresh air replaces the oxygen consumed during the wet oxidation corrosion process. Calculations show that the head space in the tall test tube is not adequate to provide all of the oxygen consumed by the corroding steel specimen in 24 hours (when corrosion inhibitors are not present in the model liquid compound).

Although the shaking period can be prolonged when slower corroding materials such as aluminum are used, it was found that steel specimens required only 24 hours for significant results. At the end of the 24 hour period, the extent of the corrosion is observed visually, and the sample is cathodically stripped of the corrosion product so that it can be weighed and metal loss determined. Samples of the inhibited model compound are generally run in triplicate. The inhibitor-free model compound is tested in a group of 10 specimens to establish a corrosion rate for the system, which is then used to calculate the degree of corrosion and inhibitor efficiency, as shown in Table 2. Ten steel samples provide corrosion rate results which proved reproducible within 3% of the mean.

TABLE 2 — Corrosion Inhibitor Performance

Degree of corrosion (DC) = $\dfrac{\text{\% Wt. loss with inhibitor}}{\text{\% Wt. loss w/o inhibitor}}$

Degree of protection (Corrosion inhib. eff.) = $1.0 - DC$

Inhibitor efficiency	
1.0	complete protection
0.9	good protection
0.0	same as with no inhibitor
−0.2	accelerates corrosion

Materials

A heavy duty water bath shaker manufactured by New Brunswick Scientific Co., Inc., Model RW-650, was used to handle as many as 20 systems or 60 tubes at one time. Steel specimens were cut from unground SAE 1010 panels purchased from Q Panel Co., cleaned with methyl ethyl ketone, dried at 110 C for 30 minutes, cooled in a desiccator, and weighed. Test tubes are of the culture rimless type available from Kimble Glass Co.

Diphenyl Oxide, perfume grade, was purchased from Dow Chemical Company; Ashland Chemical Company supplied plasticizer grade diisodecyl phthalate.

Cathodic cleaning was accomplished in 5% sulfuric acid containing 0.1% methyl quinoline using 4 amperes for 4 minutes with a graphite anode. Prior to cathodic cleaning, the steel specimens were dipped in methyl ethyl ketone to remove the model liquid compound, then flushed in distilled water. The samples were dipped in distilled water after the elecrolytic bath, then immersed in an ultrasonic bath containing 3 g/l alkaline detergent for 5 minutes to

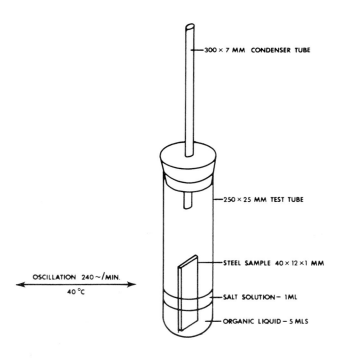

FIGURE 1 — Aerated salt corrosion test.

® Trademark Union Carbide Corporation.

TABLE 3 — Corrosion Inhibitor Efficiency of Carboxylic Acids Dissolved in DIDP/3% Salt Water

	Concentration/100 grams		
	5 mM	0.5 mM	0.05 mM
Oleic acid	1.00	.96	.93
Undecylenic acid	.99	.93	.31
Octadecenylsuccinic acid	.98	—	.99
Dodecenylsuccinic acid	.98	.39	.38
Octadecene-MA copolymer	.96	.71	.24
Di-tri-oleic acid	.99	.83	.74
Benzoic acid	.96	.32	−.04
Cyclohexane carboxylic acid	.96	.58	.01

TABLE 4 — Corrosion Inhibitor Efficiency of Amines Dissolved in Phenyl Ether/3% Salt Water

	Concentration/100 grams 5 mM
Dodecyl amine	0.18
Didodecyl amine	−0.02
N, N dimethyldodecyl amine	0.04
Octadecyl amine	0.72
Oleyl amine	0.52
N-oleyl 1, 3 diaminopropane	0.99

TABLE 5 — Corrosion Inhibitor Efficiency Effect of Functional Groups Dissolved in Phenyl Ether/3% Salt Water

	Concentration/100 grams 5 mM
Octadecyl amine	0.72
Oleyl amine	0.52
Oleyl alcohol	0.18
Oleyl nitrile	0.07
Dodecyl nitrile	0.07
Dodecyl mercaptan	0.10

TABLE 6 — Corrosion Inhibitor Study on Steel: The Effect of Hydrophobic Segment Size of Gallic Acid When Dissolved in Phenyl Ether/3% Salt Water

	Inhibitor Efficiency mM%		
	5.0	0.5	0.05
Gallic — acid	0.54		
Propyl ester	0.61		
Dodecyl ester	0.99	0.98	0.68

remove any adhering graphite particles from the anode, dipped in methyl ethyl ketone, dried on a clean ceramic top 110 C hot plate for 5 minutes, and stored in capped vials purged with nitrogen, until weighed.

Results and Discussion

Carboxylic acids of various types were dissolved in diisodecyl phthalate in concentrations shown in Table 3. Millimoles were used, rather than weight concentrations, so that the number of molecules could be held constant and, thereby, provide some correlation of structure with corrosion inhibitor performance. All of the carboxylic acids show high efficiencies at 5 mM but drop off significantly at 0.05 mM, which approximates a monomolecular layer. It is shown that 11 carbon undecylenic acid is not as effective as 18 carbon oleic acid in DIDP. Similar compounds, which have two carboxyl groups at one end, are even more efficient than their counterparts, as demonstrated by octadecenyl succinic acid and dodecenyl succinic acid. However, compounds with less symmetry such as di-tri-oleic acid and octa-decene-maleic acid copolymer are less effective than oleic acid. The millimole concentrations of the the polymeric species were adjusted to reflect the repeating unit, rather than the complete polymer. Benzoic acid and cyclohexane carboxylic acid, which find use as corrosion inhibitors, perform poorly in this system.

The inhibitor efficiency at low concentration is related to the size of the hydrophobic segment and the polarity of the molecule. The carboxyl group chemisorbs on the steel, whereas, the nonpolar segment prevents the intrusion of the salt.

Similar results are seen with amines in Table 4, except that phenyl ether is used as the model compound. Primary amines are more efficient than similar secondary amines. The longer chain and more polar amines give superior performance, as demonstrated by octadecyl amine and N-oleyl 1,3 diamino-propane. Oleyl amine, having a centrally located double bond in its hydrophobic segment, is less effective than a saturated straight chain compound in repelling the corrosive salt. Other functional groups, shown in Table 5, are less effective than the primary amines.

Some corrosion inhibitors may function by forming a complex with the iron or oxide on the steel surface. Gallic acid, 3, 4, 5 trihydroxy benzoic acid, is known for its reactivity with iron salts. Examination of Table 6 shows that an increase in the size of the hydrophobic segment on gallic acid results in significant corrosion inhibitive properties which are retained at low concentrations.

Materials not soluble in the model organic compound may be dispersed by grinding or dissolving in the salt solution. It has been found convenient to use a 125 ml Erlenmeyer flask containing 3 mm glass beads to disperse insoluble materials. The flask is placed on a rack in the water bath shaker, which is then operated for the necessary time, generally overnight. The beads act as a grinding media similar to a

TABLE 7 — Corrosion Inhibitor Efficiency of Inorganic materials Disolived in Phenyl Ether/3% Salt Water

	Concentration/100 grams		
	5.0 mM	0.5 mM	0.05 mM
Zinc phosphate	0.76		
Sodium nitrite	0.91	0.21	0.10
Sodium chromate	0.99	0.20	0.10
Sodium phosphate — dibasic	0.95	0.81	0.27
Sodium phosphate — monobasic	0.42	0.11	

TABLE 8 — Corrosion of Steel Wetted with Phenyl Ether and Exposed to 3% Salt Water

Exposure Time hours	Wt. Loss %	Salt Solution pH
0	0	6.3
1½	0.1	7.5
4	0.4	8.3
7	0.6	8.8
24	2.0	9.9

ball mill. The liquid is separated from the beads by filtering through a size 4 Coors Buchner funnel having 1 mm holes.

Several inorganic materials, which are used commercially as inhibitors, were evaluated in this manner, and are shown in Table 7. The chromate, phosphate, and nitrite proved very effective, as would be expected.

Although this model compound method examines the corrosion inhibitors under ideal diffusion conditions, it is felt that the information developed is valuable in selecting materials for further evaluations, including studies of chemisorption on steel in a non-aqueous environment and ability to diffuse through the coating to the substrate. This method approximates the boundary conditions existing when a steel coating is under salt attack. It has been reported[4] that failures in salt spray exposure are similar in type to cathodic exposure failures which result in alkali displacement of the coating. Failures of these types are common with hydrolyzable resins such as polyesters or alkyds.

Steel specimens wetted with phenyl ether and exposed to 3% salt water according to the previously described procedure produce alkali. The salt solution attains pH of 9.9 after 24 hours exposure during which time the steel specimen loses 2% weight by corrosion (Table 8). The small amount of corrosive medium in contact with the steel specimens suggests that chemical changes, such as the pH of the medium, reflect the chemistry between the organic coating and the steel during the corrosion process. Other recent work by Ritter and Kruger confirms the generation of a high pH fluid at the interface of corroding coated iron by a combination of pH measurements and ellipsometry.[5]

A similar test with zinc, instead of steel, results in a higher pH of 11.4, which, apparently, is adequate to prevent corrosion of steel. The dispersion of 5% zinc powder in phenyl ether also suffices to prevent corrosion of the steel specimen in this test, suggesting that the alkaline reaction may be part of the protective mechanism of zinc rich coatings, since a galvanic circuit cannot exist under the experimental conditions. It is well known that zinc rich coatings require resins with good alkali resistance for proper performance.

The test method discussed in this paper provides reproducible data which can be used to identify chemical structures that perform well as corrosion inhibitors. The distribution of the hydrophobic segment and functional groups of the molecule significantly effect inhibitor efficiency. Time and equipment requirements are modest, so that this technique has proven extremely useful in the screening of a wide variety of compounds. The better performing materials will be studied for chemisorption kinetics and film transport properties.

References

1. Mayne, J.E.O., *Br. Corrosion J.*, **5**, p. 106 (1970).
2. Kumins, C.A., *Official Digest*, p. 843, Aug., 1962 Kumins, C.A. and London, A., *J. Polymer Science*, **46**, p. 395 (1960).
3. Leidheiser, H., Jr., *Progress in Organic Coatings*, **7**, pp. 79-104 (1979).
4. Smith, A.G. and Dickie, R.A., Adhesion Failure Mechanism of Primers, Ind. Eng. Chem. Prod. Res. Dev., **17**, No. 1, 1978.
5. Ritter, J.J. and Kruger, J., Studies on the Subcoating Environment of Coated Iron Using Qualitative Ellipsometric and Electrochemical Techniques, this volume.

A Modified Sequential Sampling Plan for Painting Inspection

Raymond Tooke, Jr.,* Dr. Harrison M. Wadsworth**

Introduction

Background

It is the nature of the inspection task that decisions must be made continuously. Moreover, the inspection of construction work usually imposes an additional complication that units to be inspected are not presented in an orderly and uniform manner as, say, bottles of Coca Cola moving along the production line. In our trade, protective painting, the inspection operation will tend to follow the path of the painter's progress, depending in some cases on the painter's rigging for access to the surface. The surface may be variable and irregular in contour and texture and the inspector must adapt to these conditions.

In spite of the great variety of structures encountered, it is possible to classify the painting task into three general categories: (1) areas; (2) edges; (3) discontinuities. Quantitatively, areas are square measure, edges are linear, and discontinuities a numeric count. More about this later. The point is, established inspection procedures should take into account any particular conditions that may obtain on a job, and guide the inspector to assure the desired level of workmanship quality, without undue waste of time and effort. In short, the inspector should follow a rational plan.

Existing Reference Methods

Table 1 is a listing of the inspection report items for applied coatings, as taken from the Manual for Coatings Work.[1] Since the visually determined paint film characteristics permit essentially 100% inspection without undue penalties in time or cost, the problem we are now considering relates to the requirements for instrumentally determined characteristics. Of these, film thickness is primary in importance. Others, such as pencil hardness and permeability (pinholes), are usually specified only for immersion service.[1]

Surface preparation inspection is excluded from consideration here only because it is based entirely on visually determined characteristics. Surface profile measurements, if required, can reasonably be limited to spot checks.

**Micro-Metrics Company, Atlanta, Georgia*
***Georgia Institute of Technology, Atlanta, Georgia*

TABLE 1 — Inspection Report Items for Applied Coatings

Item Number	Item	Item Number	Item
1	Pin holes	9	Runs and sags
2	Blisters	10	Film thickness, dry
3	Color & gloss un'fty	11	Film thickness, wet
4	Bubbling	12	Holidays, skips
5	Fish eyes	13	Dry spray
6	Orange peel	14	Foreign contaminents
7	Mud cracking	15	Mechanical damage
8	Curing properties	16	Uniformity

In certain specialized protective coatings applications, inspection becomes a part of the production activity. McAbee covered this subject thoroughly in his paper at CORROSION 80 this year.[2] Examples of this practice would be tank interior coatings and coatings for the primary containment area of nuclear power plants.[1,3] Here the risk of a single flaw is prohibitive; thus, 100% inspection of all painted surfaces is essential. On the other hand, painting work on most structural steel, including bridges, tank exteriors, and even chemical plants, does not require the total supervisory type of inspection identified previously. Until the present time, however, painting inspection practice has not adopted formally the concepts of statistical sampling. Thus, the subjective judgment of the painting inspector, rather than a rational plan, dictates the sampling rate of observations.

The only reference method known to the writers which makes any mention of sampling is SSPC-PA 2.[4] Section three of this method calls for five "spot" measurements (each an average of three closely spaced readings) spaced evenly over 100 square feet of the structure. The overall average shall not be less than the specified thickness. "Spot" values may be no less than 80% of the specified value, and individual readings may be less (anything). Strictly speaking, while the foregoing procedure defines a sampling unit (100 square feet) and the criterion of acceptance or rejection of the unit, it does not address the subject of unit sampling at all.

In the National Association of Corrosion Engineers (NACE) manual on coatings and linings,[3] under Film

Thickness Inspection, the statement, "The required dry film thicknesses of the applied coating should be a part of the job specifications. These thicknesses should be specified to fix, as required, the maximum, minimum, and nominal," appears. No reference is made to the interval of observations. The NACE guide to paint application specifications[5] merely states, "the actual dry film thickness shall be not less than herein specified on all parts of the structure." In an earlier specification of SSPC[6] may be found the statement "No portion of the paint film shall be less than these specified film thicknesses." In his NACE publication on Industrial Painting,[7] Paul E. Weaver states "specify minimum dry film thickness per coat and total minimum dry film thickness."

Perception of the Need

If there were a majority consensus among informed professionals that all units of protective painting should receive 100% inspection in all cases, then there would be no basis for proposing the use of sequential analysis. This analysis rests first on the premise of statistical sampling, and second on the idea that the intensity of examination of a section of painting work may be guided and adjusted properly as the inspection work proceeds and information accumulates concerning the quality level. Complete 100% inspection of units is, in fact, a limiting case of the sequential analysis procedure. At the other extreme, a reduction of measurements to 25 or 30% without undue risk might be realized for work of superb quality.

Scope and Definitions

The proposed sampling plan is not intended to be exclusively applicable to any particular applied coatings test, except that it relates to measured, rather than observed, attributes of films. The plan provides efficiency of inspection tasks by utilizing accumulating test data to determine an appropriate sampling rate. The risk of accepting unsatisfactory units may be reduced to any desired low probability level. The following terms will be used, and are defined for the purposes of this paper.

Unit. The smallest area (100 sq ft) or length (40 ft) on which an acceptance judgement is made.

Job or lot. A set of not more than 100 contiguous units of painting which is accepted or rejected as a whole.

Inspection by attributes. A unit is classified as either defective or acceptable.

Sample. A single unit selected for inspection.

Sampling plan. The plan specifies the manner of inspection of units in a job and the criteria for determining the acceptability of the job.

Risk. Probability that a job containing more than "X" defective units will be accepted (beta risk), or that a job containing less than "Y" defective units (alpha risk) will be rejected.

Sequential analysis. A test in which, after each measurement, beyond the required minimum, one may either accept or reject the hypothesis (pass or fail), or request further measurements.

Concepts From the Factory Production Line

While quality control is at least as old as the pyramids of Egypt, statistical quality control had its beginnings in the Bell Telephone Laboratories in the 1920s. The primary motivation was the need for better tolerances in the manufacture of mating parts to achieve required close fits. Production completely free of all defects is the ideal, of course. But it may be unnecessary, or uneconomic, or even impossible.

Obviously, if a test is destructive, and we test every part, we will have completely destroyed our work. Also with thousands, even millions, of parts (or areas), testing each one may not be feasible. It was to meet these and related problems that the tools of statistical quality control were developed.

For various logical reasons, statistical quality control methods have gained small headway in the construction industry generally, and in the painting trade particularly. Perhaps the key explanation is simply that a total shutdown of a construction job will seldom be at stake if a paint film defect occurs; whereas, a factory may be completely closed by a single out of spec part. It is interesting to note, that in 1978, an article entitled "Acceptance Sampling of Structural Paints" appeared in a publication of the Transportation Research Board.[8] This study was applicable to liquid paint in containers, rather than the final applied film, but it would seem inevitable that we, too, shall be overtaken by a concept whose time has come.

The writers would be frankly dubious about efforts to directly apply the "classical" control chart methods to painting inspection in the field. The virtue and appeal of sequential analysis, however, is its potential to organize and simplify the inspection task. It is hoped that the discussion which follows will illuminate this virtue.

A circumstance especially favoring the use of sequential analysis is the regular or occasional employment of an "executive" or "final review" inspection team, which may be required to cover an entire project as expeditiously as possible. In such situations, overall uniformity of procedures and economy of time are of greatest importance.

The concepts of sequential analysis were born during WWII, and formalized by Abraham Wald in 1947.[9] The primary advantage over earlier methods is the economy of effort achieved by the ability to recognize immediately during testing when a quality goal is reached (accepted) or a defective lot (job) is found (rejected).

A Plan of Inspection for Painting

Purpose

The "Modified Sequential Sampling Plan" (MSSP) provides an objectively guided procedure of inspection, which may be executed by various inspectors with improved productivity and uniformity of results.

The "intensity" of the inspection procedure may be defined by the engineer to reduce the risk of occurrence of defective units in a job to any desired low level. The procedure may be especially useful for wide ranging "survey" inspections, which may be required to cover large projects rapidly, uniformly, and efficiently. It is applicable even to the smallest job, where it usually results in 100% inspection of units.

Limitations

As has been indicated, where single flaws can be catastrophic to the supporting structure, nothing less than 100% inspection can be tolerated. This represents a limiting case of the sequential analysis method. A single defective unit causes rejection of a job. At the other extreme, up to 100% rejected units might be allowed in a job (each defective unit found being corrected). In this case, the inspector is performing a test of application uniformity which should be done by the applicator. Neither of these limiting cases demonstrate the special virtue of the modified sequential analysis approach; namely, the abbreviation of the unit testing procedure, either on the basis of accumulating evidence of high or low quality, as the case may be. Thus, it becomes feasible to hasten significantly the process of approving superior application and rejecting inferior application on a job basis without undue risk of error. It is possible to compute the risks for any specified process. Before discussing risk considerations however, it will be desirable to become more familiar with the general procedure proposed.

Procedure

In brief, the modified sequential sampling inspection involves these steps.

1. Interpret specification requirements with respect to workmanship.
2. Select an Inspection Table corresponding to workmanship requirements.
3. Note status limits (defective unit limitations) in the Inspection Table.
4. Define the physical boundaries of a job, nominally 100 units. If it exceeds 125 units, subdivide into 60 to 125 unit jobs.
5. Begin inspection at a randomly chosen unit and inspect units contiguously, returning to starting point.
6. ...ing on "Full Procedure" (5 measure... marking each successive unit on the...
7. ...cessive result corresponds to the ..."Demerit Status" of the table (based on cumulative defective units observed), then change to "Abbreviated Procedure" (1 measurement per unit).
8. Continue testing units until another "status change" occurs, and proceed as required to: a) resume Full Procedure; b) reject job; or c) accept job.
9. Jobs are rejected if in "Demerit Status" after the last unit is inspected.

When heard or read for the first time, the foregoing steps may seem confusing to some. In fact, however, using the computer generated inspection tables provided in the Appendix of this paper, the inspector merely counts rejected units as inspection proceeds, records on the table the results for each unit, and notates the job status based on the tabulated status requirements for each unit. The tables provide both procedural instructions and documentation of inspection findings.

If this approach to painting inspection gains significant acceptance, much of the credit will have to be attributed to our BASIC computer program which generates inspection tables at the drop of a "byte." With the correct table at hand, the inspector is totally relieved of any computational task. For those who wish to examine the underlying statistical mathematics, the Appendix includes a portion of the BASIC program which shows all computations. Of more general interest, are the computer generated sequential sampling plots which illustrate graphically the inspection process, and could also be used as control charts if desired. In a few minutes time, the program is capable of generating CRT or hard copy of plots and tables for any desired test criteria.

Risk Decision

For the purposes of this paper, we are assuming that a 100 square foot unit of painted area or a forty foot length of "edge" work unit or forty discontinuities always correctly determined to be acceptable or not acceptable on the basis of five uniformly distributed determinations. Each determination consists of three closely spaced readings, none of which may be less than the specified minimum. This will be seen to be a modification of the SSPC-PA 2-73 procedure that is somewhat more stringent. Our analysis assumes that if this full procedure were followed with every unit, then the paint job would be flawless when approved.

Until detailed experiments can be performed under field conditions, it is not possible to make judgements with confidence about the precise level of painting workmanship corresponding to a particular set of statistical parameters. We have chosen values for our tables which have been labeled, for convenience, as standard, high, and superior.

With our Modified Sequential Sampling Program (MSSP) any set of parameters may be evaluated rapidly. In this paper, we have made no attempt to consider the full range of possible combinations, but believe our choices are conservative while providing significant opportunity for increasing the productivity and uniformity of the inspection function.

Conclusions

The problem. Much has been written about painting inspection, but sampling frequency has been barely considered, and statistical methods totally ignored. It is doubtful if any experimental facts are documented on painting inspection performance.

Optional approaches. The subject of painting inspection procedures cannot be ignored, because application is the primary key to paint system performance, and inspection is the key to application. Present "methods" are archaic, if not primitive, and do not utilize any of the technology of industrial inspection.

Engineering a method. A review of existing statistical inspection tools revealed attractive features of the sequential testing method. The "pure" sequential test appeared too "harsh" for painting inspection, so that the modified technique was evolved.

Potentials and further needs. The MSSP provides the first rational general approach to the job acceptance decision of the painting inspector. MSSP may have shortcomings, but it does provide a concrete model, against which other alternatives may be tested. Statistically designed field tests are needed for practical assessment of the consequences of various test criteria.

References

1. ASTM Committee D-1, Manual of Coating Work for Light-Water Nuclear Power Plant Primary Containment and other Safety-Related Facilities, ASTM, 1979, p. 56.
2. McAbee, P., Organizing and Conducting an Effective Inspection Program, Paper No. 253, CORROSION 80, Chicago, 1980.
3. NACE TCP Publication No. 2, Coatings and Linings for Immersion service, NACE, 1972, p. 22, 23.
4. SSPC-PA 2-73, Method of Measurement of Dry Paint Thickness with Magnetic Gages, Steel Structures Painting Council, 1973.
5. NACE Committee T-6J, Guide to the Preparation of Contracts and Specifications for the Application of Protective Coatings, NACE, 1962, p. 27.
6. SSPC-PA 1-64, Paint Application Specifications, No. 1 Shop, Field, and Maintenance Painting, Steel Structures Painting Council, 1964, paragraph 3.5.1.8.
7. Weaver, Paul E., Industrial Maintenance Painting, NACE, 1973, p. 110.
8. Law, D. A., and Anania, G. L., Acceptance Sampling of Structural Paints, Transportation Research Record 692, Transportation Research Board, 1978.
9. Wald, A., Sequential Analysis, Wiley, New York, 1945.

APPENDIX A

MSSP Computer Outputs — SUPERIOR Workmanship Level.

```
                        SAMPLING PARAMETERS
     TEST CRITERIA:                COMPUTED QUANTITIES:
     ------------------------      ----------------------------
     ALPHA(PR'DR RSK)---- .05      SLOPE,S---------- = .0397474
     BETA(CN'SRS RSK)--- .1        INTERCEPT,-H1---- =- 1.02932
     P1(ACCEPT'L- AQL )-- .01      INTERCEPT,H2----- = 1.32151
     P2(LIMITING - LQL )- .1       FINAL ORDINATE,F1 = 2.94543
     UNITS IN LOT/JOB---- 100      FINAL ORDINATE,F2 = 5.29625
     WORKMANSHIP LEVEL:            TRUNCAT'N UNITS,NT= 89.0973
       SUPERIOR
                        SEQUENTIAL SAMPLING PLOT
WORKMANSHIP LEVEL: SUPERIOR                      REJECTS LIMIT LINE => !

                     -- NUMBER OF REJECTS FOUND--
          -2        0        2        4        6        8        10
          +   +*    +*       +        +        +!  +    +        +    +

                    +        *                 !

                *+                              !

                    +*            *             !

                    + *           *             !

                    +      *      *             !

                    +             *             !

                    +      *                    !

                    +           *     *         !

                    +           *                !

          100 +++             *       *!

            100 UNITS APPROVAL LINE - REJECT LINE != 5.29625
```

APPENDIX A (cont'd.)

** SEQUENTIAL SAMPLING TABLE **

WORKMANSHIP LEVEL: SUPERIOR   JOB SIZE: 100

DATE: /   /   JOB NO.-DESCR.:

(TABULATED VALUES ARE DEFECTIVE UNITS ALLOWED OR REQUIRED.)

UNITS TESTED	MERIT STATUS	STANDARD STATUS	DEMERIT STATUS	ACCEPT LOT	REJECT LOT	CUMUL. # REJ.	JOB STATUS
0	-1	1	5	-1	6	( )	( )
1	-1	1	5	-1	6	( )	( )
2	-1	1	5	-1	6	( )	( )
3	-1	1	5	-1	6	( )	( )
4	-1	1	5	-1	6	( )	( )
5	-1	1	5	-1	6	( )	( )
6	-1	1	5	-1	6	( )	( )
7	-1	1	5	-1	6	( )	( )
8	-1	1	5	-1	6	( )	( )
9	-1	1	5	-1	6	( )	( )
10	-1	1	5	-1	6	( )	( )
11	-1	1	5	-1	6	( )	( )
12	-1	1	5	-1	6	( )	( )
13	-1	1	5	-1	6	( )	( )
14	-1	1	5	-1	6	( )	( )
15	-1	1	5	-1	6	( )	( )
16	-1	1	5	-1	6	( )	( )
17	-1	1	5	-1	6	( )	( )
18	-1	2	5	-1	6	( )	( )
19	-1	2	5	-1	6	( )	( )
20	-1	2	5	-1	6	( )	( )
21	-1	2	5	-1	6	( )	( )
22	-1	2	5	-1	6	( )	( )
23	-1	2	5	-1	6	( )	( )
24	-1	2	5	-1	6	( )	( )
25	-1	2	5	-1	6	( )	( )
26	0	2	5	-1	6	( )	( )
27	0	2	5	-1	6	( )	( )
28	0	2	5	-1	6	( )	( )
29	0	2	5	-1	6	( )	( )
30	0	2	5	-1	6	( )	( )
31	0	2	5	-1	6	( )	( )
32	0	2	5	-1	6	( )	( )
33	0	2	5	-1	6	( )	( )
34	0	2	5	-1	6	( )	( )
35	0	2	5	-1	6	( )	( )
36	0	2	5	-1	6	( )	( )
37	0	2	5	-1	6	( )	( )
38	0	2	5	-1	6	( )	( )
39	0	2	5	-1	6	( )	( )
40	0	2	5	-1	6	( )	( )
41	0	2	5	-1	6	( )	( )
42	0	2	5	-1	6	( )	( )
43	0	3	5	-1	6	( )	( )
44	0	3	5	-1	6	( )	( )
45	0	3	5	-1	6	( )	( )
46	0	3	5	-1	6	( )	( )
47	0	3	5	-1	6	( )	( )
48	0	3	5	-1	6	( )	( )
49	0	3	5	-1	6	( )	( )
50	0	3	5	-1	6	( )	( )

APPENDIX A (cont'd.)

UNITS TESTED	MERIT STATUS	STANDARD STATUS	DEMERIT STATUS	ACCEPT LOT	REJECT LOT	CUMUL. # REJ.	JOB STATUS
51	0	3	5	-1	6	( )	( )
52	1	3	5	-1	6	( )	( )
53	1	3	5	-1	6	( )	( )
54	1	3	5	-1	6	( )	( )
55	1	3	5	-1	6	( )	( )
56	1	3	5	-1	6	( )	( )
57	1	3	5	-1	6	( )	( )
58	1	3	5	-1	6	( )	( )
59	1	3	5	-1	6	( )	( )
60	1	3	5	-1	6	( )	( )
61	1	3	5	-1	6	( )	( )
62	1	3	5	-1	6	( )	( )
63	1	3	5	-1	6	( )	( )
64	1	3	5	-1	6	( )	( )
65	1	3	5	-1	6	( )	( )
66	1	3	5	-1	6	( )	( )
67	1	3	5	-1	6	( )	( )
68	1	4	5	-1	6	( )	( )
69	1	4	5	-1	6	( )	( )
70	1	4	5	-1	6	( )	( )
71	1	4	5	-1	6	( )	( )
72	1	4	5	-1	6	( )	( )
73	1	4	5	-1	6	( )	( )
74	1	4	5	-1	6	( )	( )
75	1	4	5	-1	6	( )	( )
76	1	4	5	-1	6	( )	( )
77	2	4	5	-1	6	( )	( )
78	2	4	5	-1	6	( )	( )
79	2	4	5	-1	6	( )	( )
80	2	4	5	-1	6	( )	( )
81	2	4	5	-1	6	( )	( )
82	2	4	5	-1	6	( )	( )
83	2	4	5	-1	6	( )	( )
84	2	4	5	-1	6	( )	( )
85	2	4	5	-1	6	( )	( )
86	2	4	5	-1	6	( )	( )
87	2	4	5	-1	6	( )	( )
88	2	4	5	-1	6	( )	( )
89	2	4	5	-1	6	( )	( )
90	2	4	5	-1	6	( )	( )
91	2	4	5	-1	6	( )	( )
92	2	4	5	-1	6	( )	( )
93	2	5	5	-1	6	( )	( )
94	2	5	5	-1	6	( )	( )
95	2	5	5	0	6	( )	( )
96	2	5	5	1	6	( )	( )
97	2	5	5	2	6	( )	( )
98	2	5	5	3	6	( )	( )
99	2	5	5	4	6	( )	( )
100	2	5	5	5	6	( )	( )

(c) COPYRIGHT 1980 MICRO-METRICS CO.

APPENDIX B
MSSP Computer Outputs — HIGH Workmanship Level.

SAMPLING PARAMETERS

```
TEST CRITERIA:                    COMPUTED QUANTITIES:
--------------------------        --------------------------
ALPHA(PR'DR RSK)---- .05          SLOPE,S---------- = .0711269
BETA(CN'SRS RSK)---- .1           INTERCEPT,-H1---- =- 1.40362
P1(ACCEPT'L- AQL )-- .025         INTERCEPT,H2----- = 1.80206
P2(LIMITING - LQL )- .15          FINAL ORDINATE,F1 = 5.70907
UNITS IN LOT/JOB---- 100          FINAL ORDINATE,F2 = 8.91475
WORKMANSHIP LEVEL:                TRUNCAT'N UNITS,NT= 95.7124
  HIGH
```

SEQUENTIAL SAMPLING PLOT

WORKMANSHIP LEVEL: HIGH   REJECTS LIMIT LINE => !

```
            -- NUMBER OF REJECTS FOUND--
         -2        0      2      4      6      8      10
          + * +    +  +   *+  +  +   +  r  +   +  +!  +

                  *  +       *                      !

                   +*        *                      !

                   + *       *                      !

                   +   *                            !

                   +      *                         !

                   +            *                   !

                   +         *      *               !

                   +            *      *            !

                   +               *      *  !

          100 +++                     *          *!
```

100 UNITS APPROVAL LINE - REJECT LINE != 8.91475

** SEQUENTIAL SAMPLING TABLE **

WORKMANSHIP LEVEL: HIGH   JOB SIZE: 100

DATE: /   /   JOB NO.-DESCR.:

(TABULATED VALUES ARE DEFECTIVE UNITS ALLOWED OR REQUIRED.)

UNITS TESTED	MERIT STATUS	STANDARD STATUS	DEMERIT STATUS	ACCEPT LOT	REJECT LOT	CUMUL. # REJ.	JOB STATUS
0	-1	1	8	-1	9	( )	( )
1	-1	1	8	-1	9	( )	( )
2	-1	1	8	-1	9	( )	( )
3	-1	2	8	-1	9	( )	( )
4	-1	2	8	-1	9	( )	( )
5	-1	2	8	-1	9	( )	( )
6	-1	2	8	-1	9	( )	( )
7	-1	2	8	-1	9	( )	( )

APPENDIX B (cont'd.)

UNITS TESTED	MERIT STATUS	STANDARD STATUS	DEMERIT STATUS	ACCEPT LOT	REJECT LOT	CUMUL. # REJ.	JOB STATUS
8	-1	2	8	-1	9	( )	( )
9	-1	2	8	-1	9	( )	( )
10	-1	2	8	-1	9	( )	( )
11	-1	2	8	-1	9	( )	( )
12	-1	2	8	-1	9	( )	( )
13	-1	2	8	-1	9	( )	( )
14	-1	2	8	-1	9	( )	( )
15	-1	2	8	-1	9	( )	( )
16	-1	2	8	-1	9	( )	( )
17	-1	3	8	-1	9	( )	( )
18	-1	3	8	-1	9	( )	( )
19	-1	3	8	-1	9	( )	( )
20	0	3	8	-1	9	( )	( )
21	0	3	8	-1	9	( )	( )
22	0	3	8	-1	9	( )	( )
23	0	3	8	-1	9	( )	( )
24	0	3	8	-1	9	( )	( )
25	0	3	8	-1	9	( )	( )
26	0	3	8	-1	9	( )	( )
27	0	3	8	-1	9	( )	( )
28	0	3	8	-1	9	( )	( )
29	0	3	8	-1	9	( )	( )
30	0	3	8	-1	9	( )	( )
31	0	4	8	-1	9	( )	( )
32	0	4	8	-1	9	( )	( )
33	0	4	8	-1	9	( )	( )
34	1	4	8	-1	9	( )	( )
35	1	4	8	-1	9	( )	( )
36	1	4	8	-1	9	( )	( )
37	1	4	8	-1	9	( )	( )
38	1	4	8	-1	9	( )	( )
39	1	4	8	-1	9	( )	( )
40	1	4	8	-1	9	( )	( )
41	1	4	8	-1	9	( )	( )
42	1	4	8	-1	9	( )	( )
43	1	4	8	-1	9	( )	( )
44	1	4	8	-1	9	( )	( )
45	1	5	8	-1	9	( )	( )
46	1	5	8	-1	9	( )	( )
47	1	5	8	-1	9	( )	( )
48	2	5	8	-1	9	( )	( )
49	2	5	8	-1	9	( )	( )
50	2	5	8	-1	9	( )	( )
51	2	5	8	-1	9	( )	( )
52	2	5	8	-1	9	( )	( )
53	2	5	8	-1	9	( )	( )
54	2	5	8	-1	9	( )	( )
55	2	5	8	-1	9	( )	( )
56	2	5	8	-1	9	( )	( )
57	2	5	8	-1	9	( )	( )
58	2	5	8	-1	9	( )	( )
59	2	5	8	-1	9	( )	( )
60	2	6	8	-1	9	( )	( )
61	2	6	8	-1	9	( )	( )
62	3	6	8	-1	9	( )	( )
63	3	6	8	-1	9	( )	( )
64	3	6	8	-1	9	( )	( )
65	3	6	8	-1	9	( )	( )
66	3	6	8	-1	9	( )	( )
67	3	6	8	-1	9	( )	( )
68	3	6	8	-1	9	( )	( )

APPENDIX B (cont'd.)

UNITS TESTED	MERIT STATUS	STANDARD STATUS	DEMERIT STATUS	ACCEPT LOT	REJECT LOT	CUMUL. # REJ.	JOB STATUS
69	3	6	8	-1	9	( )	( )
70	3	6	8	-1	9	( )	( )
71	3	6	8	-1	9	( )	( )
72	3	6	8	-1	9	( )	( )
73	3	6	8	-1	9	( )	( )
74	3	7	8	-1	9	( )	( )
75	3	7	8	-1	9	( )	( )
76	4	7	8	-1	9	( )	( )
77	4	7	8	-1	9	( )	( )
78	4	7	8	-1	9	( )	( )
79	4	7	8	-1	9	( )	( )
80	4	7	8	-1	9	( )	( )
81	4	7	8	-1	9	( )	( )
82	4	7	8	-1	9	( )	( )
83	4	7	8	-1	9	( )	( )
84	4	7	8	-1	9	( )	( )
85	4	7	8	-1	9	( )	( )
86	4	7	8	-1	9	( )	( )
87	4	7	8	-1	9	( )	( )
88	4	8	8	-1	9	( )	( )
89	4	8	8	-1	9	( )	( )
90	4	8	8	-1	9	( )	( )
91	5	8	8	-1	9	( )	( )
92	5	8	8	0	9	( )	( )
93	5	8	8	1	9	( )	( )
94	5	8	8	2	9	( )	( )
95	5	8	8	3	9	( )	( )
96	5	8	8	4	9	( )	( )
97	5	8	8	5	9	( )	( )
98	5	8	8	6	9	( )	( )
99	5	8	8	7	9	( )	( )
100	5	8	8	8	9	( )	( )

(c) COPYRIGHT 1980 MICRO-METRICS CO.

APPENDIX C

MSSP Computer Outputs — STANDARD Workmanship Level.

SAMPLING PARAMETERS

```
TEST CRITERIA:                      COMPUTED QUANTITIES:
------------------------            ------------------------
ALPHA(PR'DR RSK)----  .05           SLOPE,S---------- = .110292
BETA(CN'SRS RSK)----  .1            INTERCEPT,-H1---- =- 2.0248
P1(ACCEPT'L- AQL )--  .05           INTERCEPT,H2----- = 2.59959
P2(LIMITING - LQL )-  .2            FINAL ORDINATE,F1 = 9.00436
UNITS IN LOT/JOB----  100           FINAL ORDINATE,F2 = 13.6287
WORKMANSHIP LEVEL:                  TRUNCAT'N UNITS,NT= 134.103
   STANDARD
```

SEQUENTIAL SAMPLING PLOT

WORKMANSHIP LEVEL: STANDARD   REJECTS LIMIT LINE => !

```
           -- NUMBER OF REJECTS FOUND--
       -2      0      2      4      6      8      10
       +*   r   +   +   + *  +   +   +   +   +   +   +

           *   +                  *

               +*                     *

               +     *                     *

               +                              *

               +          *

               +                *                    *

               +                     *

               +                          *

               +                               *

       100 +++                                       *
```

100 UNITS APPROVAL LINE - REJECT LINE != 13.6287

** SEQUENTIAL SAMPLING TABLE **

WORKMANSHIP LEVEL: STANDARD   JOB SIZE: 100

DATE: /   /   JOB NO.-DESCR.:

(TABULATED VALUES ARE DEFECTIVE UNITS ALLOWED OR REQUIRED.)

UNITS TESTED	MERIT STATUS	STANDARD STATUS	DEMERIT STATUS	ACCEPT LOT	REJECT LOT	CUMUL. # REJ.	JOB STATUS
0	-1	2	13	-1	14	( )	( )
1	-1	2	13	-1	14	( )	( )
2	-1	2	13	-1	14	( )	( )
3	-1	2	13	-1	14	( )	( )

APPENDIX C (cont'd.)

UNITS TESTED	MERIT STATUS	STANDARD STATUS	DEMERIT STATUS	ACCEPT LOT	REJECT LOT	CUMUL. # REJ.	JOB STATUS
4	-1	3	13	-1	14	( )	( )
5	-1	3	13	-1	14	( )	( )
6	-1	3	13	-1	14	( )	( )
7	-1	3	13	-1	14	( )	( )
8	-1	3	13	-1	14	( )	( )
9	-1	3	13	-1	14	( )	( )
10	-1	3	13	-1	14	( )	( )
11	-1	3	13	-1	14	( )	( )
12	-1	3	13	-1	14	( )	( )
13	-1	4	13	-1	14	( )	( )
14	-1	4	13	-1	14	( )	( )
15	-1	4	13	-1	14	( )	( )
16	-1	4	13	-1	14	( )	( )
17	-1	4	13	-1	14	( )	( )
18	-1	4	13	-1	14	( )	( )
19	0	4	13	-1	14	( )	( )
20	0	4	13	-1	14	( )	( )
21	0	4	13	-1	14	( )	( )
22	0	5	13	-1	14	( )	( )
23	0	5	13	-1	14	( )	( )
24	0	5	13	-1	14	( )	( )
25	0	5	13	-1	14	( )	( )
26	0	5	13	-1	14	( )	( )
27	0	5	13	-1	14	( )	( )
28	1	5	13	-1	14	( )	( )
29	1	5	13	-1	14	( )	( )
30	1	5	13	-1	14	( )	( )
31	1	6	13	-1	14	( )	( )
32	1	6	13	-1	14	( )	( )
33	1	6	13	-1	14	( )	( )
34	1	6	13	-1	14	( )	( )
35	1	6	13	-1	14	( )	( )
36	1	6	13	-1	14	( )	( )
37	2	6	13	-1	14	( )	( )
38	2	6	13	-1	14	( )	( )
39	2	6	13	-1	14	( )	( )
40	2	7	13	-1	14	( )	( )
41	2	7	13	-1	14	( )	( )
42	2	7	13	-1	14	( )	( )
43	2	7	13	-1	14	( )	( )
44	2	7	13	-1	14	( )	( )
45	2	7	13	-1	14	( )	( )
46	3	7	13	-1	14	( )	( )
47	3	7	13	-1	14	( )	( )
48	3	7	13	-1	14	( )	( )
49	3	8	13	-1	14	( )	( )
50	3	8	13	-1	14	( )	( )
51	3	8	13	-1	14	( )	( )
52	3	8	13	-1	14	( )	( )
53	3	8	13	-1	14	( )	( )
54	3	8	13	-1	14	( )	( )
55	4	8	13	-1	14	( )	( )
56	4	8	13	-1	14	( )	( )
57	4	8	13	-1	14	( )	( )
58	4	8	13	-1	14	( )	( )
59	4	9	13	-1	14	( )	( )
60	4	9	13	-1	14	( )	( )
61	4	9	13	-1	14	( )	( )
62	4	9	13	-1	14	( )	( )
63	4	9	13	-1	14	( )	( )
64	5	9	13	-1	14	( )	( )

APPENDIX C (cont'd.)

UNITS TESTED	MERIT STATUS	STANDARD STATUS	DEMERIT STATUS	ACCEPT LOT	REJECT LOT	CUMUL. # REJ.	JOB STATUS
65	5	9	13	-1	14	( )	( )
66	5	9	13	-1	14	( )	( )
67	5	9	13	-1	14	( )	( )
68	5	10	13	-1	14	( )	( )
69	5	10	13	-1	14	( )	( )
70	5	10	13	-1	14	( )	( )
71	5	10	13	-1	14	( )	( )
72	5	10	13	-1	14	( )	( )
73	6	10	13	-1	14	( )	( )
74	6	10	13	-1	14	( )	( )
75	6	10	13	-1	14	( )	( )
76	6	10	13	-1	14	( )	( )
77	6	11	13	-1	14	( )	( )
78	6	11	13	-1	14	( )	( )
79	6	11	13	-1	14	( )	( )
80	6	11	13	-1	14	( )	( )
81	6	11	13	-1	14	( )	( )
82	7	11	13	-1	14	( )	( )
83	7	11	13	-1	14	( )	( )
84	7	11	13	-1	14	( )	( )
85	7	11	13	-1	14	( )	( )
86	7	12	13	-1	14	( )	( )
87	7	12	13	0	14	( )	( )
88	7	12	13	1	14	( )	( )
89	7	12	13	2	14	( )	( )
90	7	12	13	3	14	( )	( )
91	8	12	13	4	14	( )	( )
92	8	12	13	5	14	( )	( )
93	8	12	13	6	14	( )	( )
94	8	12	13	7	14	( )	( )
95	8	13	13	8	14	( )	( )
96	8	13	13	9	14	( )	( )
97	8	13	13	10	14	( )	( )
98	8	13	13	11	14	( )	( )
99	8	13	13	12	14	( )	( )
100	9	13	13	13	14	( )	( )

(c) COPYRIGHT 1980 MICRO-METRICS CO.

APPENDIX D
BASIC Coding — Formulas For Computed Quantities.

```
1000 '----------sAMPLING PARAMETERS
1010 CLS:Z$=""
1020 PRINT:PRINT:PRINT"ENTER SELECTED VALUES FOR THE FOLLOWING
1030 PRINT"SAMPLING PARAMETERS
1040 PRINT
1050 INPUT"ALPHA (PRODUCER'S RISK)-------------";AL
1060 INPUT"BETA (CONSUMER'S RISK)--------------";BE
1070 INPUT"P1 (ACCEPTABLE QUALITY LEVEL)------";P1
1080 INPUT"P2 (LIMITING   QUALITY LEVEL)------";P2
1090 INPUT"LOT (JOB) SIZE (UNITS)-------------";NU
1095 INPUT"WORKMANSHIP LEVEL(STD, HIGH, SUPER)-";WO$
1100 PRINT:INPUT"<C>ORRECTION OR <EN>TER IF OK";Z$
1110 IF Z$="C" GOTO 1010
1120 RETURN
1130 GOTO200
2000 '         -- COMPUTED QUANTITIES
2010 S = LOG((1-P1)/(1-P2))/(LOG((P2*(1-P1))/(P1*(1-P2))))
2020 H1 =   LOG((1-AL)/BE)/LOG(P2*(1-P1)/P1*(1-P2))
2030 H2 = LOG((1-BE)/AL)/LOG(P2*(1-P1)/P1*(1-P2))
2040 F1 = - H1 + NU*S
2050 F2 = H2 + NU*S
2060 NT =   2.5*H1*H2/(S*(1-S))
2070 Y1 = -H1 + X*S
2080 Y2 = H2 + X*S
2090 CLS
2100 PRINT:PRINT"GIVEN THE PARAMETERS:":PRINT
2110 PRINT"ALPHA =";AL;"BETA =";BE;"P1 =";P1;"P2 =";P2;"SIZE =";NU
2120 PRINT:PRINT"THE FOLLOWING WERE COMPUTED:"
2130 PRINT
2140 PRINT"SLOPE,S---------------------- =";S
2150 PRINT"INTERCEPT, - H1------------- = -";H1
2160 PRINT"INTERCEPT,H2---------------- =";H2
2170 PRINT"FINAL ORDINATE,F1---------- =";F1
2180 PRINT"FINAL ORDINATE,F2---------- =";F2
2190 PRINT"TRUNCATION UNITS,NT-------- =";NT
```

High Temperature, Long Term Performance of Temperature Indicating Paints

*J. F. Delahunt**

Introduction

Evaluations have been carried out concerned with temperature indicating (TI) paints which are frequently employed within refineries and petrochemical plants, not for aesthetics or atmospheric corrosion control, but as a process aid. These programs were carried out to establish the long term color stability of TI paints when applied to process equipment operating at elevated temperatures. It was discovered that, in fact, modern TI paints have well defined color stability curves, which may be used to predict long term stable performance for the usable period prior to maintaining the system.

TI Paints Designed in 1955

In 1955, it was foreseen that there would be a need to design a process aid which would function as an onstream inspection tool that could be applied as easily as maintenance paints to refinery and petrochemical process equipment. TI paints resulted from a program carried out under contract with an independent polymer laboratory. Applied to process equipment, TI paint gives early warning (by means of a color change) of vessel overheating due to hot gas bypassing or failure of the internal refractory lining. Normally, the color change is from the original color to snow white. The program resulted in the development and the use of two TI paints: one royal blue that changed to white at 575 F; one a second generation green that changed to white at 750 F operation temperature.

Hydrogen Attack Accelerated TI Paint Development

During the 1950s, various hydrocarbon processes were developed to remove sulfur from motor gasoline feedstocks and to increase octane ratings of motor gasoline feedstocks by formation of aromatic hydrocarbons. These process steps are carried out frequently in fixed bed catalytic reaction vessels at high temperature and high pressure in a hydrocarbon rich atmosphere. Such conditions are conducive for hydrogen attack of the reaction pressure vessel.

Carbon or low alloy steels are subject to hydrogen attack when exposed to hydrogen at temperatures above 260 C (500 F), and at hydrogen partial pressures above 690 kPa (100 psi). Under such conditions, attack will result in loss of strength and ductility of the material, rather than by any metal loss. Exposure to the high temperature, high pressure hydrogen results in diffusion of atomic hydrogen (from thermally dissociated) into the steel. Reaction of this hydrogen with iron carbide in the steel forms methane internally, at the grain boundaries, leading to the generation of intergranular fissures as a result of the methane pressure. Metallographically, the attack is evidenced by the presence of intergranular fissures and associated decarburization of the steel, as shown by Figure 1.

Hydrogen attack differs from corrosion in that the damage can occur throughout the thickness of the metal without metal loss. Therefore, the use of corrosion allowance added to the vessel is of little protective value for this type of attack. Similarly, stainless cladding or weld overlay, which are immune to hydrogen attack, will not provide protection to the base metal underneath, since the atomic hydrogen will diffuse through the stainless portions and cause hydrogen attack. Therefore, whenever stainless cladding or overlay is required for corrosion protection in an H_2S/H_2 environment it is still necessary to provide a base metal of suitable alloy composition which will resist hydrogen attack. In addition, once attack has occurred, the metal cannot be welded or repaired normally; therefore, it must be replaced.

With the advent of these new process designs in 1955, it was deemed imperative to have available a process aid, such as a TI paint, to warn of overheating of the process vessel and avoid the possibility of hydrogen attack and potential catastrophic failure of the reaction vessel.

Ti Paints Developed for Use within Exxon

In 1955, the first TI paints were developed, and a number of applications were made to process equipment. In their first application, a blue TI paint turned

---

*Exxon Research & Engineering Co., Florham Park, New Jersey

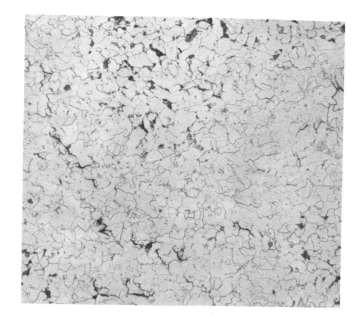

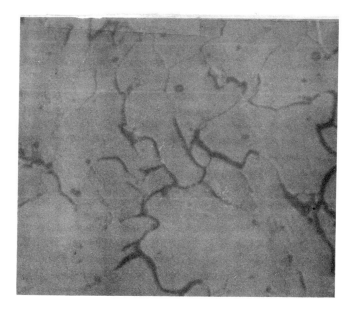

FIGURE 1 — In upper view, extensive fissuring and decarburization due to hydrogen attack are evident throughout carbon steel. At higher magnification, grain boundary separation and localized decarburization are shown in lower view.

from a royal blue to snow white, indicating early warning of vessel overheating and potential hydrogen attack. The process was immediately shut down, the vessel opened, and the necessary repairs and modifications made to prevent hydrogen rich gas, high temperature contact of the carbon steel shell. Such performance has continued from that point, and the overall value of the TI paints as a process aid and early warning tool has far out distanced the few thousand dollars it cost to be developed.

Two TI paints have been used frequently: one royal blue, and one green. The TI color change points were specified as nearly immediate at temperatures of 550 F and 750 F, respectively. The TI blue paint is applied to carbon steel vessels; the green is applied to vessels fabricated from low chrome alloys such as 1½ to 2½ chrome steels. In recent years, however, various affiliates have reported that, at times, the TI paints apparently changed colors at temperatures much lower than those specified. This improper color change may cause unwarranted unit shutdowns, and the expenditure of valuable manpower to discover the source of color change.

Because of the overall concern due to the insiduous nature of hydrogen attack, it was decided to evaluate long term high temperature exposure TI paint, and its effect on color change point.

Houston Test Stand Selected for Evaluation

In order to evaluate TI paints completely at elevated temperatures, a number of methods were evaluated, and the Houston high temperature paint test stand selected. This apparatus was developed in the late 1940s and early 1950s. It is completely described in the literature of that period. The following is a short description of the apparatus, its operation, and the quality of its results.

The apparatus is shown schematically in Figure 2. It consists of four component parts

1. Core Pipe — carbon steel 1½ inch NPS, standard wall thickness

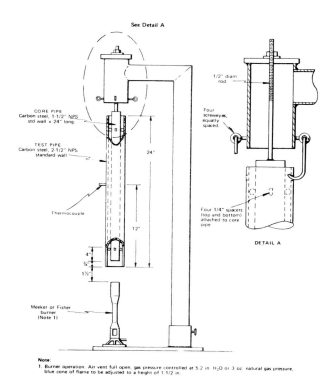

FIGURE 2 — Schematic of high temperature coating test apparatus used during investigations of temperature indicating paint.

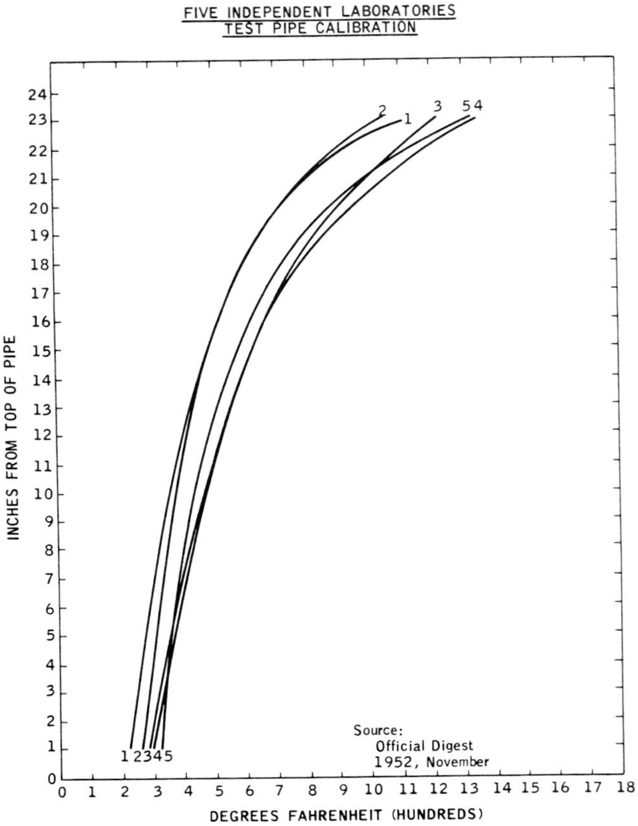

In preparation for evaluating high temperature coating tests, it is first necessary to prepare a test stand calibration curve. A pipe test coupon is fitted with suitable thermocouples, which are peened into the surface of the pipe along the length of the pipe at one inch intervals. The test coupon is mounted in the test stand, and heated until equilibrium conditions are established; that is, until temperature readings are consistant with no significant shifting. At that point, a temperature profile curve as shown in Figure 3 may be constructed. Table 1 shows data from the "Official Digest," November 1952, displaying the temperature profile data obtained at that time. The pipe length is measured from the bottom of the pipe nearest the heat source. Accuracy reported is ± 25 F when the equipment was first developed.

In addition, Figure 3 shows curves for pipe calibrations developed approximately 25 years ago by five different operators in various laboratories. These again demonstrate the amazingly good correlation between different test apparatus.

Temperature Indicating Paint Test Procedures

In the past, color change points of TI paints were often investigated by exposure of test panels in high temperature ovens at static temperatures. The Hous-

FIGURE 3 — Calibration curves for heat resistant paint test apparatus developed approximately 30 years ago.

TABLE 1 — Test Pipe Temperature Data

Pipe Length (in)[1]	Temperature (F)
1	995
5	710
10	490
15	375
20	300
22	275

[1] Measured from end nearest heat source

2. Test Pipe — carbon steel 2½ inch NPS, standard wall thickness
3. Gas Burner — Fisher Burner with calibrated gas supply
4. Test Stand — carbon steel shielded by suitable barriers for personnel protection and to reduce unsuitable air currents

This equipment was built and used in 1955 and '56 for testing high temperature paints and coatings. A second apparatus was built to carry out the objective of this investigation.

FIGURE 4 — View of TI paint suspended from test stand with midpoint thermocouple in position.

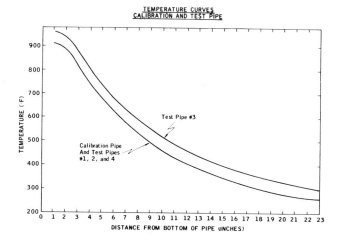

FIGURE 5 — Curves depicting temperature curves for calibration pipe as well as test pipes used during this investigation.

TABLE 2 — TI Paints Investigated

Manufacturer[1]	Color Change	Data Sheet Color Stability Point (F)
AB[2]	Blue-White	560 - 585
AG[2]	Green-White	725 - 750
BB[2]	Blue-White	560 - 585
BG[2]	Green-White	725 - 750
BM	Multiple	Green 400 / Blue 600
CB	Blue-White	500
DM	Multiple	—

[1] First letter denotes manufacturer; second letter establishes color (B = Blue, G = Green, M = Multiple).
[2] Presumably same formulation from different manufacturer.

ton test stand was used, but only for short exposure periods of 24 hours or less. Because of an increasing number of reports from manufacturing plants, it was decided to investigate color stability in the laboratory by monitoring its stability over 800 to 1000 hours of exposure.

Test specimens (24 inch long, 2½ inch NPS standard wall carbon steel pipe) were prepared by applying test paint systems to each longitudinal half, or to the entire pipe. In addition, a thermocouple is peened to the midpoint of the test pipe (Figure 4). The temperature profile for each test pipe was determined by measuring the midpoint temperature, and comparing it to the calibration curve. Figure 5 shows the temperature profiles for the test pipes. The color change temperature was determined periodically by measuring the distance from the bottom of the pipe to the point where the paint changed color from white. This distance was compared to the temperature profile for that test pipe to obtain the temperature at that point. This was the color change temperature for that time period. Table 2 indicates the systems tested.

It is of interest to note that Paint AB and AG were tested approximately two years before BB and BG. They were made by a different manufacturer, but reportedly from the same TI paint formulation.

In addition to these variations, there was modification of the paint system. In the field, Exxon affiliates often specify that two coats of TI paint be applied to commercially sandblasted reaction vessels. However, other affiliates and manufacturers have specified one coat of zinc pigmented organic primers, or one coat of inorganic zinc rich primers. Therefore, test work including these general primers was also undertaken.

Color Stability Defined

When high temperature stability tests were conducted for 1600 hours for blue TI paint from time temperature curves, it was determined that accurate depiction of such a curve could be obtained after 850 hours (one month) test. After concluding the evaluation, it is possible to establish a color stability point after 20,000 hours of service. This period represents 2½ years of field exposure, at which time recoating of the equipment is specified at a normally scheduled turnaround.

Figure 6 illustrates the performance of blue TI paint applied: in two coats; as a topcoat over zinc rich primer; as a topcoat over inorganic zinc rich primer (IZRC). After 20,000 hours, it is expected that the color stability point shall lie between 470 and 440 F after 2½ years of service. It was concluded that the performance does not depend upon the primer coat. The performance is nearly equivalent when applied to itself and organic or IZRC prime coat.

Figure 7 depicts color stability behavior of green TI paint AG_1, after a similar period of time. It is concluded that for stable color retention, AG_1 should not be applied to equipment operating at temperatures exceeding 600 F. Again, selection of prime coat made little difference in long term paint TI color stability performance.

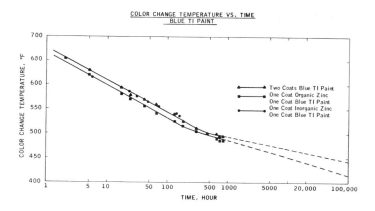

FIGURE 6 — Blue temperature indicating paint applied in two coats, compared to its performance applied to zinc rich primers.

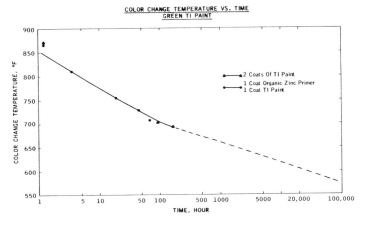

FIGURE 7 — Curve illustrating color stability of green temperature, indicating paint applied in two coats and to an organic zinc rich primer.

TABLE 3 — Temperature Indicating Paint Color Stability Characteristics

Manufacturer[1]	20,000 hours life (F)	
	White	Original
AB₁[2]	440	360
AG₁[2]	600	—
BB₁[2]	380	300
CB	360	315
BG[2]	510	440
BM	380	310
DM	280	235

[1] First letter establishes manufacturer; second letter establishes color (B = blue; G = green; M = multiple)
[2] Presumably same formulation, different manufacturers.

Effect of Various Manufacturers

In the last few years, additional tests of TI paints were undertaken on paints obtained from different manufacturers than those described previously. These results are shown in Table 3. As shown, at times there was almost 100 to 150 F difference between the point at which the white terminated and the original color was retained.

It was determined that, due to the change of manufacturers, color stable points for paints had changed by about 60 F. Therefore, the minimum surface temperature allowable, prior to suspecting reaction vessel over heating is reduced to 300 F. A second finding was that multiple color change paints manufactured in the United States and United Kingdom were not practical for use in a process plant; color changes were not sharp, and could not be delineated easily. Therefore, these paints are not recommended. It was discovered, with one brand of TI paint, that after 120 hours exposure the original green color of this paint had changed to blue that extended to the upper, or coolest, end of the test pipe specimen.

Other Conclusions Reached as A Result of Testing

The selection of a TI paint must be made first for color stability combined with the operating temperature of the process equipment. A TI paint can also be selected for sharpness of color change, as defined by the fade zone temperature range, and the brillance of color difference.

Recommendations

Refinery and petrochemical plant reports that TI paint color conversion could occur at temperatures approaching 300 F were confirmed by test work. However, this point would be reached only after many thousands of hours of service. Field performance of TI paints can be influenced by atmospheric weather. Probably more important, it can be influenced by minor process upsets, or minor internal lining deterioration, that could result in higher than expected shell temperatures. Therefore, it is recommended that TI paints be selected as a process aid based upon a long term (2½ year) field exposure, considering normal shell steel operating temperatures plus 100 F. This should resolve satisfactorily the problems that have been experienced in the past.

Painting "O"-Ring Sealing Surfaces to Prevent Corrosion

*Colin J. Sandwith**

Introduction

This report summarizes the procedures and results of a one year test program undertaken by this laboratory for the Naval Undersea Warfare Engineering Station, Keyport, Washington, to investigate painting the "O"-ring sealing surfaces of Mk 48 torpedoes to prevent corrosion. Although the technique of painting "O"-ring sealing surfaces to prevent corrosion is new for the Mk 48 torpedo, it has been used successfully since 1966 on at least one other submersible shell (Exercise Head Mk 78 Mod 2). In that application, epoxy plastic primer 3-0-11 and plastic coating 3-0-1 were applied to a thickness of 0.002 inch on the "O"-ring sealing surfaces to prevent rapid corrosion of the shell, which was composed of a magnesium alloy, per Note 19, BUWEPS Drawing No. 1874180, U.S. Naval Underwater Weapons Research and Engineering Station, Newport, Rhode Island (now Naval Underwater Systems Center).

The tests were limited to paints accepted currently for use on the Mk 48. The specific objectives were to determine the pressure keeping capabilities of the "O"-rings, the compatibility of the "O"-ring/paint/Otto Fuel interface, the chemical compatibility between the paints and the "O"-rings, and paint creep characteristics. For practical and safety reasons, the Mk 48 "O"-ring sealing surfaces and conditions were simulated, using a specially designed apparatus.

Test Setup

The hazards involved in maintaining a large volume at high pressure for a long period indicated that the pressurized volume should be minimized. Structural stability and safety considerations suggested that the design utilize a cylinder/piston arrangement, as opposed to parallel pressurized plates, even though the "O"-rings used in the Mk 48 are face type seals. A piston type seal requires closer tolerances in machining and painting and more effort in assembly and disassembly; however, it makes up for these differences in increased safety and by allowing a greater number of "O"-rings to be tested at one time. To test the hypothesis that the high contact pressures from

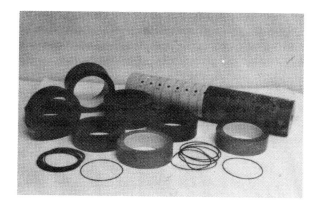

FIGURE 1 — Disassembled test apparatus showing piston, cylinders, and "O"-rings.

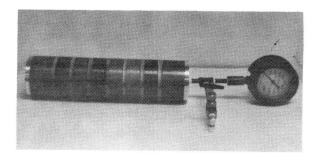

FIGURE 2 — Assembled test apparatus.

the "O"-rings would produce creep in the paints, the smallest practical "O"-ring cross section (0.07 inch) was used at the highest practical pressure to simulate worst case conditions. The stresses produced by the contact between the "O"-ring and the sealing surfaces are similar for large diameter to cross section ratios in both piston and face type seals.

Figure 1 is a picture of the disassembled apparatus showing the hollow piston with its "O"-ring grooves, the "O"-rings, and the cylindrical rings with inside sealing surfaces that were positioned over the piston and "O"-rings. Between these cylinders were 0.5 inch thick, clear plastic spacers which served as view ports for detecting leaks. The 16 grooves on the piston held 8 pairs of "O"-rings. When one of the cylinders was

---

*Applied Physics Laboratory, University of Washington, Seattle, Washington

TABLE 1 — Paint Systems Used in Tests

Designation	Color	Components
LP (epoxy primers)	Yellow[1]	Wash primer (Phos-Pho-Neal 31-G-6) Chromate primer (Dexter Corp. 4-G-14)
LT (epoxy topcoat)	Light green[2]	Wash primer (Phos-Pho-Neal 31-G-6) Chromate primer (Dexter Corp. 4-G-14) Topcoat, Class I (Dexter Corp. 34108)
PP (polyurethane primers)	Pale green	Wash primer (Phos-Pho-Neal 31-G-6) Primer (Hughson TS3236-26 Part A plus Hardener TS3236-26 Part B)
PT (polyurethane topcoat)	Dark green (glossy)	Wash primer (Phos-Pho-Neal 31-G-6) Primer (Hughson TS3236-26 Part A plus Hardener TS3236-26 Part B) Topcoat (Hughson elastomeric TS3236-23A plus Hardener TS3236-23B)

[1] Manufacturer's "yellow-green"
[2] Manufacturer's "flat dark green"

positioned over a pair of "O"-rings, it produced a control volume between the cylinder and the piston.

The assembled apparatus is shown in Figure 2. The test fluid entered through the valve, was distributed throughout the 1 inch diameter hollow center of the piston, and entered the volumes formed between the piston and the cylinders through ports in the piston wall. Each pair of "O"-ring grooves on the piston and each cylinder were numbered for identification.

The sealing surfaces of the cylinders and the piston, including the grooves, were painted with various combinations of epoxy or polyurethane paint and/or its primer. Half the piston was painted with the primers for the epoxy paint system; the other half was painted with the primers for the polyurethane paint system. Because of the piston type seal and the small cross section of the "O"-rings, only relatively thin coatings of paint could be applied. This was not a compromise, since only thin coatings would be permitted in actual use, owing to the viscoelastic behavior of the paint. Based on an analysis of the various tolerances involved ("O"-ring dimensions, machined dimensions, anodizing thicknesses, paint buildup, etc.), the paint thickness was limited to a range of 0.0015 to 0.002 inch. The paint systems used in the tests are listed in Table 1.

Test Procedure

Each of the paint/primer combinations was subjected to a 1 or 2 week exposure to 4 simulated environments to check paint deterioration (bubbling, peeling, creep, cracking, or tearing). Saturation of the paints and interfaces by the pressurized fluids should be complete in this time; therefore unbonding of the paint due to the saturated interfaces should have been detectable. However, corrosion damage in the form of oxidation or pitting would not be expected to occur during the short test time. Thus, our criterion for predicting anticorrosion effectiveness is: if the paint remains intact on the sealing surface, then corrosion will be reduced or prevented.

The 4 test environments were: 1) fresh water (with the inadvertent addition of several percent of isopropyl alcohol) at 1500 psig for 1 week; 2) artificial sea water (also with several percent of isopropyl alcohol) at 1500 psig for about 1 day; 3) artificial sea water at 1500 psig for 3 weeks; 4) Otto Fuel II at approximately 1000 psig for about 2 weeks.

Because the first 2 test environments were altered by the inadvertent addition of small quantities of isopropyl alcohol, and since the test procedure was complicated by this event, it may be more clear to describe briefly the complete sequence.

TABLE 2 — Paint Systems Used in One Week Fresh Water Test Beginning on April 18, 1979[1]

System	Cylinder No.	Piston No.
LP (epoxy primers)	1, 2	1, 2, 3, 4
LT (epoxy topcoat)	3, 4	None
PP (polyurethane primers)	7, 8	5, 6, 7, 8
PT (polyurethane topcoat)	5, 6	None

[1] The cylinder/piston combinations used for the freshwater test are shown in Table 4.

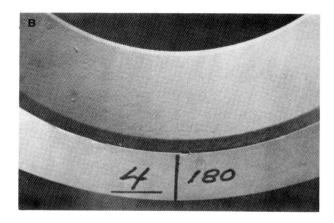

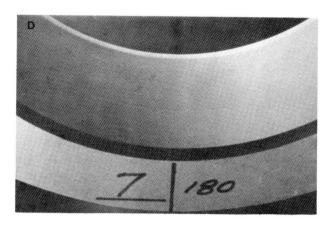
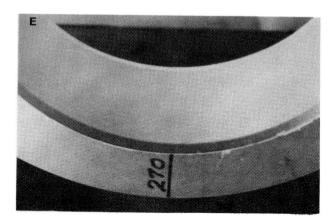

FIGURE 3 — Typical condition of painted surfaces before the freshwater test. A) Epoxy primer; B) epoxy topcoat; C) polyurethane topcoat; D) polyurethane primer; E) fingerprint with a roughness of 200 rms on cylinder 2 at 270° before freshwater test.

Test 1
1. The piston and cylinders were machined, anodized, cleaned, and checked dimensionally.
2. The piston and cylinders were painted with the first combinations of paint.
3. Before test photographs were taken of each quadrant of the cylinders.
4. Before test roughness measurements were taken.
5. The test apparatus was assembled, and the "O"-rings installed with copious quantities of "O"-ring grease.
6. The system was pressurized to 1500 psig with fresh water (contaminated with isopropyl alcohol).
7. The pressurization was maintained for 1 week.
8. The apparatus was disassembled, cleaned, and inspected for damage to the paint and "O"-rings.
9. The post test photographs were taken.
10. The post test roughness was measured.

Test 2
1. The paint was removed from the cylinders.

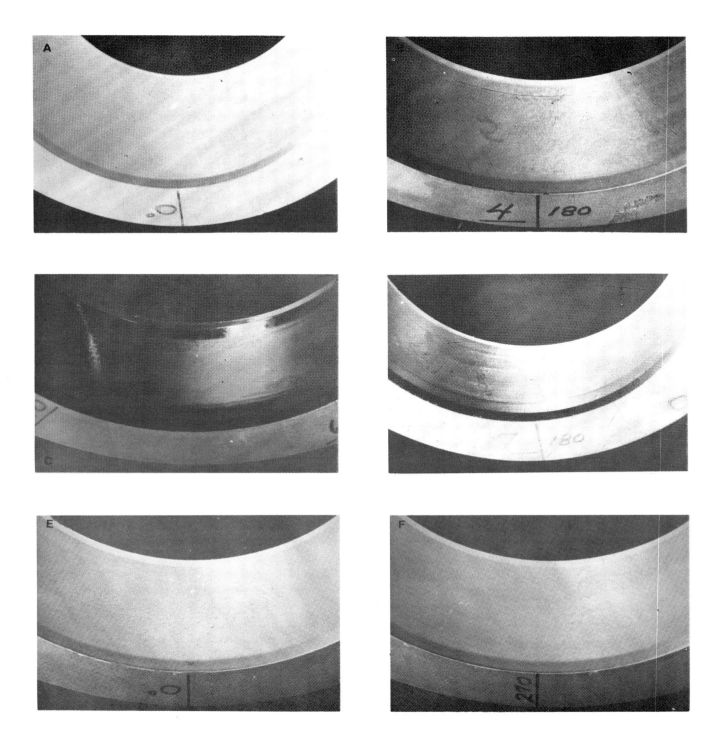

FIGURE 4 — Examples of the painted surfaces after the freshwater test. A) Cylinder 1 at 0°; B) cylinder 4 at 180°; C) cylinder 6 at 230°; D) cylinder 7 at 180°; E) cylinder 2 at 0°; F) cylinder 2 at 270°.

2. The cylinders were painted with the next combination of paints.
3. The before test roughness was measured.
4. The apparatus was reassembled using new "O"-rings and large quantities of "O"-ring grease.
5. The apparatus was pressurized to 1500 psig with sea water (slightly contaminated with isopropyl alcohol).
6. After 1 day of exposure, the alcohol was detected; immediately, the apparatus was disassembled and the surfaces were cleaned completely.
7. The surfaces were inspected, and all effects due to exposure to alcohol were noted.

Test 3

1. The apparatus was reassembled and pressur-

TABLE 3 — Comparison of Surface Roughness (RMS) Before and After One Week Fresh Water Test Beginning April 18, 1979 (instrument: Cleveland Roughness Meter B 6103 Power Track BK 3911)

Cylinder	Angle	Maximum Before	Maximum After[1]	Minimum Before	Minimum After	Average Before	Average After	Extreme Before	Extreme After
1 yellow	0	65	71	40	32	55	49	70	87
	90	70	57	40	32	55	41	85	87
	180	75	55	50	30	60	44	75	80
	270	75	61	45	28	55	43	100	93
	Average	71	61	44	30	56	45	83	87
2 yellow	0	120	93	45	34	75	49	120	300
	90	100	100	40	30	75	62	110	300
	180	70	87	40	23	65	42	75	180
	270	80	88	40	24	60	37	85	112
	260[2]	150	161	50	27	200	53	240	300
	Average	93	93	41	28	69	48	97	223
3 light green	0	140	116	70	38	100	75	180	196
	90	110	131	60	49	95	90	130	261
	180	120	125	60	30	90	61	180	225
	270	110	79	30	33	90	86	120	196
	Average	120	113	55	38	94	78	153	220
4 light green	0	120	110	70	20	90	61	120	173
	90	150	143	60	22	100	80	180	195
	180	120	126	50	29	80	74	140	176
	270	120	167	60	26	100	79	120	329
	Average	127	138	60	25	92	74	140	220
5 dark green	0	30	31	10	11	25	19	50	42
	90	35	31	15	10	20	19	40	45
	180	30	29	12	10	20	19	35	31
	270	25	18	10	10	20	14	80	22
	Average	30	27	12	10	21	18	51	35
6 dark green	0	20	21	8	10	12	13	30	128
	90	20	38	10	15	15	21	22	63
	180	15	19	8	6	12	9	ND	19
	270	25	19	15	8	20	11	30	27
	Average	20	25	11	10	15	14	28	60
7 pale green	0	45	53	25	40	30	48	ND	57
	90	45	53	25	40	30	44	ND	67
	180	45	57	25	38	30	45	ND	93
	270	60	63	30	42	45	51	ND	80
	Average	48	57	27	39	34	47	ND	75
8 pale green	0	50	57	30	12	35	30	60	69
	90	40	44	30	13	35	31	45	64
	180	50	53	35	15	40	32	60	64
	270	50	81	35	17	40	32	60	155
	Average	48	59	32	15	37	32	56	88

After 500A Emery Paper Was Used to Take High Points Off the Paint

Cylinder	Angle	Maximum	Minimum	Average	Extreme
3	0	130	55	90	140
	90	140	80	100	150
	180	120	35	95	140
	270	120	10	80	130
4	0	90	55	80	100
	90	110	50	90	140
	180	100	40	80	110
	270	90	55	80	100

TABLE 3, cont.

Outside Circumference Roughness (Anodic Finish Only)

Cylinder	Angle	Maximum Before	Maximum After	Minimum Before	Minimum After	Average Before	Average After	Extreme Before	Extreme After
3	0	65	112	55	38	60	78	70	210
	90	65	ND	55	ND	60	ND	70	ND
	180	65	ND	55	ND	60	ND	70	ND
	270	75	ND	60	ND	65	ND	75	ND
	Average	67	112	56	38	61	78	71	210
4	0	80		60		70		90	
	90	75		60		65		ND	
	180	80		55		70		ND	
	270	75		60		70		ND	
	Average	77		59		69		90	
	0	—	80	—	35	—	55	—	93
	0	—	132	—	55	—	74	—	164
	0	—	55	—	84	—	73	—	136
	0	—	102	—	48	—	70	—	148

(1) Roughness readings were recorded on a Brush 280 Strip chart recorder in all measurements after 4/18/79. Settings: Roughness meter, 0.03, 1000 scale; Recorder, damping circuit, 5 mm/second, 50 × 1 Sensitivity, Sensitivity dial full ccw.
(2) Finger print. Not included when calculating roughness average.

TABLE 4 — Data Reduction and Observations

Cylinder No.	Paint Sys.	Piston No.	Paint Sys.	Average B	Average A	ΔAverage (A_B-A_A)	Range (max/min) B	Range (max/min) A	ΔRange (R_B-R_A)	Extreme B	Extreme A	ΔExtreme (E_B-E_A)
1	LP	1	LP	56	45	11	27	31	− 4	83	87	− 4
2	LP	7	PP	69	49	20	52	64	− 12	97	300	− 203
3	LT	4	LP	94	78	16	65	85	− 20	147	169	− 22
4	LT	5	PP	92	78	14	67	113	− 46	140	220	− 80
5	PT	2	LP	21	18	3	18	17	1	51	35	+ 16
6	PT	6	PP	15	14	1	9	15	− 6	97	300	− 203
7	PP	3	LP	34	47	− 13	21	18	3	ND	75	− 75
8	PP	8	PP	37	32	5	16	44	− 28	56	88	− 32

Cylinder	Observations After Test	Cylinder	Performance Rating
1	No reaction, no bubbles, slight ridge at 90 & 180°	1	Acceptable
2	Bubbles conc. at 0°, diffusion at 180 & 90°; no grooves	2	Acceptable
3	No reaction; detectable ridges at 180°	3	Good
4	No reaction; severe ridges at 270 and 180°	4	Acceptable
5	Grease discoloration conc. at 180°; minor groove	5	Good
6	Bubbles, local 180 to 270°; discoloration and minor ridge at 180°	6	Good
7	Reaction, H_2O severe at 180°; obvious depression	7	Unacceptable
8	Reaction, H_2O severe; obvious depression at 180°	8	Unacceptable

B = Before
A = After
A_B = Average Before
A_A = Average After
ND = no data
R_B = Range Before
R_A = Range After
E_B = Extreme Before
E_A = Extreme After

ized to 1500 psig with artificial sea water for 3 weeks.
2. The system was disassembled, cleaned, and inspected for defects.
3. The post test photographs were taken.
4. The post test roughness was measured.

Test 4
1. The test apparatus was reassembled with new "O"-rings using copious quantities of "O"-ring grease.
2. The system was pressurized to 1000 psig with Otto Fuel II for approximately 2 weeks.

TABLE 5 — Paint System Used in Sea Water and Otto Fuel II Tests[1]

Paint System	Cylinder No.	Piston No.
LP (epoxy primers)	None	1, 2, 3, 4 [2]
LT (epoxy topcoat)	1, 3, 4	None
PP (polyurethane primers)	None	5, 6, 7, 8
PT (polyurethane topcoat)	2, 7, 8, **5, 6**	None

[1] For the sea water and Otto Fuel II tests, the cylinder/piston configuration was as follows: 1/1, 2/5, 3/7, 4/3, 5/4, 6/6, 7/2, 8/8.
[2] Bold indicates that the designated cylinder or piston was not stripped and repainted.

3. The system was disassembled, cleaned, and inspected.
4. The post test photographs were taken.
5. The post test roughness was measured.

Each time the apparatus was assembled before testing, new "O"-rings were inserted with copious quantities of silicone "O"-ring lubricant.

Results

The paint systems used in the freshwater test beginning April 18, 1979, are listed in Table 2. Figure 3 shows 5 of the 36 photographs of the painted cylinders taken before the test. (In addition to the 32 quadrants of the cylinders, special areas were photographed to observe defects such as fingerprints and particles.) Figure 3a (cylinder 1 at 0°) shows the typical condition of the primer system for the epoxy. This system exhibited an average roughness of 55

TABLE 6 — Comparison of Surface Roughness (RMS) of Painted "O"-Ring Sealing Surfaces Before and After the Three Week Sea Water Test Beginning June 29, 1979

Cylinder	Angle	Minimum B	Minimum A	Average B	Average A	Maximum B	Maximum A	Extreme B	Extreme A	Grooves (RMS)	Comments
1	0	30	50	67	125	120	210	205	>250	—	
	90	44	38	78	100	130	220	215	"	—	Noise too high to identify
	180	40	38	78	115	138	210	220	"	—	grooves
	270	25	40	80	120	150	220	240	"	—	
2	0	0	8	5	20	15	50	25	60	I	Bubbles
	90	0	8	9	15	22	30	43	50	I	Dent at 90°
	180	0	10	3	16	12	25	30	90	>250	Grooves and Bubbles
	270	0	10	6	18	12	30	18	45	180	Grooves
	190	—	—					>250	>250		Flattened bubbles
3	0	42	40	85	130	126	220	210	>250	I	
	90	27	30	86	95	165	210	235	"	"	Noise too high to measure
	180	42	30	95	98	168	220	225	"	"	height of grooves
	270	42	35	95	90	155	210	208	"	"	
4	0	30	50	96	120	163	210	225	>250	—	
	90	38	50	92	120	180	210	210	"	—	Noise too high to identify
	180	45	45	95	105	155	210	210	"	—	grooves
	270	26	35	65	85	102	210	198	"	—	
5	0	0	10	7	20	25	40	42	55	60	
	90	0	10	12	20	22	40	35	55	60	
	180	0	10	12	22	27	25	42	45	NI	
	270	0	8	5	12	10	20	15	25	NI	
6	0	0	10	3	15	6	25	10	40	NI	
	90	0	8	4	15	12	25	18	40	NI	
	180	0	6	4	15	12	25	16	38	NI	
	270	0	6	6	18	16	35	28	45	>250	Grooves
7	0	0	10	6	18	15	30	28	65	120	Definitely grooves
	90	0	10	10	15	20	35	30	50	100	Definitely grooves
	180	0	10	5	12	12	22	18	35	140	Definitely grooves
	270	0	10	5	15	10	22	15	35	250	Definitely grooves
8	0	0	10	0	15	18	33	55	40	150	Grooves
	90	12	10	0	15	42	40	78	55	140	Grooves
	180	25	10	10	20	25	32	45	40	140	Grooves
	270	0	15	12	20	30	40	50	50	>250	Grooves

I = Indication of grooving, but noise too high to measure height.
NI = No indication of grooving.

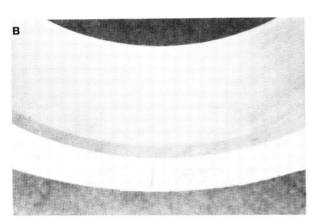

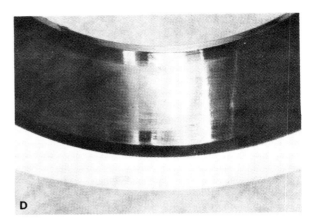
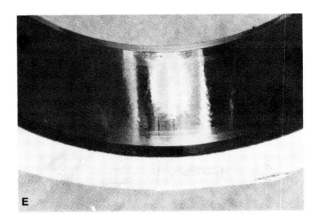

FIGURE 5 — Examples of the painted surfaces after the sea water test. A) Cylinder 2 at 180°; B) cylinder 4 at 180°; C) cylinder 6 at 270°; E) cylinder 7 at 180°.

rms. Figures 3b, 3c, and 3d show, in turn, typical conditions of the epoxy topcoat, the polyurethane topcoat, and the polyurethane primer systems. These surfaces exhibited roughnesses of 80, 19, and 30 rms, respectively. Figure 3e shows a fingerprint with a roughness of 200 rms on cylinder 2 at 270°. The roughness of the painted surfaces before and after the freshwater test are listed in detail in Table 3.

Figure 4 shows selected photographs from the post test series. Cylinder 1 at 0° (Figure 4a) shows a slight grooving near the center. Cylinder 4 at 180° (Figure 4b) shows grooving and a transfer of the number 5 from the piston to the cylinder by diffusion. Cylinder 6 at 230° (Figure 4c), exhibits bubbles under and in the paint. Cylinder 7 at 180° (Figure 4d), exhibits some grooving and marks which were left by the profilometer stylus. Cylinder 2 at 0° (Figure 4e), exhibits a bubble pattern, and cylinder 2 at 270° (Figure 4f), still exhibits the fingerprint. Table 4 shows the reduced data.

Table 5 shows the paints applied to the cylinders for the sea water and Otto Fuel II tests. Before the application of the paints listed in Table 5, all paints and greases were stripped as completely as possible from the cylinders to be repainted. Cylinders 5 and 6 were not stripped or repainted.

Table 6 compares the surface roughness of the cylinders before and after testing in sea water for 3

TABLE 7 — Comparison of Surface Roughness (RMS) Before and After One Week Otto Fuel II Test Beginning August 1, 1979

Cylinder	Angle	Minimum B	Minimum A	Average B	Average A	Maximum B	Maximum A	Extreme B	Extreme A	Comments
1	0	50	48	125	90	210	150	>250	190	
	90	38	45	100	70	220	125	"	185	
	180	38	45	115	90	210	155	"	195	
	270	40	35	120	90	220	155	"	190	
2	0	8	6	20	15	50	30	60	85	
	90	8	5	15	14	30	30	50	85	
	180	10	5	16	10	25	25	90	120	
	270	10	5	18	10	30	25	45	—	
	190		5		10		25	>250	110	Flattened bubbles
3	0	40	48	130	120	220	200	>250	>200	
	90	30	35	95	90	210	190	"	"	
	180	30	38	98	105	220	190	"	"	
	270	35	55	90	95	210	190	"	"	
4	0	50	50	120	95	210	175	>250	>200	
	90	50	35	130	90	210	185	"	"	
	180	45	40	105	75	210	130	"	"	
	220	35	50	85	95	210	130	"	"	
5	0	10	8	20	18	40	40	55	60	
	90	10	8	20	15	40	38	55	55	
	180	10	8	22	15	35	35	45	60	
	270	8	5	12	10	20	28	25	38	
6	0	10	5	15	10	25	22	40	42	
	90	8	5	15	10	25	22	40	42	
	180	6	8	15	10	25	20	38	50	
	270	6	5	18	10	35	25	45	50	
	280	8	5	15	8	25	—	>300	>200	Flattened bubbles
7	0	10	10	18	25	30	35	65	50	
	90	10	10	15	20	35	42	50	70	
	180	10	10	12	20	22	35	35	50	
	270	10	10	15	20	22	45	35	75	
8	0	10	5	15	15	33	35	40	100	
	90	10	5	15	10	40	30	55	125	
	180	10	5	20	10	32	25	40	45	
	270	15	5	20	10	40	30	50	185	Grooves and bubbles
	50		5		10		30		75	Paint peeling

weeks. Figure 5 shows examples selected from the post test series to show both extreme and typical behavior. Cylinder 2 at 180° (Figure 5a), shows grooving and bubbles. Cylinder 4 (Figure 5b) shows relatively good resistance to the environment. At 270° (Figure 5c), cylinder 6 shows an extreme concentration of bubbles; at 180° (Figure 5d), it shows defects and painted over particles. Cylinder 7 (Figure 5e) shows good resistance to the test environment. All of the paint systems not shown in the figure had reasonably good resistance to the environment.

Table 7 compares the surface roughness before and after the Otto Fuel II tests. Figure 6 shows selected post test photographs. Cylinder 2 at 180° (Figure 6a), exhibits bubbles and grooves. Cylinder 4 (Figure 6b), exhibits remarkably good resistance. Cylinder 6 at 270° (Figure 6c), exhibits extreme bubbling and ridging. Cylinder 8 (Figure 6d), exhibits relatively good resistance.

Conclusions

None of the 64 "O"-rings leaked at any time. Under the test conditions, painting the sealing surfaces does not produce leak paths past the "O"-rings. Under the test conditions, all of the paints and paint combinations showed some susceptibility to deterioration by isopropyl alcohol. Hardened topcoats of the epoxy paint showed the best resistance, whereas the polyurethane primers and topcoats showed the least resistance.

Apparently, deterioration by alcohol contamination is exhibited by bubbling and weakening of the paint, thinning of the topcoats, and dissolving of the primer. The deterioration due to the alcohol did not produce leak paths past the "O"-rings.

The typical surface roughness of the polyurethane topcoats was approximately 5 to 25 rms, in microinches, whereas the surface roughness of the epoxy topcoats ranged between 40 and 125 rms, in micro-

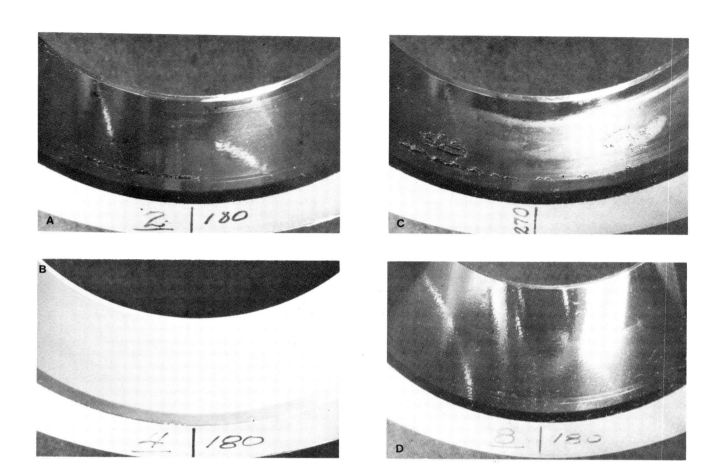

FIGURE 6 — Examples of the painted surfaces after the Otto Fuel test. A) Cylinder 2 at 180°; B) cylinder 4 at 180°; C) cylinder 6 at 270°; D) cylinder 8 at 180°.

inches. The differences in surface roughness apparently have little effect on the seal produced by the contact between the "O"-ring and the paint.

Over time, the pressure of the "O"-rings produces creep in all of the paints tested. The polyurethane topcoat was most susceptible to grooving by creep, whereas the epoxy topcoat was least susceptible to grooving. The depth of the grooves averaged up to approximately 50% of the thickness of the paint. The grooving produced a ridge about 20% higher than the average height of the paint. Sharper ridges were produced in the polyurethane than in the epoxy paint system. Creep or grooving did not cause exposure of bare metal.

Exposure to artificial sea water alone produced no indications of deterioration to the paint or "O"-ring sealing surfaces. Exposure to artificial sea water for 3 weeks increased the average roughness by 10 to 20 rms.

The exposure to Otto Fuel II produced some indications of interaction with the paint, but none that were serious. The fuel produced a characteristic staining in the epoxy paints and the primers, but no adverse effects. The fuel did produce distortion, wrinkling, and color changes in the polyurethane topcoat, but apparently the reaction was limited to areas of previous damage due to contamination by alcohol or foreign matter (dust particles) entrapped in the paint. Exposure to Otto Fuel II for 2 weeks decreased the average roughness by about 5 rms. However, this value is within the calibration error of ±5 rms.

Acknowledgment

This work was supported by the Naval Undersea Warfare Engineering Station, Keyport, Washington, under contract N00406-77-C-0778.

Subject Index

A

Abrasives, profile height resulting from selected, 222
Absorption, RH, temp. info. on water, 25
Absorption, water: primed steel, 259

ACIDS
acetic, generation by dryers in coatings, 165
aliphatic, evolution by dryers, 168
hydrochloric, generation by vinyl coatings, 169
propionic, generation by dryers in coatings, 165

Acidic: leachates from thermal insulation, 161
Activation energy: polymer delamination in chlorides, 76

ACRYLICS
emulsions, 186
+ Epoxy systems, 191
steel surface profile affecting, 223
water proofing for concrete, 164

ADHESION
acid-base interaction factors, 1
application internal influencing, 174
chlorinated rubber to steel, chemical reaction, 238
coatings failures in waste water, coal scrubber environments due to, 126
coatings performance, significance in, 87
coatings, shrinkage causing failure, 125
epoxy to Al, 75
epoxy single, two component formulations, 120
failure due to application error, 224
mechanism of loss, 103
polar sulfonates, 145
polybutadiene to Al, 75
six coatings to steel, including steel surface profile, 224
surface prep. effects, 227

Adsorption: solvent isotherms, 1
Adsorption effects: water displacing compounds, 152

ALKALI
delamination from interface, 70
disbonding effects, 108
leaching from thermal insulation, 161
sulfonate complexes containing, 145

ALKYDS
acid gas evolution from, 167
Arabian Gulf performance, 138
atmospheric effects, 186
+ methacrylate reactivity, 193

polarization tests, 56
+ soya, linseed, DCO, 24
steel surface profile affecting, 223

Aluminum, coatings delamination from, 70
American Society for Testing and Materials: salt fog test, 247
Ammonium chloride delamination effects, 71
Anion exchange properties test, 21
Arabian American Oil Co., coatings performance data, 138
Arabian Gulf, coatings performance data, 138

ASPHALT
accelerated test vs SO_2, 182
aqueduct, 320 microns inside, 172
sea water vs, 105

Atlas Test Cell, 181
Atmospheric effects: coal tar, chlorinated rubber and epoxies vs, 140
Automobile, coatings tested for, 255
Automobile, sulfonate hot melt for, 147

B

Bernard cells, 245
Biological effects: barnacles vs chlorinated rubber, 140
Bitumens, 206

BLISTERS
alkali cations, effect on, 107
anions, effect on, 107
ASTM D714 method, evaluation of, 249
coal tar epoxy, 200 micron film on steel, 54
coatings, evaluation factor, 174
coatings on flue gas condensate, 178
electrochemical reactions in, 104
epoxy, high solids, 123
mechanism, 93, 97, 103
organic coatings in salt fog, 223
polyester vs SO_2, 179
rating standard modification, 250
sodium chloride effects, 106
sodium hydroxide in sea marine, 105
temperature gradient, effect on, 105

Bonds, coatings tests, 82
Bonds, coatings to steel, six organic, one inorganic, 224
Butyl, vapor barrier for thermal insulation, 164

C

CALCULATIONS
accelerated tests, coordination of, 258

coatings, adhesion correlation of data, 226
delamination rate, 71
heats of adsorption interpretation, 2
mass transfer through coatings, 62
neutral species flux through polymer membrane, 63
rust, extent of, ASTM D 610 standard evaluation, 251
salt fog test data correlation, 249
sampling analysis: formulas for computing quantities, 279
Zn phosphate, data correlations on performance, 216

Capacitance, coal tar epoxy tests, 52
Capacitance, dielectric film, effects of adsorbed water in, 34
Carbon, organic in flue gas condensate, 182

CASE HISTORIES
coatings failure, 123
coatings failure and repair in aqueduct, 172
insulation, thermal; attacks under, 158

Castor oil, 24

CATHODIC
electrodeposition, 111
polarization, coated steel panels, 259
polarization epoxy vs NaCl under, 108

CATHODIC PROTECTION
alkali at interface from, 70
coatings disbondment by, 78
film impedance vs, 154
offshore petroleum structures, 140
pipelines, influence vs SCC, 203
thermally insulated pipe, galvanic, 164

Cations, alkali, influence on blistering, 107
Cations, delamination influence, 74
Cement, corrosion resistant vs SO_2, accelerated test, 182
Cesium chloride delamination effects, 71
Charge transfer factors vs attack rate, 36

CHLORIDES
ion vs water vapor permeation rates, 26
L vs D area penetration, 14
thermal insulation, concentration under, 161

Chlorinated rubber, 52, 139, 140
Chlorine, ionic permeation characterization, 136
Chromate, coatings containing vs pipeline SCC, 203

Cleaning, steel, high strength, low carbon, 213
Clusters, in polyethylene, 4

COAL TAR
enamel, 140
sea water vs, 105
steel surface profile affecting, 223

COAL TAR EPOXY
accelerated test vs SO_2, 182
coal tar epoxy, 51, 105, 130, 139, 164, 182
pipe coating, 164

COATINGS, APPLICATION
anodic deposition, 111
cathodic electrodeposition, 111
importance, 122
review, 128

COATINGS, CONVERSION
phosphate: impedance probe tests on epoxy coatings, 45
phosphate: morphology and kinetics, 236
Zn phosphate, 201
Zn phosphate on high strength steel, 211
Zn phosphate on steel, oxygen reduction tests, 214
Zn phosphate weight vs surface prep. method, 234

Coatings, inorganic, accelerated test vs SO_2, 182

COATINGS, METAL LOADED (Zn)
Arabian Gulf experience, 138
impedance probe scan influence, 48
inorganic under thermal insulation, 164
Na silicate, others vs sea water, 106
organic coatings, delamination from, 70
rich primer failure, 124

COATINGS, ORGANIC
also generic names
automobile, tests for, 255
carbon interface on cold rolled steel, 211
chlorinated rubber on steel, 238
dryers, vehicles attacking metals, 165
emulsion binders, 186
extruded polymeric, 206
formulation variables, 128
heat indicating, 286
high solids solvent based, 192
immersion service, 139
latex evaluation, 252
latex for plaster, concrete, wood, 140
"O" ring coating, 285
pigment particle convection in films, 245
rusting rating in salt fog, 223
steel aqueducts, performance in, 172
sulfonate base, 144
test methods, 263
vapor barrier for cold thermal insulation, 164
water proofing for hydraulic concrete, 163

Coatings thickness, disbonding effects, 110

Cobalt, chloride delamination effects, 73
Concrete, hydraulic attack under, 158
Concrete, hydraulic formulation requisites, 163
Condensation, exhaust gas stacks, 179

CONDUCTIVITY
DC contribution to adhesion loss, 103
epoxy films changes vs electrolyte, 40
leachates from organic foams, 162

Crevice attack at organic holidays on Al, steel, 93
Crosslinking, ion permeation, effects of, 16
Curing mode effect in epoxy-polyamine delamination, 75

D

DCO, 24
Defects, impedance tests to identify, 42
Defects, impedance tests, probe indication of film, 45

DELAMINATION
also disbondment
delamination, 70
oxidation mechanism, 30

Design, controls limiting attack under thermal insulation, 163
Design, errors causing attack under thermal insulation, 163
Dielectric properties, polymers, 4
Dielectric properties, tests, 19

DIFFUSION
impedance measurements characterizing, 35
polymeric coatings, properties analysis, 62
water through exhaust stack linings, 179

Dioctylphthalate, 24

DISBONDMENT
also delamination
anodic, 93
cathodic, 70, 87
cathodic by alkalis, 108
cationic effects on, 109
coatings in combustion gas stacks, 180

Dispersion, iron oxide pigment effects, 199
Dryers, acid gas generation by, 165

E

ELECTRICAL
charge phenomena in permeable dielectric film, 34
coatings, properties measurements, 38
effects, AC frequency on coal tar epoxy capacitance, 52
effects, AC impedance tests of attack through polymers, 45
parameters, electrodeposition of polymers, 112
resistance, coating effects vs blistering, 106
resistance, DC, influence of primer on topcoat, 105

Electroosmosis, 97, 104
Electroosmosis, anodic and cathodic blistering due to, 104
Endosmosis, hydrophilic effect of aggressive species, 89
Endosmosis, linings in exhaust gas stacks, 179

EPOXY
amine accelerated test vs SO_2, 182
Arabian Gulf performance, 138
capacitance and impedance measurements, 38
cathodic delamination, 70
cathodic electrodeposition, 112, 118
characteristics of electrodeposited films, 118
coal tar, see coal tar epoxy
coal tar, in North Sea, 105
emulsion films, solvent resistance, 190
high solids, ketimine, blisters in, 123
inhibited thin films vs SCC, 206
"O" ring coatings, 286
phenolic vs K chloride, 92
polyamide, accelerated test vs SO_2, 182
polyamide, DC resistance test, 12
powder, hot curing, 80
topcoats, resistance, 191
splash zone, sea marine, for, 140
steel interface analysis in NaCl, 46
steel surface profile affecting, 223
water base, 188

Erosion, water displacing compound vs, 153

F

FAILURE ANALYSIS
coatings in NaCl solution, 133
failure analysis, 122, 133, 158
thermal insulation, 158

Ferrites, 199
Filiform attack mechanism, 101

FILMS
AC frequency dependent electrical properties, 41
analysis of surfaces after preparation, 229
condensed water identification in, 43
crystal reinforcement of sulfonate, 145
D and I type properties, 12
D area origin and characteristics, 12, 15
dehydration method, 20
dielectric polarization study, 56
dielectric properties vs AC frequencies, 43
electrical resistance vs permeation rate, 33
ellipsometric characterization, 28
emulsion agglomeration, 187
impedance measurements, 38, 51
ion exchange properties, 18
membrane reaction in diffusion cell, 63
nitrogen, influence in, 18
organic impedance tests, 51
pigmented, L and D area influence, 16
pigmented, water permeation, 198

polar groups in organic, 58
polymeric, emulsion properties, 190
polymeric, inhomogeneity effects, 12
polymeric, species transport, properties of, 62
thickness, characteristics, 12
thickness, effects, 13
thickness, permeation, influence on, 17
water displacing compounds forming dry, 153

Filters, flue gas scrubber linings, 178
Filters, scrubbers, coatings for, 126
Fireproofing materials, attack under, 158
Flash rusting, 186
Flocculation, pigment: effects on water permeation, 200
Fluorocarbons, accelerated test vs SO_2, 182
Fluoropolymer, combustion gas vs, 183

G

Galvanic effects, bare vs coated steel, 106
Galvanized steel, also coatings, metal loaded (Zn)
Gas, combustion, scrubber linings, 178
Gas, exhaust, coatings vs, 126
Glass, closed cell foam block vs SO_2, 182
Glass, foam properties, 162
Grain effects, high strength steels, influence annealing, 212

H

Halides, delamination effects, 7
Heat treatment, steels, high strength, 212
Holidays, effects of, 88

HUMIDITY
emulsion coatings, influence of, 187
interface water activity, influence of, 88
metal salts, liquefaction vs relative, 101

Hydroxyl ions, formation at coatings defects, 74

I

IMPEDANCE
cathodic protection influence on film impedance, 54
coating films in NaCl solutions, 136
tests, 45, 51

INHIBITORS
amines in coatings vs salt water, 265
coatings vs NaCl solutions, 130
coatings on steel pipelines vs SCC, 203, 206
coatings, test methods, 264
eight rated for steel vs salt water, 265
ferrite effect in primers, 201
ferrites, mechanism in coatings, 201
phosphate and sodium compounds in coatings vs salt water, 266
phosphate, Zn, strontium chromates vs NaCl, 28
soil additions vs pipeline SCC, 204

water displacing compounds containing, 153

Insulation, thermal coatings under, 158
Insulation, thermal organic foams, 161

INSPECTION
coatings, sampling program, 267
coatings, sequential sampling table, 271
density determination, statistical, 269
sequential analysis, 268, 271

INTERFACE
coatings, metal, pH effect on attack, 266
current flow in electrolyte at, 88
hydrophilic effect of aggressive species at, 89
ionic activities at, 91, 95
pH effects at, 95
phenomena, pH measurement, 28
water absorption at, 261

INTERFACE EFFECTS
alkali from cathodic reactions, 70
chlorinated rubber on steel, 241
coatings bond to steel, 78
impedance tests for, 45
sodium chloride on steel, 54

Ion activity, coatings disbondment in, 85
Ion exchange, coatings properties, 62
Ionic effects, transport phenomena at interface, 91

IRON OXIDE PIGMENTS
color range, particle sizes, 197
micronized: properties, 199
volume concentration, particle distance effects, 199

Isocyanates, polarization tests, 56

L

Light, coatings discoloration due to, 238
Linseed oils, 24
Lithium chloride delamination effects, 71
Localized attack: pitting, exhaust gas stacks, 180

M

Mass transfer, blister formation due to, 104
Melamine, 24
Metal salts, liquefaction vs humidity, 101
Methacrylate reactivity, 193
Mill oils, Zn phosphate reactions vs, 219
Miscibility, 152
Monel sheathing for splash zone protection, 140

N

NACE, protective coatings immersion test standard, 181

NITROGEN
films, properties, influence of, 18
oxide attack in flue gases, 178
salt spray test results, influence of, 22

O

Osmosis, blisters due to, 104
Osmosis, linings in flue gas stacks, 179
Osmotic effects, 97
Osmotic pressure, 42
Oxidation, DCPEMA (methacrylate), 193
Oxidation, interface measurement, 29
Oxide, films after surface prep., 231

OXYGEN
cells at coating to base metal interface, 30
diffusion rates, 95
interface activity, 92
permeation through polymer coatings, 74

P

Packaging, parameters conducive to acid gas generation in, 169
Penthaerythritol alkyd, Dc resistance test, 12
Perforation, exhaust gas stacks, 180

PERMEATION
electrical currents through coatings, 88
film, electrical, resistance vs rate, 33
inorganic anions, 97
ionic, 24, 103
ionic, influence on adhesion, 103
ions through coatings, thickness effects, 90
oxygen through polymer coatings, 74
water, flocculation of pigment affecting, 200
water through coal tar epoxies, 130
water through coatings vs corrosion rate, 24
water through pigmented films, 198
water vapor: N in films vs, 18

Permselectivity, urethanes, 22
Petroleum production: offshore structures, coatings, 140
Petroleum refineries, thermal insulation, attack under, 158

pH
alkaline for water base coatings, 189
condensate, blister fluids in exhaust gas stacks, 178
disbonding, effects of, 110
flue gas condensate, 179
interface effects and measurement, 28
iron corrosive interface in salt water, 266
water extract of ferrites, 201

Phase separation effects, 98
Phenol formaldehyde tung oil, DC resistance test, 12
Phenolics, acid gas evolution from, 167
Phenolics, para varnishes with tung, linseed, soya oils, 24

PHOSPHATE
inhibitors in soil vs SCC of pipelines, 204
-izing, + chromate, laboratory tests, 28
organic, effects in coatings, 146

Phosphorus, phosphate porosity affected by surface, 220

PIGMENTS
active and inactive effects, 201
calcium ferrite, 200
chromates affecting L and D films, 16
dielectrics, effects on, 58
emulsion coatings, influence of, 187
mechanism of influence on corrosion rates, 200
oxides affecting L and D films, 16
primers, effects in, 197
Zn ferrite, 200

Piting, also Localized attack
Pitting, locus at holidays, 105
Pipes, aqueduct, coatings failure inside, 172

PIPELINES
inhibited coating vs SCC, 203
thermally insulated failure, 160
underground, coating for thermal insulation, 164

Plasticity, sulfonate water based coatings retention, 148
Polarization, resistance measuring ionic resistance, 34
Polyacrylates, polarization tests, 56
Polybutadiene, cathodic delamination, 70
Polybutadiene, hot curing, 80
Polycarbonate, dielectric properties, 4

POLYESTERS
accelerated test vs SO_2, 182
blistering in SO_2, 179
glass filled tank lining failure, 125

POLYETHYLENE
dielectric properties, 4
disbonding, bare edge and other, 109
pipe coating, extruded jacket, 164

POLYMERIC MATERIALS
also generic names
inhomogeneity effects, 12
solubilized electrodeposition of, 111

Polysulfone, bound water effect, 8
Polysulfone, dielectric properties, 4
Polyvinyl acetate, bound water effects, 9
Pores, magnitude effects, 100
Porosity, Zn phosphate vs steel, 215

POTASSIUM
chloride delamination effects, 71
chloride polymers vs, 14
chromate Fe treatment with, 28
fluoride delamination effects, 71

Preferential adsorption: compounds exhibiting vs water, 150

PRIMERS
automobile tests, 255
coatings, DC resistance affected by, 105
latex, 195
phosphate, tests, 255
synthetic iron oxide pigments in, 197
surface on steel affecting coatings, 222, 225

Q

Quality control, statistical, 268

R

Relative humidity, water permeation, effects of, 25
Residential coatings, 140
Resistivity, ionic, tests, 21
Rust, ASTM evaluation of rate, 251
Rust, characteristics under blisters, 99

S

Sampling: parameters, high workmanship level, 273
Sampling: parameters, standard workmanship level, 276
Sand, abrasion by blowing, 139
Saponification, pigments affecting, 200
Scrubbers, see Filters
Shelf life, coatings in Arabian Gulf area, 138
Silicates, pipeline coating inhibitors, 209
Silicon, carbide: chlorinated rubber reactions with, 241

SODIUM CHLORIDE
AC influence on activity, 33
coal tar epoxy vs, 51
delamination effects, 71
steel contamination effects vs coating impedance, 54
epoxy amide vs, 47
epoxy versamid vs, 40
Fe vs, 30
hydroxide Fe vs, 30

SOLVENTS
adsorption isotherms, 1
epoxy emulsion films vs, 190
interface effects in blistering, 98
residual effects at interface, 107
water displacement compounds containing, 153

Soya oils, 24
Spalling, concrete from rusted steel, 159
Splash zone protection, 140

STEEL, CARBON
chlorinated rubber adhesion to, 238
coatings delamination from, 70
high strength cold rolled, coatings on, 211
surface profile affecting bonding to, 224
thermal insulation, attack under, 158

STANDARDS
ASTM blistering, 182
NACE, coatings immersion test, 181
German, salt fog coating, 255

Statistics, coatings inspection sampling, 268
Stress corrosion cracking, chromium nickel steel insulation jackets, 164
Stress corrosion cracking, steel pipelines, inhibited coatings vs, 203, 208
Sulfates, flue gas content, 179
Sulfonates, 144
Sulfonates, petroleum: adsorption enhancers, 152

SULFUR
concentration under thermal insulation, 161
dioxide, polymeric materials accelerated test vs, 182
dioxide, scrubber coatings, 178

Sunlight, coal tar coatings vs, 140

SURFACE EFFECTS
carbon on cold rolled steel from mill oils, 211, 213
elements on low carbon steels, 213
energetics related to surface prep., 231
film analysis after surface prep., 228
mapping of steel after surface prep., 233

SURFACE PREPARATION
acid pickling, 228
adhesion effects, 227
alkaline derusting, 228
anodic etching, 228
bonding strength effects, 225
chemical smoothing, 228
coatings for inadequate, 138
degreasing, 228
delamination, influence on, 70
disbondment tests, 78
electropolishing, 228
epoxy glass reinforced, importance for, 140
hot alkaline, 228
passivation, 228
phosphoric acid, 228
pickling, 228
pipeline, 78
review, 128
salt fog test, 249
sandblasting, 228
solvent cleaning, 227
steel, influence of profile on coatings performance, 222
sulfuric dichromate, 228
US Federal, British code of practice pec., 227

SURFACE ROUGHNESS
steel in fresh water causing, 289
steel in Otto fuel, 293
steel in sea water causing, 291

Surface tension, compounds with low, 152

T

Tanks, thermal insulation, attack under, 160

TEMPERATURE
25 C, water vs polycarbonate, 7

epoxy electrodeposits, failure, effects of, 118
film properties, effects, 42
gradient effects across coatings, 105
paints, indicating, 288
permittivity effects, 57
stack lining attack, influence of, 178
sulfonate coatings resistance, 148
thermal insulation, influence on attack under, 159
water permeation, influence of, 25

TEMPERATURE, LOW
−140 C, water vs polycarbonate, 7
−105 to −5 C, condensed water in films at, 43
0 to 50 C, potassium chloride vs polymers, 14
water vs polymers, 7

TESTS, ACCELERATED
atmospheric: exhaust stack linings, 181
atmospheric: sulfonates, 146
automobile coatings, 255
coatings, surface profile, influence of, 222
emulsion coatings, 188
epoxy emulsion topcoats, 191
iron oxide pigmented films, 198
organic coatings, 263
early rusting, latex 253
salt fog evaluation, 247
salt spray, N contents affecting results of, 22
scribe procedure auto coating, 256

TEST EQUIPMENT
AC and DC impedance probe, 45
cathodic detachment, 90
cathodic electrodeposition, 112
coatings inhibitors, 264
direct current, 32
electrochemical, automobile coating, 258
electrochemical, ellipsometric, 28
impedance for organic coatings, 38
orientation polarization, 56
temperature indicating coatings, 281

TESTS, FIELD
long term, problems with, 141
pipelines, inhibited coatings on, 207
sea marine, iron oxide pigmented primers, 199

TESTS, LABORATORY
acid gases, chromatographic of, 166
acid gas evolution from drying coatings, 165
activation energies of alkyds, polyacrylates, isocyanates, 60
alkyd, oleoresinous, coal tar epoxy and epoxy coatings vs NaCl, 130
anion exchange, 21
cathodic currents vs organic membranes, 90
chlorinated rubber on steel, 238
dielectric, mechanical properties of films, 59
dielectric properties, 19

dielectric spectroscopic, condensed water, 43
disbonding to qualify coatings, 110
discoloration, chlorinated rubber, 239
electrochemical, 28
electrochemical, AC and DC impedance of polymer to metal interface, 45
electrochemical, AC frequency effects, 32
electrochemical, automobile coatings, 257
electrochemical, orientation polarization, 56
electron micrographic, epoxy electrodeposited, 118
ellipsometric, 28
ellipsometric, steel surfaces, 233
flame ionization tests of acid gases, 167
flue gas condensate vs steel, 183
high strength steel, annealing influence, 212
ionic diffusion through urethane, 62
impedance, organic coatings, 38
leaching, organic foam insulation, 162
radiotracer, ion diffusion effects of, 63
scanning electron microscope, disbonded coatings, 84
steel surface profile effects, 222
water displacing compounds vs sea water, 154
water vapor permeability, 18
Zn phosphate on steel, 215

TEST METHODS
coating vs fresh and salt water, 286
coating thickness vs drying, application intervals, 172
disbonded coating characteristics, 83
DC resistance of detached films, 12
drying process, 172
inhibitors vs SCC, 205, 206, 208
iodine/methanol oxide stripping, 82
organic coatings, 263
salt fog for coating, 247
temperature indicating coatings, 281
water cluster analysis, 4
XPS spectra analysis, 78

Thickness, specifications, 268
Thixotropic sulfonate based coatings, 144

TIME
coating changes in electrolyte vs, 40
conductivity, diffusivity, effects of, 66
curing, Zn rich ethyl silicate primers, 125
drying, acid gas generation, influence on, 170
drying tests vs, 176
film properties, influence of, 58
inhibitors in pipelines persistence vs time, 208
salt fog effects exposure, 251

Titanium dioxide in alkyds, melamine, nitrocellulose, vinyl, 24
Tradenames, 195
Tung oils, 24
Tung oils, acid vapors from drying, 170

U

Underfilm attack, 256

URETHANE
AC effects on vs NaCl, 33
chloride, sodium ion diffusion through, 62
D constituents in, 16
electrochemical properties, 18
ionic capacity test, 63
"O" ring coating, 286
permeability coefficients, 64
thermal insulation attack under, 160

V

VINYLS
accelerated test vs SO_2, 182
ester failure in exhaust gas, 126
hydrochloric acid evolved from drying, 169
latex: steel surface profile affecting, 223
mastic, 139
sodium chloride vs, 92
thin film reactivity, 192
water proofing for concrete, 164
+ Zn chromate in steel aqueduct, 172
Zn loaded: inorganic, surface profile affecting, 223

Voltage, + current, time effects on cathodic electrodeposition, 112

W

WATER
absorption: primers on steel, 260
Al to coating interface, volume at, 93
bound and free, calculation of amounts, 20
capacitance change to calculate uptake, 41
coating resistance affected by bound, 67
displacing compounds, 150
interface activity of, 88
migration through thermal insulation, 164
organic coatings absorption rates, 97
permeation pigmented films, 198
pigment to binder interface accumulation, 199
polyethylene, measurement in, 5
polymers, condensed in, identification of,
polymer uptake rates, 75
salt: amines tested as coatings inhibitors in, 265
sea, blisters on steel in, 105
sea, coatings for use in and near, 140

XYZ

ZINC
acid gases from dryers vs, 169
chromate: Fe treatment with, 29
inhibitors in coatings, 206

Author Index

Anderson, E.W. 4	Joslin, T. 1	Ruggeri, R. T. 62
Bair, H. E. 4	Jullien, H. 18	Salensky, G. A. 263
Beck, T. R. 62	Kargol, J. A. 211	Sandwith, C. J. 285
Berger, D. M. 178	Kielmanson, Z. 165	Scantlebury, J. D. 51
Berry, W. E. 203	Koehler, E. L. 87	Schmidt, H. J. Jr. 128
Brooman, E. W. 203	Koesters, E. 255	Schwab, L. K. 222
Castle, J. E. 78	Kresse, P. 197	Schwenk, W. 103
Cech, L. S. 144	Kruger, J. 28	Sekine, I. 130
Delahunt, J. F. 158, 280	Leidheiser, H. Jr. 38, 70	Smelt, D. H. 238
Devay, J. 56	Letai, P. 165	Standish, J. V. 38
Drisko, R. W. 222	Lineman, D. M. 203	Sun, C-Y 1
El-Aasser, M. S. 111	Liu, T. 247	Sussex, G. A. M. 51
Fessler, R. R. 203	Lumsden, J. B. 227	Szadkowski, V. 197
Forsberg, J. L. 144	Mansfeld, F. 227	Tator, K. B. 122
Fowkes, F. M. 1	Mayne, J. E. O. 12	Tooke, R. Jr. 267
Funke, W. 97	Meszaros, L. 56	Trewella, R. J. 178
Goering, W. 255	Mills, D. J. 12	Tsai, S. 227
Hegedus, C. R. 150	Moore, E. M. Jr. 138	Vanderhoff, J. W. 111
Higgins, W. A. 144	Muenster, R. 255	Verbist, R. 32
Ho, C. C. 111	Odenthal, R. H. 197	Washburne, R. N. 186
Hughes, M. C. 45	Parks, J. M. 45	Wang, W. 70
Hymayun, A. 111	Piens, M. 32	Watts, J. F. 78
Janaszik, F. 56	Rhodes, K. I. 138	White, J. H. 165, 172
Jeanjaquet, S. L. 227	Ritter, J. J. 28	Wummer, C. J. 178
Johnson, G. E. 4	Rothschild, W. 172	Yaseen, M. 24
Jordan, D. L. 211		